国家出版基金资助项目
现代数学中的著名定理纵横谈丛书
丛书主编　王梓坤

TSCHEBYSCHEFF APPROXIMATION THEOREM

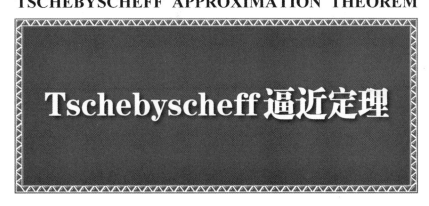

佩捷　等　编著

哈尔滨工业大学出版社
HARBIN INSTITUTE OF TECHNOLOGY PRESS

内容简介

本书详细介绍了 Tschebyscheff 逼近问题的相关知识及应用.全书共 21 章,读者可以较全面地了解 Tschebyscheff 这一类问题的实质,并且还可以认识到它在其他学科中的应用.

本书适合数学专业的本科生和研究生以及数学爱好者阅读和收藏.

图书在版编目(CIP)数据

Tschebyscheff 逼近定理/佩捷等编著.—哈尔滨:哈尔滨工业大学出版社,2016.6
(现代数学中的著名定理纵横谈丛书)
ISBN 978-7-5603-5779-9

Ⅰ.①T… Ⅱ.①佩… Ⅲ.①切比雪夫逼近
Ⅳ.①O174.41

中国版本图书馆 CIP 数据核字(2016)第 003391 号

策划编辑	刘培杰 张永芹
责任编辑	张永芹 聂兆慈
封面设计	孙茵艾
出版发行	哈尔滨工业大学出版社
社　　址	哈尔滨市南岗区复华四道街 10 号　邮编 150006
传　　真	0451-86414749
网　　址	http://hitpress.hit.edu.cn
印　　刷	哈尔滨市石桥印务有限公司
开　　本	787mm×960mm 1/16 印张 33.25 字数 356 千字
版　　次	2016 年 6 月第 1 版　2016 年 6 月第 1 次印刷
书　　号	ISBN 978-7-5603-5779-9
定　　价	88.00 元

(如因印装质量问题影响阅读,我社负责调换)

◦ 代序

读书的乐趣

你最喜爱什么——书籍.
你经常去哪里——书店.
你最大的乐趣是什么——读书.

这是友人提出的问题和我的回答.真的,我这一辈子算是和书籍,特别是好书结下了不解之缘.有人说,读书要费那么大的劲,又发不了财,读它做什么？我却至今不悔,不仅不悔,反而情趣越来越浓.想当年,我也曾爱打球,也曾爱下棋,对操琴也有兴趣,还登台伴奏过.但后来却都一一断交,"终身不复鼓琴".那原因便是怕花费时间,玩物丧志,误了我的大事——求学.这当然过激了一些.剩下来唯有读书一事,自幼至今,无日少废,谓之书痴也可,谓之书橱也可,管它呢,人各有志,不可相强.我的一生大志,便是教书,而当教师,不多读书是不行的.

读好书是一种乐趣,一种情操；一种向全世界古往今来的伟人和名人求教的方法,一种和他们展开讨论的方式;一封出席各种社会、体验各种生活、结识各种人物的邀请信;一张迈进科学宫殿和未知世界的入场券;一股改造自己、丰富自己的强大力量.书籍是全人类有史以来共同创造的财富,是永不枯竭的智慧的源泉.失意时读书,可以使人重整旗鼓;得意时读书,可以使人头脑清醒;疑难时读书,可以得到解答或启示;年轻人读书,可明奋进之道;年老人读书,能知健神之理.浩浩乎!洋洋乎!如临大海,或波涛汹涌,或清风微拂,取之不尽,用之不竭.吾于读书,无疑义矣,三日不读,则头脑麻木,心摇摇无主.

潜能需要激发

我和书籍结缘,开始于一次非常偶然的机会.大概是八九岁吧,家里穷得揭不开锅,我每天从早到晚都要去田园里帮工.一天,偶然从旧木柜阴湿的角落里,找到一本蜡光纸的小书,自然很破了.屋内光线暗淡,又是黄昏时分,只好拿到大门外去看.封面已经脱落,扉页上写的是《薛仁贵征东》.管它呢,且往下看.第一回的标题已忘记,只是那首开卷诗不知为什么至今仍记忆犹新:

日出遥遥一点红,飘飘四海影无踪.

三岁孩童千两价,保主跨海去征东.

第一句指山东,二、三两句分别点出薛仁贵(雪、人贵).那时识字很少,半看半猜,居然引起了我极大的兴趣,同时也教我认识了许多生字.这是我有生以来独立看的第一本书.尝到甜头以后,我便千方百计去找书,向小朋友借,到亲友家找,居然断断续续看了《薛丁山征西》《彭公案》《二度梅》等,樊梨花便成了我心中的女英雄.我真入迷了.从此,放牛也罢,车水也罢,我总要带一本书,还练出了边走田间小路边读书的本领,读得津津有味,不知人间别有他事.

当我们安静下来回想往事时,往往会发现一些偶然的小事却影响了自己的一生.如果不是找到那本《薛仁贵征东》,我的好学心也许激发不起来.我这一生,也许会走另一条路.人的潜能,好比一座汽油库,星星之火,可以使它雷声隆隆、光照天地;但若少了这粒火星,它便会成为一潭死水,永归沉寂.

抄,总抄得起

好不容易上了中学,做完功课还有点时间,便常光顾图书馆.好书借了实在舍不得还,但买不到也买不起,便下决心动手抄书.抄,总抄得起.我抄过林语堂写的《高级英文法》,抄过英文的《英文典大全》,还抄过《孙子兵法》,这本书实在爱得狠了,竟一口气抄了两份.人们虽知抄书之苦,未知抄书之益,抄完毫末俱见,一览无余,胜读十遍.

始于精于一,返于精于博

关于康有为的教学法,他的弟子梁启超说:"康先生之教,专标专精、涉猎二条,无专精则不能成,无涉猎则不能通也."可见康有为强烈要求学生把专精和广博(即"涉猎")相结合.

在先后次序上,我认为要从精于一开始.首先应集中精力学好专业,并在专业的科研中做出成绩,然后逐步扩大领域,力求多方面的精.年轻时,我曾精读杜布(J. L. Doob)的《随机过程论》,哈尔莫斯(P. R. Halmos)的《测度论》等世界数学名著,使我终身受益.简言之,即"始于精于一,返于精于博".正如中国革命一样,必须先有一块根据地,站稳后再开创几块,最后连成一片.

丰富我文采,澡雪我精神

辛苦了一周,人相当疲劳了,每到星期六,我便到旧书店走走,这已成为生活中的一部分,多年如此.一次,偶然看到一套《纲鉴易知录》,编者之一便是选编《古文观止》的吴楚材.这部书提纲挈领地讲中国历史,上自盘古氏,直到明末,记事简明,文字古雅,又富于故事性,便把这部书从头到尾读了一遍.从此启发了我读史书的兴趣.

我爱读中国的古典小说,例如《三国演义》和《东周列国志》.我常对人说,这两部书简直是世界上政治阴谋诡计大全.即以近年来极时髦的人质问题(伊朗人质、劫机人质等),这些书中早就有了,秦始皇的父亲便是受害者,堪称"人质之父".

《庄子》超尘绝俗,不屑于名利.其中"秋水""解牛"诸篇,诚绝唱也.《论语》束身严谨,勇于面世,"己所不欲,勿施于人",有长者之风.司马迁的《报任少卿书》,读之我心两伤,既伤少卿,又伤司马;我不知道少卿是否收到这封信,希望有人做点研究.我也爱读鲁迅的杂文,果戈理、梅里美的小说.我非常敬重文天祥、秋瑾的人品,常记他们的诗句:"人生自古谁无死,留取丹心照汗青""谁言女子非英物,夜夜龙泉壁上鸣".唐诗、宋词、《西厢记》《牡丹亭》,丰富我文采,澡雪我精神,其中精粹,实是人间神品.

读了邓拓的《燕山夜话》,既叹服其广博,也使我动了写《科学发现纵横谈》的心.不料这本小册子竟给我招来了上千封鼓励信.以后人们便写出了许许多多的"纵横谈".

从学生时代起,我就喜读方法论方面的论著.我想,做什么事情都要讲究方法,追求效率、效果和效益,方法好能事半而功倍.我很留心一些著名科学家、文学家写的心得体会和经验.我曾惊讶为什么巴尔扎克在51年短短的一生中能写出上百本书,并从他的传记中去寻找答案.文史哲和科学的海洋无边无际,先哲们的明智之光沐浴着人们的心灵,我衷心感谢他们的恩惠.

读书的另一面

以上我谈了读书的好处,现在要回过头来说说事情的另一面.

读书要选择.世上有各种各样的书:有的不值一看,有的只值看20分钟,有的可看5年,有的可保存一辈子,有的将永远不朽.即使是不朽的超级名著,由于我们的精力与时间有限,也必须加以选择.决不要看坏书,对一般书,要学会速读.

读书要多思考.应该想想,作者说得对吗?完全吗?适合今天的情况吗?从书本中迅速获得效果的好办法是有的放矢地读书,带着问题去读,或偏重某一方面去读.这时我们的思维处于主动寻找的地位,就像猎人追找猎物一样主动,很快就能找到答案,或者发现书中的问题.

有的书浏览即止,有的要读出声来,有的要心头记住,有的要笔头记录.对重要的专业书或名著,要勤做笔记,"不动笔墨不读书".动脑加动手,手脑并用,既可加深理解,又可避忘备查,特别是自己的灵感,更要及时抓住.清代章学诚在《文史通义》中说:"札记之功必不可少,如不札记,则无穷妙绪如雨珠落大海矣."

许多大事业、大作品,都是长期积累和短期突击相结合的产物.涓涓不息,将成江河;无此涓涓,何来江河?

爱好读书是许多伟人的共同特性,不仅学者专家如此,一些大政治家、大军事家也如此.曹操、康熙、拿破仑、毛泽东都是手不释卷,嗜书如命的人.他们的巨大成就与毕生刻苦自学密切相关.

王梓坤

目录

第 0 章　引言　//1

第 1 章　Tschebyscheff 小传　//9

第 2 章　什么是逼近　//22

第 3 章　Tschebyscheff 多项式　//41

第 4 章　多项式动力学和 Fermat 小定理的一个证明　//51

  4.1　引言　//51

  4.2　Tschebyscheff 多项式　//52

  4.3　结论　//54

第 5 章　最佳逼近多项式的特征　//58

第 6 章　Tschebyscheff 多项式的三角形式在几何中的应用　//64

  6.1　第一型 Tschebyscheff 多项式　//66

  6.2　第二型 Tschebyscheff 多项式　//70

第 7 章　Tschebyscheff 多项式的三角形式不等式　//76

第 8 章　Tschebyscheff 多项式的拉格朗日
　　　　形式　//81
第 9 章　再谈最佳逼近多项式　//86
第 10 章　最小偏差多项式　//94
第 11 章　高次 Tschebyscheff 逼近　//98
　11.1　一道集训队试题　//98
　11.2　П. Л. Tschebyscheff 定理　//100
第 12 章　Tschebyscheff 多项式与不等式　//136
第 13 章　Tschebyscheff 多项式与马尔可夫
　　　　定理　//143
　13.1　多项式与三角多项式的导数增长的阶　//144
　13.2　函数的可微性质的表征　//148
第 14 章　多元逼近　//153
第 15 章　多元逼近问题中的未解决问题　//155
第 16 章　非线性 Tschebyscheff 逼近　//158
第 17 章　巴拿赫空间中的 Tschebyscheff
　　　　多项式　//161
第 18 章　FIR 数字滤波器设计的 Tschebyscheff
　　　　逼近法　//163
　18.1　Tschebyscheff 最佳一致逼近原理　//165
　18.2　利用 Tschebyscheff 逼近理论设计 FIR 数字
　　　　滤波器　//166
　18.3　误差函数 $E(\omega)$ 的极值特性　//172
第 19 章　苏格兰咖啡馆的大本子　//176
第 20 章　逼近论中的伯恩斯坦猜测　//184
　20.1　引言　//184

20.2　高精度计算 $\{2nE_{2n}(|x|)\}_{n=1}^{52}$　//189

20.3　计算伯恩斯坦常数 β 的上界　//192

20.4　计算伯恩斯坦常数 β 的下界　//202

20.5　数 $\{2n\sum_{2n}(|x|)\}_{n=1}^{52}$ 的理查森外插　//205

20.6　某些未解决的问题　//207

20.7　$|x|$ 在 $[-1,+1]$ 上的有理逼近　//210

附录 I　关于非线性 Tschebyscheff 逼近的几点注记　//219

附录 II　几个多项式问题　//225

　　1. 全 k 次方值蕴涵 k 次方式　//225

　　2. Tschebyscheff 多项式引申出的几个问题　//227

　　3. 二次函数的几个问题 //234

附录 III　离散逼近论　//240

　　1. Banach 空间的离散逼近 //241

　　2. 闭算子的离散逼近 //244

附录 IV　Tschebyscheff 正交多项式问题　//246

　　1. Tschebyscheff 正交多项式 //246

　　2. 用 Tschebyscheff 方法逼近函数 //253

附录 V　联合最佳 L_p 逼近　//258

　　1. 引言 //258

　　2. 存在定理 //260

　　3. L_1 逼近的特征定理 //262

　　4. L_p 逼近的特征定理 //268

附录 VI　多元函数的三角多项式逼近　//276

　　1. 引论 //276

　　2. 定理 6.2 的证明 //282

  3. 定理 6.3 的证明 //286

  4. 定理 6.3 的另一证明 //290

附录Ⅶ 多元周期函数的非整数次积分与三角多项式逼近 //297

  1. 多元周期函数的非整数次积分 //297

  2. 非整数次积分的性质 //304

  3. 三角多项式逼近 //314

附录Ⅷ 在具有基的 Banach 空间中的最佳逼近问题 //325

附录Ⅸ С. Н. Мергелян 定理的推广 //346

附录Ⅹ 平方逼近 //362

  1. 函数按最小二乘法的逼近 //362

  2. 周期函数借助于三角多项式的平方逼近 //369

  3. 借助于线性无关函数组的逼近表示 //374

  4. 平方逼近的 Tschebyscheff 公式 //378

  5. 非线性的依从于一个或几个参数的函数的逼近 //388

  6. 分段连续函数的逼近 //390

  7. 用以确定平方逼近的系数的方程组 //394

  8. 平方误差的计算 //397

  9. 多个自变量函数的平方逼近 //399

参考文献 //403
编辑手记 //515

引 言

第 0 章

先看一道清华大学金秋营试题.

例1 求方程 $x^5+10x^3+20x-4=0$ 的所有根.

解法1 设 $x=z-\dfrac{2}{z}$,则方程 $x^5+10x^3+20x-4=0$ 可化为

$$z^5-\dfrac{32}{z^5}-4=0$$

于是 $z^5=8$ 或 $z^5=-4$,故 $z=\sqrt[5]{8}\,\mathrm{e}^{\frac{2k\pi}{5}\mathrm{i}}$ 或 $z=-\sqrt[5]{4}\,\mathrm{e}^{-\frac{2k\pi}{5}\mathrm{i}}$(其中 $k=0,1,2,3,4$).

解法2 设

$$x=\lambda\left(t-\dfrac{1}{t}\right)$$

Tschebyscheff 逼近定理

则原方程变为

$$\lambda^5\left(t-\frac{1}{t}\right)^5+10\lambda^3\left(t-\frac{1}{t}\right)^3+20\lambda\left(t-\frac{1}{t}\right)=4$$

即

$$\lambda^5\left(t^5-\frac{1}{t^5}\right)-5\lambda^2(\lambda^2-2)\left(t^2-\frac{1}{t^2}\right)+$$

$$10(\lambda^2-1)(\lambda^2-2)\left(t-\frac{1}{t}\right)=\frac{4}{\lambda}$$

令 $\lambda=\sqrt{2}$,则

$$t^5-\frac{1}{t^5}=\frac{1}{\sqrt{2}}=\sqrt{2}-\frac{1}{\sqrt{2}}$$

由单调性可知

$$t^5=\sqrt{2}$$

从而

$$t=\sqrt[10]{2}\,\mathrm{e}^{\mathrm{i}\frac{2k\pi}{5}}\quad(k=0,1,2,3,4)$$

因此

$$x=\sqrt{2}\left(t-\frac{1}{t}\right)=\sqrt[5]{8}\,\mathrm{e}^{\mathrm{i}\frac{2k\pi}{5}}-\sqrt[5]{4}\,\mathrm{e}^{-\mathrm{i}\frac{2k\pi}{5}}\quad(k=0,1,2,3,4)$$

解法 3 首先介绍一类特殊的一元五次方程

$$x^5+px^3+\frac{p^2}{5}x+q=0$$

的解法.

令 $x=u+v$,则

$$(u+v)^5+p(u+v)^3+\frac{p^2}{5}(u+v)+q=0$$

而

$$(u+v)^5=u^5+v^5+5uv(u^3+v^3)+10u^2v^2(u+v)=$$
$$u^5+v^5+5uv(u+v)^3-5u^2v^2(u+v)$$

故
$$u^5+v^5+(5uv+p)(u+v)^3-\left(5u^2v^2-\frac{p^2}{5}\right)(u+v)+q=0$$

令 $uv=-\dfrac{p}{5}$，则
$$u^5+v^5+q=0$$

由
$$\begin{cases} uv=-\dfrac{p}{5} \\ u^5+v^5=-q \end{cases}$$

可解得 u,v.

回到原题，$p=10$，$q=-4$，有
$$\begin{cases} uv=-2 \\ u^5+v^5=4 \end{cases}$$

解得
$$\begin{cases} u_1=\sqrt[5]{4}\,\mathrm{e}^{\mathrm{i}\frac{\pi}{5}} \\ v_1=\sqrt[5]{8}\,\mathrm{e}^{\mathrm{i}\frac{4\pi}{5}} \end{cases}$$

$$\begin{cases} u_2=\sqrt[5]{4}\,\mathrm{e}^{\mathrm{i}\frac{3\pi}{5}} \\ v_2=\sqrt[5]{8}\,\mathrm{e}^{\mathrm{i}\frac{2\pi}{5}} \end{cases}$$

$$\begin{cases} u_3=\sqrt[5]{4}\,\mathrm{e}^{\mathrm{i}\pi}=-\sqrt[5]{4} \\ v_3=\sqrt[5]{8}\,\mathrm{e}^{\mathrm{i}0}=\sqrt[5]{8} \end{cases}$$

$$\begin{cases} u_4=\sqrt[5]{4}\,\mathrm{e}^{\mathrm{i}\frac{7\pi}{5}} \\ v_4=\sqrt[5]{8}\,\mathrm{e}^{\mathrm{i}\frac{8\pi}{5}} \end{cases}$$

$$\begin{cases} u_5=\sqrt[5]{4}\,\mathrm{e}^{\mathrm{i}\frac{9\pi}{5}} \\ v_5=\sqrt[5]{8}\,\mathrm{e}^{\mathrm{i}\frac{6\pi}{5}} \end{cases}$$

Tschebyscheff 逼近定理

因此原方程的五个根为
$$x_i = u_i + v_i$$
其中 $i=1,2,\cdots,5$.

解法 4 注意到满足 $f(2\sqrt{2}\sinh t) = 8\sqrt{2}\cosh 5t$ 的多项式是
$$f(x) = x^5 + 10x^3 + 20x$$
即
$$8\sqrt{2} \cdot \frac{e^{5t} - e^{-5t}}{2} = 4$$

所以 $e^{5t} = \sqrt{2}$ 或 $-\frac{\sqrt{2}}{2}$.

于是 $e^t = \sqrt[10]{2}\, e^{\frac{2k\pi}{5}i}$ 或 $-\frac{1}{\sqrt[10]{2}} e^{\frac{2k\pi}{5}i}$（其中 $i=0,1,2,3,4$），

故 $x = \sqrt{2}(e^t - e^{-t}) = \sqrt[5]{8}\, e^{\frac{2k\pi}{5}i} - \sqrt[5]{4}\, e^{-\frac{2k\pi}{5}i}$（其中 $k=0,1,2,3,4$）.

评注 本题的背景是 Tschebyscheff（切比雪夫）多项式，若不熟悉 Tschebyscheff 多项式，要做出此题，难度异常之大.

例 2 对于 $-1 \leqslant x \leqslant 1$，令
$$T_n(x) = \frac{1}{2^n}\left[(x+\sqrt{1-x^2}\,i)^n + (x-\sqrt{1-x^2}\,i)^n\right] \quad (n \in \mathbf{N})$$

(1) 求证：对于 $-1 \leqslant x \leqslant 1$，$T_n(x)$ 是 x 的首项系数为 1 的 n 次多项式，且 $T_n(x)$ 的最大值是 $\frac{1}{2^{n-1}}$.

(2) 假设 $p(x) = x^n + a_{n-1}x^{n-1} + \cdots + a_1 x + a_0$ 是首项系数为 1 的实系数多项式，使得对在 $-1 \leqslant x \leqslant 1$ 内的所有 x，$p(x) > -\frac{1}{2^{n-1}}$. 求证：$[-1,1]$ 内存在 x^*，使得 $p(x^*) \geqslant \frac{1}{2^{n-1}}$.（1994 年中国台北数学奥林匹克试题）

证明 （1）令
$$T^*(x) = 2^{n-1}T_n(x) \qquad ①$$
于是，有
$$T_1^*(x) = x, \quad T_2^*(x) = 2x^2 - 1 \qquad ②$$
$$2xT_{n-1}^*(x) - T_{n-2}^*(x) =$$
$$2^{n-1}xT_{n-1}(x) - 2^{n-3}T_{n-2}(x) =$$
$$x[(x+\sqrt{1-x^2}\,\mathrm{i})^{n-1} + (x-\sqrt{1-x^2}\,\mathrm{i})^{n-1}] -$$
$$\frac{1}{2}[(x+\sqrt{1-x^2}\,\mathrm{i})^{n-2} + (x-\sqrt{1-x^2}\,\mathrm{i})^{n-2}] =$$
$$(x+\sqrt{1-x^2}\,\mathrm{i})^{n-2}[x(x+\sqrt{1-x^2}\,\mathrm{i}) - \frac{1}{2}] +$$
$$(x-\sqrt{1-x^2}\,\mathrm{i})^{n-2}[x(x-\sqrt{1-x^2}\,\mathrm{i}) - \frac{1}{2}] \qquad ③$$

这里 $n \in \mathbf{N}, n \geq 3$.

由于
$$\frac{1}{2}(x+\sqrt{1-x^2}\,\mathrm{i})^2 =$$
$$\frac{1}{2}(x^2 + 2x\sqrt{1-x^2}\,\mathrm{i} - (1-x^2)) =$$
$$x(x+\sqrt{1-x^2}\,\mathrm{i}) - \frac{1}{2} \qquad ④$$
$$\frac{1}{2}(x-\sqrt{1-x^2}\,\mathrm{i})^2 = x(x-\sqrt{1-x^2}\,\mathrm{i}) - \frac{1}{2} \qquad ⑤$$

将式④，⑤代入式③，有
$$2xT_{n-1}^*(x) - T_{n-2}^*(x) = \frac{1}{2}[(x+\sqrt{1+x^2}\,\mathrm{i})^n +$$
$$(x-\sqrt{1-x^2}\,\mathrm{i})^n] =$$
$$2^{n-1}T_n(x) = T_n^*(x) \qquad ⑥$$

由式⑥知 $T_n^*(x)$ 恰是 n 次 Tschebyscheff 多项式，

Tschebyscheff 逼近定理

首项系数是 2^{n-1},由 Tschebyscheff 多项式的性质知,对任意 $x\in[-1,1]$,有

$$|T_n^*(x)|\leqslant 1 \Rightarrow |T_n(x)|=\frac{1}{2^{n-1}}|T_n^*(x)|\leqslant \frac{1}{2^{n-1}}$$

故 $\max\limits_{x\in[-1,1]}|T_n(x)|=\frac{1}{2^{n-1}}$.

(2) 由于 $T_n(x)(-1\leqslant x\leqslant 1)$ 是 x 的首项系数为 1 的 n 次多项式,则对 $[-1,1]$ 上任一个首项系数为 1 的 n 次实系数多项式 $p(x)$,一定存在 n 个实数 $b_1,b_2,\cdots,b_n$,使得对任意 $x\in[-1,1]$,有

$$p(x)=T_n(x)+b_1T_{n-1}(x)+\cdots+b_{n-1}T_1(x)+b_n \qquad ⑦$$

取

$$x_k=\cos\frac{2k+1}{n}\pi \quad (k=0,1,2,\cdots,n-1) \qquad ⑧$$

对于 $j=1,2,\cdots,n$,由 Tschebyscheff 多项式的性质可知

$$T_j(x_k)=\frac{1}{2^{j-1}}\cos\frac{(2k+1)j}{n}\pi \qquad ⑨$$

则

$$\sum_{k=0}^{n-1}p(x_k)=\sum_{k=0}^{n-1}T_n(x_k)+b_1\sum_{k=0}^{n-1}T_{n-1}(x_k)+\cdots+b_{n-1}\sum_{k=0}^{n-1}T_1(x_k)+nb_n$$

注意到

$$\sum_{k=0}^{n-1}T_j(x_k)=\frac{1}{2^{j-1}}\sum_{k=0}^{n-1}\cos\frac{(2k+1)j}{n}\pi=$$

$$\frac{1}{2^j\sin\frac{j\pi}{n}}\sum_{k=0}^{n-1}\left[\sin\frac{2(k+1)j}{n}\pi-\sin\frac{2kj}{n}\pi\right]=$$

$$\frac{1}{2^j \sin\frac{j\pi}{n}} \sin 2j\pi = 0 \qquad ⑩$$

对 $j=1,2,\cdots,n-1$ 都成立;对 $j=n$,有

$$\sum_{k=0}^{n-1} T_n(x_k) = \frac{1}{2^{n-1}} \sum_{k=0}^{n-1} \cos(2k+1)\pi = -\frac{n}{2^{n-1}} \qquad ⑪$$

故

$$\sum_{k=0}^{n-1} p(x_k) = -\frac{n}{2^{n-1}} + nb_n \qquad ⑫$$

由已知及式 ⑧ 可知

$$p(x_k) > -\frac{1}{2^{n-1}} \Rightarrow \sum_{k=0}^{n-1} p(x_k) > -\frac{n}{2^{n-1}} \qquad ⑬$$

由式⑫,⑬ $\Rightarrow b_n > 0$. 再令

$$x_k^* = \cos\frac{2k\pi}{n} \quad (k=0,1,2,\cdots,n-1) \qquad ⑭$$

类似的,有

$$\sum_{k=0}^{n-1} p(x_k^*) = \sum_{k=0}^{n-1} T_n(x_k^*) + b_1 \sum_{k=0}^{n-1} T_{n-1}(x_k^*) + \cdots + \\ b_{n-1} \sum_{k=0}^{n-1} T_1(x_k^*) + nb_n \qquad ⑮$$

对于 $j=1,2,\cdots,n-1$,有

$$\sum_{k=0}^{n-1} T_j(x_k^*) = \frac{1}{2^{j-1}} \sum_{k=0}^{n-1} \cos\frac{2kj}{n}\pi =$$

$$\frac{1}{2^j \sin\frac{j\pi}{n}} \cdot \sum_{k=0}^{n-1} \Big[\sin\frac{(2k+1)j}{n}\pi -$$

$$\sin\frac{(2k-1)j}{n}\pi\Big] =$$

$$\frac{1}{2^j \sin\frac{j\pi}{n}} \Big[\sin\frac{(2k-1)j}{n}\pi + \sin\frac{j\pi}{n}\Big] = 0 \quad ⑯$$

Tschebyscheff 逼近定理

利用式⑭,有

$$T_n(x_k^*) = \frac{1}{2^{n-1}}\cos(2k\pi) = \frac{1}{2^{n-1}} \qquad ⑰$$

将式⑯,⑰代入式⑮中,得

$$\sum_{k=0}^{n-1} p(x_k^*) = \frac{n}{2^{n-1}} + nb_n > \frac{n}{2^{n-1}}$$

故 n 个实数 $p(x_0^*), p(x_1^*), \cdots, p(x_{n-1}^*)$ 中必有一数 $p(x_l^*)$,满足

$$p(x_l^*) > \frac{1}{2^{n-1}}$$

注意到式⑭显然有 $|x_l^*| = \left|\cos\dfrac{2l\pi}{n}\right| \leq 1.$

Tschebyscheff 小传

第1章

对于像 П. Л. Tschebyscheff 这样世界级的数学大师,自然会有许多传记材料,为了保证材料准确,我们选择两个反映 Tschebyscheff 不同侧面的传记文章,一篇是陈惟勇先生译自前苏联 Г·И·林科夫所著《中学数学的课外活动》一书. 它侧重于介绍 Tschebyscheff 的生平及贡献.

П. Л. Tschebyscheff 在 1821 年 5 月 26 日生于卡鲁士斯基港的欧卡托夫村. 他被人们认为是 19 世纪最伟大的数学家和应用力学家之一,以及俄国自立的数学学校和俄国机械科学的奠基人.

Tschebyscheff 逼近定理

入莫斯科大学前的初等教育,他是在家里受到的. 教育家,一本著名课本的作者伯格烈利斯基,是 Tschebyscheff 的教师之一.

关于伯格烈利斯基的《代数学》,Tschebyscheff 曾说:"它以自己的简明性而言是俄国书籍中最好的一本."

Tschebyscheff 在童年时就对机械和机械玩具产生了兴趣,并喜欢亲手制作它们.

在几何课堂上,他注意了这门科学跟他的玩具之间的联系,并且热心地从事这种联系的研究.

Tschebyscheff 于 1837 年入大学. 在从一年级升入二年级的时候,他写出了"论方程的根之计算",为此他获得了银质奖章. Tschebyscheff 以自己的热心和成就,使著名的大学教授 Н·Д·布拉什曼的注意力转向了自己,布拉什曼教授开始帮助 Tschebyscheff 研究,劝导他将自己献给数学. Tschebyscheff 对自己的老师,永远怀着最深切的敬意.

大学毕业后又过了五年,Tschebyscheff 在莫斯科大学以"概率论的初等分析的经验"论文被授予了硕士学位.

过了三年,他在彼得堡大学以"同余式论"论文被授予了博士学位.

1853 年,Tschebyscheff 被选为研究生,且在 1859 年被选为科学院的普通院士.

Tschebyscheff 的教授生涯大部分是在彼得堡大学里度过的,他在这里度过了 35 年——从 1847 年到 1882 年. 1882 年,Tschebyscheff 停止了讲课,全力投入到科学研究工作中.

为了参观各种工厂和研究实用力学,在 1852 年,Tschebyscheff 曾出差到国外去,他同欧洲各国的一些著名的学者们交谈,出席了一些学术会议.

从 1856 年到 1873 年,Tschebyscheff 是学校总局学术委员会(后更名为人民教育部学术委员会)的会员.

Tschebyscheff 在数学和应用方面卓越的成就给他带来了作为世界上最伟大的学者之一的荣誉.他被数个外国科学院:法国的、意大利的以及英国的皇家协会等选为院士.他还是俄国的以及外国的许多大学学术团体的名誉会员.在费拉捷里费亚和契卡果的两次国际展览会上,Tschebyscheff 以自己的机械模型获得了奖状.

Tschebyscheff 在 1894 年 12 月 8 日在彼得堡逝世,被埋葬在欧卡托夫村,享年 73 岁.

Tschebyscheff 的著作,是经 A·A·马尔柯夫和 H·Я·索宁两位院士编纂后由科学院出版的,它分为两卷且印成两种文字:俄文和法文.

Tschebyscheff 的科学兴趣所及,几乎包括了数学的一切主要分支.

他对数论和概率论方面进行了研究,他解决了积分学上的许多复杂问题,他研究了用连分数的插补问题,特别是用最小二乘法研究了插补问题.

Tschebyscheff 的数学研究不仅在成就上是出色的,而且那些用来获得这些成就的新颖方法的深奥和优美也是出色的.

Tschebyscheff 逼近定理

力求以最少限度的一些最简单的方法获得一些成就,这也是 Tschebyscheff 的特点.

Tschebyscheff 在数论中的一些主要成就,是关于质数在自然数列中的分布问题. 这是一个直到今天尚未彻底解决的问题,它有好几个世纪之久了. 很早,欧几里得就证明了:质数的集合是无限的. 后来经验证得出:质数在自然数列中的分布是不均匀的. 例如,在第一个一百万中质数的个数是 78 499,而在第二个一百万中质数的个数是 70 433,但是质数分布的精确定律的建立却未成功.

贝特朗提出了一个命题:在两个自然数 x 和 $2x-2(x>3)$ 之间含有一个质数.

法国数学家列日安得尔以及其他的一些数学家们在经验所知的基础上,试图建立表示不超过一已知数 x 的质数个数的一些近似公式.

Tschebyscheff 证明了贝特朗命题的正确性,从而建立了关于质数在整个自然数列中的分布定律的第一个意义重大的命题.

Tschebyscheff 进一步证明了:若 $\pi(x)$ 是表示小于 x 的质数个数的函数,则 $0.921\ 29 \leqslant \dfrac{\pi(x)}{\dfrac{x}{\ln x}} \leqslant 1.105\ 55$,

并且,若当 $x \to +\infty$ 时比式 $\dfrac{\pi(x)}{\dfrac{x}{\ln x}}$ 的极限存在,则这个极限应等于 1. 因而,对于充分大的 x 值,有近似等式 $\pi(x) = \dfrac{x}{\ln x}.$

这些著作——С·Н·别尔什金院士指出——一

第 1 章　Tschebyscheff 小传

下子就把 Tschebyscheff 在全世界人们的面前列入了现代最伟大的数学家的行列中,而这是很自然的:在 Tschebyscheff 建立了科学基础并获得了最初的一切巨大成果以后,罗巴切夫斯基就解决了几个世纪以来的平行问题,它①是如此的古老,或许比质数问题还要难些.

Tschebyscheff 在概率论方面的研究,首先是关于它的基本原理——大数法则与对于独立事件的量之和的极限定理.

在相当广泛的一些假定之上的这些定理的准确公式和严谨的数学证明,都是 Tschebyscheff 做的.

Tschebyscheff 发展了最良近似函数理论和正交多项式(Tschebyscheff 多项式)理论,后者在近代数学分析以及其邻近的各科中起着重要的作用. 在 Tschebyscheff 的一本数学分析的著作中,他阐述的无理函数积分的研究也占据着一个主要的位置. 在这里他大大地发展了在他以前由挪威数学家阿贝尔所获得的一些结果.

Tschebyscheff 的求问于实践而使其满足的方针,是他的科学著作的一个最大的特点.

建立一种为适于最有效地利用自然力的技术,发展一些为保障达到此目的的数学方法——这就是 Tschebyscheff 的基本思想,他在生活中坚持不懈地贯彻这种思想,并在自己的一些言论中强调它.

Tschebyscheff 说过:"理论跟实践的接近产生了一

---

① 指平行问题.

Tschebyscheff 逼近定理

些最良好的结果,且不唯独是实践,因此有所收获.而科学本身在它的影响下发展起来,它给科学发现了一些新的研究对象或早已知道的事物中的一些新方面.因此一种理论从旧理论的新应用中或从旧理论的新发展中获得很多."

Tschebyscheff 从事了所谓机械平行四边形理论的研究,它是专供蒸汽转动机用的. 燃料的节约以及机器的坚固都是以这些机械的合理制造为转移. 正是这些研究,将 Tschebyscheff 引向了一个新的、极其有用的数学理论——与零成最小规避的函数理论. Tschebyscheff 在机械学上的研究是有其具体的体现的:他制作了许多的机械模型. 这些模型被保存在苏联科学院里.

人们公正地认为, Tschebyscheff 是俄国数学的导师,俄国自立的数学学校①的创建者. 对于这些,俄国科学界不仅应当感谢 Tschebyscheff 的天才,而且也应当感谢他的惊人的教育才能和智慧. Tschebyscheff 的一个令人永记难忘的事迹是:他在一些学术谈话中,以自己的著作和指示将自己的学生们引向探求本质的一些有用的问题上,并使他们注意这样的一些问题,这对他们的研究总会或多或少地得出一些重要的结果.

例如,他曾向自己最亲近的一个学生李雅普诺夫(后来是科学院院士)提出了一个关系着绕轴旋转的同类液体质量形态研究的问题. 李雅普诺夫在这个问题上研究了整个一生,并获得了一些卓越的成就,这些成就大大地超过了在这方面由著名的法国数学家 A·

---

① 原文并无"数学"一词,而与前文相对照可看出是应该有的.

第 1 章　Tschebyscheff 小传

普安卡尔所做出的一切.

　　Tschebyscheff 关于各种数学问题的讲演,都利用了通俗、生动而又诱人的叙述,这是他讲演的特色,它们是穿插着许多有趣的见解,即关于这些或另一些问题,或一些科学方法的意义和重要性.

　　1944 年 12 月,苏联人民委员会发布了一项《关于纪念科学院院士 П · Л · Tschebyscheff 逝世 50 周年的措施》的决议.

　　前苏联人民委员会决定:

　　(1)同意苏联科学院关于出版科学院院士 П · Л · Tschebyscheff 的全集的决议.

　　(2)以科学院院士 П · Л · Tschebyscheff 命名设立两项为数 20 000 卢布的奖金,它们是每三年奖给在数学方面以及在机械学和机器理论方面有优秀著作的人.

　　(3)以科学院院士 П · Л · Tschebyscheff 命名在莫斯科大学和列宁格勒大学为研究生和候补博士们设立的若干奖学金,而在苏联科学院数学研究所里是以斯切科洛夫命名的.

　　(4)在列宁格勒大学的校舍前立起一尊 Tschebyscheff 的半身铜像.

　　另外一篇是在《Tschebyscheff 全集》(第一卷)数论部分中前苏联著名数学家维诺格拉朵夫为 Tschebyscheff 写的一个小传,以数学家的视角对 Tschebyscheff 进行了评介:

　　Tschebyscheff 于 1821 年 5 月 26 日(俄历 5 月 14 日)诞生于卡鲁士斯基港的欧卡托夫村(自己)父亲的

15

Tschebyscheff 逼近定理

领地上. Tschebyscheff 的家庭是属于古老的贵族之一,他父亲在那时是个有教养并富有的人.

Tschebyscheff 初期所受的教育——即入大学前的教育——是在家中获得的. 他的母亲奥格拉芬娜·伊万诺芙娜教他读书认字,其他别的几门功课,包括算术和法文,则是她的堂姊妹苏哈列娃——一位颇有教养而且对 Tschebyscheff 起着重作用的人教授的.

1832 年,为了准备送 Tschebyscheff 和他的哥哥上大学,Tschebyscheff 举家迁居莫斯科. 为了达到这个目的,家里还给他们聘请了莫斯科卓越的教师,其中有当时著名的教育家、数学教师伯格烈利斯基. 关于他的《代数学》一书,Tschebyscheff 后来谈起过,认为这本书是那时俄文版本中最好的一本,因为它阐述得最简明. 在这个时期,Tschebyscheff 对数学的才华已经显示出来了. 因而,他后来给自己选择了数学物理系.

Tschebyscheff 于 1837 年考入了莫斯科大学. 当他升到二年级的时候,他已经完成了"论方程的根之计算"一文的写作,这篇论文使他获得了银质奖章. 他对功课孜孜不倦,认真钻研的精神引起了卓有声望的布拉什曼教授的注意. 教授已经预见到自己的这位新学生未来在数学上的光辉前途,因而特意地指导他的学业,特别地谆谆善诱他把自己贡献给数学. 对于这位学者,Tschebyscheff 始终保持着最深切的敬意. 他把很晚才得到的布拉什曼的照片当作圣物似的一直保藏到他去世. 1841 年 Tschebyscheff 念完了大学的课程.

1840 年,俄国大部分地区闹饥荒,许多地主,包括 Tschebyscheff 的双亲在内都遭到了破产,他们不得不

阖家迁往乡村作久居之计,这时 Tschebyscheff 的父亲已经无力帮助儿子. 为了不受贫困的威胁,这个刚出校门的青年人看来必须去工作或者做个诲人不倦的教书匠,但是,由于他正确地考虑到这样做将会使他离开心爱的数学研究,因此他宁可忍受贫困的生活也不愿撇下自己的事业. 在这个时候,他只分到父亲在柏列兹斯清卡房产的一所住宅,他就在那儿安居下来,同时还和两个弟弟及另外两个青年同住. 这两个青年和他的弟弟一起准备投考学校. Tschebyscheff 还抽出一些时间为他们试教数学,但不久就停止了. 据 Tschebyscheff 自己承认,他是一个没有"涵养"的教育家,因为他常常生气和大声叱责自己的学生.

这段时期,Tschebyscheff 的第一批学术著作出版了. 1846 年,他的"概率论的初等分析的经验"学位论文也写完了.

Tschebyscheff 在他的"对数积分论"学位论文答辩后,获得了彼得堡大学教授助理的职位,接替离职的安库多维奇教授的位置. 所以,他于 1847 年迁居到彼得堡. 这篇论文的内容出现在 Tschebyscheff 以后的著作里,这篇论文的手稿连同他的当选词的手稿还保存下来,其中的原理部分已刊印出来了. 这篇论文使 Tschebyscheff 获得在大学讲课的权利, 1849 年, Tschebyscheff 因著名的学位论文"同余式论"而荣获博士学位.

与 Tschebyscheff 几乎同时,以常任教授身份进入彼得堡大学的布尼雅柯夫斯基是俄国最有名的数学家之一. 那时他已经是科学院的常任院士了. 另外一位是

Tschebyscheff 逼近定理

应用数学教研室的非常任教授,当时极为著名的数学家沙莫夫. 这两位学术昭著、品德高尚的杰出的学者是 Tschebyscheff 迁居彼得堡后最先接触并且意气相投的良友. 他跟他们之间的友谊一直保持到他俩去世为止. 他跟布尼雅柯夫斯基尤为亲密,布尼雅柯夫斯基赏识到 Tschebyscheff 巨大的科学潜力,很快就把他引进科学院开始作为自己出版欧拉名著的助手,Tschebyscheff 随后升为正式会员(1853 年,Tschebyscheff 被选为助理员,1859 年已经是科学院的常任院士了).

Tschebyscheff 居住彼得堡后的物质生活极为恶劣. 这个时期他家里破产得更厉害,唯一维持生活的就是靠他那份不多的教授助理的薪金.

贫乏的物质条件使他成为一个俭朴自持的人. 他从不吝惜金钱的唯一的东西就是制造他所发明的机械模型;为了制造模型,他曾花了数以千百计的卢布. 从童年时代起 Tschebyscheff 就喜爱制造各种各样的仪器:从最初用小摺刀做一些玩具开始,一直发展到后来构造复杂的计算机(该机保存在巴黎技术保存所). 他所设计的许多仪器现在还保存在圣彼得堡大学和科学院里.

Tschebyscheff 所结交的朋友为数不多. 他多半爱到布尼雅柯夫斯基的家里去,那儿常有许多数学家聚谈,著名的数学家奥斯特罗格拉茨基也是其中之一.

在青年时代,Tschebyscheff 常常爱到沙莫夫那儿去叙谈自己的发现,借助这位前辈的渊博知识,他了解这个发现从前是否已经有人完成,听从院士的指示,结果有时也正与事实相符合. 当然,这是对 Tschebyscheff

第 1 章 Tschebyscheff 小传

尚未开始研究的那一类在他以前还没有人涉及过的数学问题的活动时期而言,这一类问题是不可能遭遇到有人抢先完成的危险的. 必须指出,比起研究其他数学家的著作,特别是当代的著作,Tschebyscheff 更为喜爱自己独立钻研. 他深入地研究欧拉、拉格朗日、高斯、阿贝尔等伟大数学家的名著. Tschebyscheff 对于研读当前的数学著作并不予以特殊的重视,他认为过多的钻研别人的作品会对自己工作的独立性产生不良的影响.

到国外旅行是 Tschebyscheff 喜爱的娱乐之一. 在学术活动初期,Tschebyscheff 并不把这些旅行用于休养,而是用之于研究重大的科学. 在他 1852 年的出国报告书中可以找到叙述他的旅行和以研究实用力学为目的的对各个工厂的访问;他常出席学术界的会议,与各国著名的学者交谈科学上的各种问题. 到了晚年,Tschebyscheff 虽然仍不忽视自己探求科学的旅行目的,但由于他的学识不断的增长而把旅行用之于休息了. 他最爱到法国去,那里有许多他熟悉的科学家. 他常列席国际学术会议,并在会上报告自己的发现. Tschebyscheff 最初几次访问巴黎时,由于他那种俭朴的特性使然,他常下榻于最简陋的旅馆,在最经济的饭店用餐,出入只乘坐公共马车. 这样一直到了晚年,他才稍稍放纵一下自己的习惯,竟也设宴招待他的法国朋友们. 当 Tschebyscheff 在国内度假的时候,他常常喜欢住在烈维尔附近的耶卡切林涅达尔.

从 1847 年到 1882 年,Tschebyscheff 的全部学术活动满 35 年(1847~1853 年为教授助理,1853~1857

Tschebyscheff 逼近定理

年为非常任教授,1857 年起为常任教授). 如果在亚历山大洛夫国立中学短期任教力学的时间不算在内的话,Tschebyscheff 的全部活动毫无例外是在彼得堡大学度过的. 他在各个时期里讲授过解析几何、高等代数、数论、积分学、概率论、有限差分法、椭圆函数论和定积分论等多门课程.

Tschebyscheff 对待自己的教学工作是严谨的, 他几乎从来没有缺过课, 也没有迟到过, 但下课铃响后他也从不在课室里多留一分钟, 虽然这样有时会使他的讲解中断. 对于课堂上尚未得出结论的课题, 只要所讲的并不是紧接着前一课的话, 他常常在下一次从头再讲起. 对于所有较为复杂的算题, 他总是先阐明它的目的和解决的一般步骤, 然后才把过程在黑板上默默地计算出来. 同时他要求学生用眼睛去看, 而不是用耳朵听. 他的演算极为迅速详尽, 使学生看起来很容易接受. 在讲授的过程中, Tschebyscheff 常系统地阐述本课程, 说出自己的观点和与此课程有关的别的数学家交谈的意见, 然后把这些论点加以比较, 指出数学中各个不同问题之间的相互联系. 这种论述使他的讲授生色不少, 使注意力紧张的听众得到休息, 引起他们在更为广泛的范围内去研究该门科学的兴趣.

按范围而论, Tschebyscheff 所任教的课程是不多的, 但是这些课程内容丰富, 易于使人接受和理解. 对于考试, Tschebyscheff 既不要求过严, 也不失之过宽, 他的态度是极为谨慎而客气的. 在学位论文辩论会上他所提出的相反意见常伴随着巧妙与机智, 因为它所牵涉到的并不是该文的细枝末节, 而往往是与该文论

第 1 章　Tschebyscheff 小传

旨有关的一般问题.

　　Tschebyscheff 作为教授的功绩永远保留在有幸获得他教诲的人们的记忆里. 他常把自己的学生一直教到他们大学毕业为止.

　　Tschebyscheff 度过了自己学术活动的 25 及 50 周年纪念日,但他自己没有一次庆祝过. 他的门徒和崇敬他的人当时曾打算用社会上最流行的方式为他祝贺,但都遭到了他坚决地拒绝.

　　在 Tschebyscheff 临终前的几天里,大家只知道他染上了轻微的流行性感冒. 虽然他自己感到十分不适,但仍不愿躺下. 去世前夕,他还照常接见宾客,那时候谁也没有想到他的终期已经这么近了. 1894 年 12 月 8 日(即俄历 11 月 26 日)早晨,Tschebyscheff 坐在书桌旁进茶的当儿,他突然感到眩晕,经过短时的弥留状态之后,即因心脏麻痹而逝世.

什么是逼近

第 2 章

先看一个例子.

笔者为浙江省桐乡高级中学教师张晓东.

例 3 2015 年 1 月的浙江省普通高中学业水平考试中数学卷的压轴题最后一问,学生觉得很新、很难,那么我们以前是否碰到过类似问题以及这类问题到底该如何解决? 我们来看题目:

(2015 年 1 月浙江省普通高中学业水平考试数学试题第 34 题)设函数 $f(x)=|\sqrt{x}-ax-b|, a,b \in \mathbf{R}$.

(Ⅰ)当 $a=0, b=1$ 时,写出函数 $f(x)$ 的单调区间;

第 2 章 什么是逼近

（Ⅱ）当 $a=\dfrac{1}{2}$ 时,记函数 $f(x)$ 在 $[0,4]$ 上的最大值为 $g(b)$,在 b 变化时,求 $g(b)$ 的最小值;

（Ⅲ）若对任意实数 a,b,总存在实数 $x_0 \in [0,4]$,使得不等式 $f(x_0) \geqslant m$ 成立,求实数 m 的取值范围.

本书只考虑第（Ⅲ）小题,下面是命题者提供的参考答案:

设 $H(t)=|at^2-t+b|$,$t\in[0,2]$,原命题等价于:对任意实数 a,b,$H(t)_{\max} \geqslant m$.

记函数 $H(t)$ 在 $[0,2]$ 上的最大值为 $G(b)$,只要 $G(b)_{\min} \geqslant m$.

（1）当 $a=0$ 时,$G(b)=\max\{H(0),H(2)\}=\max\{|b|,|b-2|\}$,此时,$G(b)$ 的最小值为 1,所以 $m\leqslant 1$;

（2）当 $a<0$ 时,$G(b)=\max\{H(0),H(2)\}=\max\{|b|,|b+4a-2|\}$,此时,$G(b)_{\min}=1-2a>1$,所以 $m\leqslant 1$;

（3）当 $\dfrac{1}{2a}\geqslant 2$,即 $0<a\leqslant\dfrac{1}{4}$ 时,$G(b)=\max\{H(0),H(2)\}=\max\{|b|,|b+4a-2|\}$,此时,$G(b)_{\min}=1-2a\geqslant\dfrac{1}{2}$,所以 $m\leqslant\dfrac{1}{2}$;

（4）当 $1\leqslant\dfrac{1}{2a}<2$,即 $\dfrac{1}{4}<a\leqslant\dfrac{1}{2}$ 时,$G(b)=\max\{H(0),H(\dfrac{1}{2a})\}=\max\{|b|,|b-\dfrac{1}{4a}|\}$,此时,$G(b)_{\min}=\dfrac{1}{8a}\geqslant\dfrac{1}{4}\geqslant 1$,所以 $m\leqslant\dfrac{1}{4}$;

Tschebyscheff 逼近定理

(5) 当 $0 < \frac{1}{2a} < 1$,即 $a > \frac{1}{2}$ 时,$G(b) = \max\{H(2),$ $H(\frac{1}{2a})\} = \max\{|b+4a-2|,|b-\frac{1}{4a}|\}$,此时,$G(b)_{\min} = \frac{1}{8a}+2a-1$,而 $\frac{1}{8a}+2a-1$ 在 $(\frac{1}{2},+\infty)$ 上递增,所以 $G(b)_{\min} \geq \frac{1}{4}$,于是 $m \leq \frac{1}{4}$.

综上,实数 m 的取值范围为 $(-\infty,\frac{1}{4}]$.

这道题是怎么命制出来的呢？我们认为,这道试题是由下面的问题 1 改编而成.

问题 1 $f(x) = |x^2 - ax - b|(x \in [0,2])$,设 $f(x)_{\max} = M$,求 M 的最小值.

由于 a,b 是任意实数,所以它等价于:

问题 2 $f(x) = |x^2 + ax + b|(x \in [-1,1])$,设 $f(x)_{\max} = M$,求 M 的最小值.

而对于定义在 $[-1,1]$,首项系数为 1 的一元 n 次多项式函数,我们有熟悉的结论:

$f(x) = x^n + a_{n-1}x^{n-1} + \cdots + a_1 x + a_0, a_0, a_1, \cdots, a_{n-1} \in \mathbf{R}$,若 $|f(x)|$ 在 $x \in [-1,1]$ 上的最大值为 M,则 M 的最小值为 $\frac{1}{2^{n-1}}$,当 $2^{n-1} \cdot f(x)$ 为 Tschebyscheff 多项式时取到最小值.

(令 $\cos x = t \in [-1,1]$,记 $T_0(t) = 1, T_1(t) = t$, $T_2(t) = 2t^2 - 1, \cdots, T_{n+1}(t) = 2t \cdot T_n(t) - T_{n-1}(t), T_n(t)$ 称为 Tschebyscheff 多项式.)

该题的精彩之处在于命题者把 x^2 改成 $\sqrt{x}$ 之后,首项系数为 1 的二次函数便成了一次项系数为 1 的二次函数,于是 Tschebyscheff 多项式不适用.

但并不是说两个问题没有相通之处,有些方法还是可以通用的.

原题等价于:

问题 3 设 $h(x) = ax^2 - x + b, x \in [0,2]$,设 $|h(x)|$ 的最大值为 M,求 $M_{\min}$.

解法一:转化为最大值减最小值.

因为 $|h(x)| \leqslant M$,所以 $h(x)_{\max} - h(x)_{\min} \leqslant 2M$.

(1) 当 $a \leqslant 0$ 时,$h(x)_{\max} = h(0) = b$,$h(x)_{\min} = h(2) = 4a - 2 + b$,所以 $2M \geqslant b - (4a - 2 + b) = 2 - 4a \geqslant 2$,所以 $M \geqslant 1$;

(2) 当 $\dfrac{1}{2a} \geqslant 2$ 时,即 $0 < a \leqslant \dfrac{1}{4}$ 时,$h(x)_{\max} = h(0) = b$,$h(x)_{\min} = h(2) = 4a - 2 + b$,所以 $2M \geqslant b - (4a - 2 + b) = 2 - 4a \geqslant 1$,所以 $M \geqslant \dfrac{1}{2}$;

(3) 当 $1 \leqslant \dfrac{1}{2a} < 2$ 时,即 $\dfrac{1}{4} < a \leqslant \dfrac{1}{2}$ 时,$h(x)_{\max} = h(0) = b$,$h(x)_{\min} = h\left(\dfrac{1}{2a}\right) = b - \dfrac{1}{4a}$,$2M \geqslant b - \left(b - \dfrac{1}{4a}\right) = \dfrac{1}{4a} \geqslant \dfrac{1}{2}$,所以 $M \geqslant \dfrac{1}{4}$,且当 $a = \dfrac{1}{2}, b = \dfrac{1}{4}$ 时,$M = \dfrac{1}{4}$;

(4) 当 $0 < \dfrac{1}{2a} < 1$ 时,即 $a > \dfrac{1}{2}$ 时,$h(x)_{\max} = h(2) = 4a - 2 + b$,$h(x)_{\min} = h\left(\dfrac{1}{2a}\right) = b - \dfrac{1}{4a}$,$2M \geqslant (4a - 2 + b) - \left(b - \dfrac{1}{4a}\right) = 4a + \dfrac{1}{4a} - 2 > \dfrac{1}{2}$,所以 $M > \dfrac{1}{4}$.

综上,$M_{\min} = \dfrac{1}{4}$.

Tschebyscheff 逼近定理

此法通过最大值减最小值消去 b 把二参转化为一参解决问题,比参考答案累次求和简单,而且是两类问题的通法.

解法二:利用绝对值三角不等式.

由 $h(x)=ax^2-x+b,x\in[0,2]$,得 $h(0)=b,h(1)=a-1+b,h(2)=4a-2+b$,消去 a,b 得 $2=3h(0)+h(2)-4h(1)$,于是

$$2=3h(0)+h(2)-4h(1)=$$
$$|3h(0)+h(2)-4h(1)|\leqslant$$
$$|3h(0)|+|h(2)|+4|h(1)|\leqslant 8M$$

所以 $M\geqslant\dfrac{1}{4}$,当 $h(0)=h(2)=\dfrac{1}{4},h(1)=-\dfrac{1}{4}$,即 $a=\dfrac{1}{2},b=\dfrac{1}{4}$ 时,$M=\dfrac{1}{4}$.

所以 $M_{\min}=\dfrac{1}{4}$

此法通过绝对值三角不等式直接求出了 M 的最小值,比解法一分类讨论简捷,也是两类问题的通法.

解法三:利用拉格朗日插值多项式

$$h(x)=\dfrac{(x-1)(x-2)}{(0-1)(0-2)}h(0)+\dfrac{(x-0)(x-2)}{(1-0)(1-2)}h(1)+$$
$$\dfrac{(x-0)(x-1)}{(2-0)(2-1)}h(2)$$

其中 x 的系数为 $-\dfrac{3h(0)}{2}+2h(1)-\dfrac{h(2)}{2}$,所以 $-\dfrac{3h(0)}{2}+2h(1)-\dfrac{h(2)}{2}=-1$,由绝对值三角不等式得 $M_{\min}=\dfrac{1}{4}$.(同解法二)

此法通过多项式恒等系数一一对应得到 $-\dfrac{3h(0)}{2}+2h(1)-\dfrac{h(2)}{2}=-1$,比解法二消去 a,b 更自然,也是两类问题的通法.

最后我们再来分析一下此题的几何背景.

设 $g(x)=\sqrt{x}$ ($x\in[0,4]$),$h(x)=ax+b$ ($x\in[0,4]$),则 $f(x)=|g(x)-h(x)|$,并设 $A(4,2)$.

(1)当 $a=\dfrac{1}{2}$ 时,如图1,先让 $h(x)$ 的图像过 O,A 两点,此时 $g(x)\geqslant h(x)$,且 $g(x)-h(x)=\sqrt{x}-\dfrac{1}{2}x=-\dfrac{1}{2}(\sqrt{x}-1)^2+\dfrac{1}{2}\leqslant\dfrac{1}{2}$,当 $x=1$ 时取最大值. 再将 $h(x)$ 的图像向上平移 $\dfrac{1}{4}$,得 $f(x)_{\max}=\dfrac{1}{4}$.

(2)当 $a>\dfrac{1}{2}$ 时,如图2,先让 $h(x)$ 的图像过点 A,此时 $g(x)\geqslant h(x)$,且 $[g(x)-h(x)]_{\max}\geqslant g(1)-h(1)>\dfrac{1}{2}$. 再将 $h(x)$ 的图像向上平移 $\dfrac{g(1)-h(1)}{2}$,得 $f(x)_{\max}>\dfrac{1}{4}$.

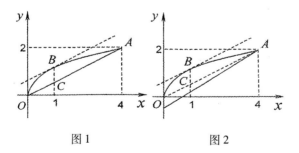

图1　　　　图2

Tschebyscheff 逼近定理

(3) 当 $a < \dfrac{1}{2}$ 时,如图3,先让 $h(x)$ 的图像过点 O,此时 $g(x) \geqslant h(x)$,且 $[g(x)-h(x)]_{\max} \geqslant g(1)-h(1) > \dfrac{1}{2}$.再将 $h(x)$ 的图像向上平移 $\dfrac{g(1)-h(1)}{2}$,得 $f(x)_{\max} > \dfrac{1}{4}$.

综上,$f(x)$ 的最大值 M 的最小值为 $\dfrac{1}{4}$,所以实数 m 的取值范围是 $\left(-\infty, \dfrac{1}{4}\right]$.

基于几何背景的理解,这类问题可以拓展为:

如图4,设 $f(x) = |g(x)-(ax+b)|$,$x \in [m,n]$,其中 $g(x)$ 在区间 (m,n) 上下凸(或下凹),对任意的 a, $b \in \mathbf{R}$,设 $f(x)$ 的最大值为 M,则

$$M_{\min} = \dfrac{|g(x_0)-(Ax_0+B)|}{2}$$

其中 $A = \dfrac{g(n)-g(m)}{n-m}$,$B = \dfrac{ng(m)-mg(n)}{n-m}$,$x_0$ 满足 $g'(x_0) = \dfrac{g(n)-g(m)}{n-m}$.

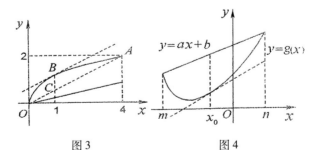

图3　　　　图4

例4　设函数 $f(x) = |\cos x + \alpha\cos 2x + \beta\cos 3x|$,$\alpha$,

第 2 章　什么是逼近

$\beta \in \mathbf{R}$. 求 $M = \min\limits_{\alpha,\beta} \max\limits_{x} f(x)$. （第 49 届莫斯科数学奥林匹克试题，1986 年）

解　显然 $f(\frac{\pi}{6}) = \left|\frac{\sqrt{3}}{2} + \frac{\alpha}{2}\right|, f(\frac{5}{6}\pi) = \left|-\frac{\sqrt{3}}{2} + \frac{\alpha}{2}\right|$.
从而

$$\max f(x) \geq \frac{1}{2}\left(\left|\frac{\sqrt{3}}{2} + \frac{\alpha}{2}\right| + \left|-\frac{\sqrt{3}}{2} + \frac{\alpha}{2}\right|\right) \geq$$

$$\frac{1}{2}\left[\left(\frac{\sqrt{3}}{2} + \frac{\alpha}{2}\right) - \left(-\frac{\sqrt{3}}{2} + \frac{\alpha}{2}\right)\right] = \frac{\sqrt{3}}{2}$$

于是得到

$$M \geq \frac{\sqrt{3}}{2} \qquad ⑱$$

另一方面，令 $\alpha = 0, \beta = -\frac{1}{6}$，则

$$f(x) = \left|\cos x - \frac{1}{6}\cos 3x\right| = \left|\frac{3}{2}\cos x - \frac{2}{3}\cos^3 x\right|$$

易知

$$\max_{x} f(x) = \max_{-1 \leq y \leq 1} |g(y)| = \max_{0 \leq y \leq 1} g(y)$$

其中 $g(y) = \frac{3}{2}y - \frac{2}{3}y^3$，因为

$$g(y) - g(\frac{\sqrt{3}}{2}) = (y - \frac{\sqrt{3}}{2})\left[\frac{3}{2} - \frac{2}{3}(y^2 + \frac{\sqrt{3}}{2}y + \frac{3}{4})\right]$$

所以，当 $0 \leq y \leq \frac{\sqrt{3}}{2}$ 时，由 $y^2 + \frac{\sqrt{3}}{2}y + \frac{3}{4} \leq \frac{9}{4}$，可得

$$g(y) - g(\frac{\sqrt{3}}{2}) \leq 0$$

当 $\frac{\sqrt{3}}{2} \leq y \leq 1$ 时，由 $y^2 + \frac{\sqrt{3}}{2}y + \frac{3}{4} \geq \frac{9}{4}$，可得

Tschebyscheff 逼近定理

$$g(y) - g(\frac{\sqrt{3}}{2}) \leq 0$$

于是得到 $\max\limits_{x} f(x) = \max\limits_{0 \leq y \leq 1} g(y) = g(\frac{\sqrt{3}}{2}) = \frac{\sqrt{3}}{2}$,由此可得

$$M \leq \frac{\sqrt{3}}{2} \qquad ⑲$$

综合式⑱和式⑲可知 $M = \frac{\sqrt{3}}{2}$.

下面我们介绍线性最佳逼近多项式的求法以及 Tschebyscheff 多项式在函数逼近中的一些应用:

(1) 线性最佳逼近多项式 $p_1^*(x)$ 的构造

若 $f(x)$ 在 $[a,b]$ 上连续,则 $f(x)$ 在 $[a,b]$ 上的零次最佳逼近多项式为

$$p_0^*(x) = \frac{1}{2}[\min\limits_{a \leq x \leq b} f(x) + \max\limits_{a \leq x \leq b} f(x)]$$

即函数 $f(x)$ 在区间 $[a,b]$ 上的零次最佳一致逼近多项式是其在该区间上最大值与最小值的算术平均值,这是显然的.

一次最佳逼近多项式 $p_1^*(x) = a_0 + a_1 x$ 的求法:

设 $f(x)$ 在 $[a,b]$ 上具有二阶导数,且 $f''(x)$ 在 $[a,b]$ 上不变号,则根据 Tschebyscheff 定理,存在点 $a \leq x_1 < x_2 < x_3 \leq b$,使

$$f(x_k) - p_1^*(x_k) = (-1)^k \sigma \cdot \max\limits_{a \leq x \leq b} |f(x) - p_1^*(x)|$$

其中 $\sigma = \pm 1, k = 1, 2, 3$. 由于 $f''(x)$ 在 $[a,b]$ 上不变号,则由式⑲可知区间 $[a,b]$ 的两个端点 a, b 都属于 $f(x) - p_1^*(x)$ 的交错点组,即有 $x_1 = a, x_3 = b$. 而另一交错点 x_2 必位于 $[a,b]$ 的内部,且它是函数 $f(x) - p_1^*(x)$ 的极值点. 故

第 2 章 什么是逼近

$$f'(x_2)-p_1^{*\prime}(x_2)=f'(x_2)-a_1=0$$

即 $f'(x_2)=a_1$. 又因为 $f''(x)$ 在 $[a,b]$ 上不变号,所以 $f'(x)$ 是 $[a,b]$ 上的严格单调函数. 从而 $f(x)-p_1^*(x)$ 在 (a,b) 内有且仅有一个极值点 x_2. 由此可得

$$f(a)-p_1^*(a)=-[f(x_2)-p_1^*(x_2)]=f(b)-p_1^*(b)$$

即

$$\begin{cases}f(a)-p_1^*(a)=f(b)-p_1^*(b)\\ f(a)-p_1^*(a)=-[f(x_2)-p_1^*(x_2)]\end{cases}$$

由 $p_1^*(x)=a_0+a_1x$,从上述方程组可解得

$$\begin{cases}a_1=\dfrac{f(b)-f(a)}{b-a}\\ a_0=\dfrac{1}{2}[f(a)+f(x_2)]-\dfrac{a+x_2}{2}\cdot\dfrac{f(b)-f(a)}{b-a}\end{cases}$$

其中 x_2 由 $f'(x_2)=a_1$ 解得. 因此 $f(x)$ 在 $[a,b]$ 上的线性最佳逼近多项式为

$$p_1^*(x)=\dfrac{f(a)+f(x_2)}{2}+\dfrac{f(b)-f(a)}{2}\left(x-\dfrac{a+x_2}{2}\right)$$

线性最佳逼近的几何意义是:直线 $Y=p_1^*(x)$ 一定与过两点 $(a,f(a))$ 与 $(b,f(b))$ 的直线平行.

例 5 求 $f(x)=\arctan x$ 在 $[0,1]$ 上的线性最佳逼近多项式.

解 由于 $f''(x)=\dfrac{-2x}{(1+x^2)^2}<0$,故 $x_1=0,x_3=1$,有

$$a_1=\dfrac{f(b)-f(a)}{b-a}=\dfrac{\arctan 1-\arctan 0}{1-0}=\dfrac{\pi}{4}\approx 0.7854$$

由 $f'(x_2)=\dfrac{1}{1+x_2^2}=\dfrac{\pi}{4}$,得 $x_2=\sqrt{\dfrac{4}{\pi}-1}\approx 0.5227$.

Tschebyscheff 逼近定理

又 $f(x_2) = \arctan 0.52277 \approx 0.48166$,故

$$a_0 = \frac{1}{2}[f(0)+f(x_2)] - a_1\frac{0+x_2}{2} =$$
$$\frac{1}{2}[0.48166 - 0.7854 \times 0.5227] \approx 0.0356$$

故 $f(x) = \arctan x$ 在 $[0,1]$ 上的线性最佳逼近多项式为

$$p_1^*(x) = 0.0356 + 0.7854x$$

(2) Tschebyscheff 多项式的应用

我们知道 Tschebyscheff 多项式 $T_n(x)$ 的最高次项 x^n 的系数为 $2^{n-1}(n=1,2,\cdots)$. 令

$$\widetilde{T}_n(x) = \frac{1}{2^{n-1}}T_n(x)$$

则 $\widetilde{T}_n(x)$ 的最高次项系数为 1. 由 Tschebyscheff 零点与极点性质可知当

$$x_k = \cos\frac{k\pi}{n} \quad (k=0,1,2,\cdots,n)$$

时有

$$\widetilde{T}_n(x_k) = \frac{(-1)^k}{2^{n-1}}$$

用 $\widetilde{P}_n(x)$ 表示最高次项系数为 1 的 n 次多项式的全体,我们有如下的 Tschebyscheff 多项式极性定理.

定理 1 在区间 $[-1,1]$ 上,最高次项系数为 1 的一切 n 次多项式集合 $\widetilde{P}_n(x)$ 中,$\widetilde{T}_n(x) = \frac{1}{2^{n-1}}T_n(x)$ 与零的偏差最小,且其偏差为 $\frac{1}{2^{n-1}}$. 即对于任何 $p(x) \in \widetilde{P}_n(x)$,有

第 2 章　什么是逼近

$$\frac{1}{2^{n-1}} = \max_{-1 \leqslant x \leqslant 1} |\widetilde{T}_n(x) - 0| \leqslant \max_{-1 \leqslant x \leqslant 1} |p(x) - 0|$$

证明　反证法. 若存在最高次项系数为 1 的另一个 n 次多项式 $p(x) \neq \widetilde{T}_n(x)$, 它与零的偏差比 $\widetilde{T}_n(x)$ 与零的偏差还小, 即有

$$\max_{-1 \leqslant x \leqslant 1} |p(x) - 0| < \max_{-1 \leqslant x \leqslant 1} |\widetilde{T}_n(x) - 0| = \frac{1}{2^{n-1}}$$

令 $q(x) = \widetilde{T}_n(x) - p(x)$. 由于 $\widetilde{T}_n(x), p(x)$ 均属于 $\widetilde{P}_n(x)$, 则 $q(x)$ 是一个次数不超过 $n-1$ 的多项式.

在 $T_n(x)$ 的交错点组 $x_k = \cos\frac{k\pi}{n}$ ($k = 0, 1, 2, \cdots, n$) 处, 由于 $|p(x_k)| \leqslant \max_{-1 \leqslant x \leqslant 1} |p(x_k)| < \frac{1}{2^{n-1}}$, 即

$$-\frac{1}{2^{n-1}} < p(x_k) < \frac{1}{2^{n-1}}$$

$\widetilde{T}_n(x)$ 在 x_k 处轮流取到 $(-1)^k \frac{1}{2^{n-1}}$, 因此

$$q(x_k) = \frac{(-1)^k}{2^{n-1}} - p(x_k)$$

显然, $q(x)$ 在 $n+1$ 个交错点处轮流取正负值. 由连续函数的介值性质知 $q(x)$ 应当具有 n 个零点, 而 $q(x)$ 至多是 $n-1$ 次多项式, 所以 $q(x) = 0, p(x) \equiv \widetilde{T}_n(x)$, 矛盾. 故定理结论成立.

此定理表明, 任何最高次项系数为 1 的 n 次多项式在区间 $[-1,1]$ 上的最大值都满足 $\max_{-1 \leqslant x \leqslant 1} |p(x)| \geqslant \frac{1}{2^{n-1}}$. 而 Tschebyscheff 正交多项式 $\widetilde{T}_n(x)$ 是最大值最小

Tschebyscheff 逼近定理

的多项式,因此 $\tilde{T}_n(x)$ 成为逼近其他函数的一种重要多项式,有很广泛的应用. 下面介绍多项式插值余项的极小化.

设在 $[-1,1]$ 上给定 $n+1$ 个互异的插值节点: x_0, $x_1, \cdots, x_n$, 函数 $f(x)$ 在 $[-1,1]$ 上具有 $n+1$ 阶连续导数 $f^{(n+1)}(x)$, 对 $f(x)$ 作插值多项式, 由拉格朗日插值余项表示式

$$R_n(x) = f(x) - L_n(x) = \frac{f^{(n+1)}(\xi)}{(n+1)!} \prod_{i=0}^{n} (x - x_i)$$

其中 ξ 在 $x_0, x_1, \cdots, x_n$ 之间. 若

$$\max_{-1 \leq x \leq 1} |f^{(n+1)}(x)| = M_{n+1}$$

则有

$$|R_n(x)| \leq \frac{M_{n+1}}{(n+1)!} \prod_{i=0}^{n} |x - x_i|$$

显然余项 $|R_n(x)|$ 的大小取决于 $\prod_{i=0}^{n} |x - x_i|$ 的大小. 问题是如何在 $[-1,1]$ 内选取节点 $x_0, x_1, \cdots, x_n$, 使 $\prod_{i=0}^{n} (x - x_i)$ 最小.

由于 $\prod_{i=0}^{n} (x - x_i)$ 是一个 $n+1$ 次且最高次系数为 1 的多项式, 由 Tschebyscheff 极性定理知, 当 x_i 满足

$$(x-x_0) \cdot (x-x_1) \cdot \cdots \cdot (x-x_n) = \frac{1}{2^n} T_{n+1}(x)$$

时, $\max_{-1 \leq x \leq 1} |(x-x_0) \cdot (x-x_1) \cdot \cdots \cdot (x-x_n)| = \frac{1}{2^n}$ 取得极小值. 也就是只要插值节点 x_i 取成 $n+1$ 次 Tschebyscheff 多项式的零点

$$x_k = \cos(2k+1)\frac{\pi}{2(n+1)} \quad (k=0,1,2,\cdots,n)$$

则插值余项在全区间$[-1,1]$上的最大绝对值极小,且有

$$\max_{-1\leq x\leq 1}|R_n(x)| \leq \frac{M_{n+1}}{(n+1)!}\max_{-1\leq x\leq 1}\left|\prod_{i=0}^{n}(x-x_i)\right| =$$

$$\frac{M_{n+1}}{(n+1)!}\max_{-1\leq x\leq 1}\frac{|T_{n+1}(x)|}{2^n} =$$

$$\frac{M_{n+1}}{(n+1)!\ 2^n}$$

此式表明用 Tschebyscheff 多项式 $T_{n+1}(x)$ 的零点

$$x_k = \cos\frac{2k+1}{2(n+1)}\pi \quad (k=0,1,2,\cdots,n)$$

作为插值节点的插值多项式 $L_n(x)$,其插值余项具有极小化性质,即 $L_n(x)$ 可作为 $f(x)$ 的近似最佳一致逼近多项式.

对于一般区间 $[a,b]$ 上的函数 $f(x)$,可通过变换 $x=\frac{a+b}{2}+\frac{b-a}{2}t$ 把函数变换成

$$f(x) = f\left(\frac{a+b}{2}+\frac{b-a}{2}t\right) \xlongequal{\triangle} g(t)$$

其中 $-1\leq t\leq 1$. 即将定义在 $[a,b]$ 上的函数 $f(x)$ 化成定义在 $[-1,1]$ 上的函数 $g(t)$. 因而,应选取

$$t_k = \cos\frac{2k+1}{2(n+1)}\pi \quad (k=0,1,2,\cdots,n)$$

即插值节点为

$$x_k = \frac{a+b}{2}+\frac{b-a}{2}\cdot\cos\frac{2k+1}{2(n+1)}\pi \quad (k=0,1,2,\cdots,n)$$

可使 $\max_{a\leq x\leq b}|R_n(x)|$ 达到极小,并有

Tschebyscheff 逼近定理

$$\max_{a\leq x\leq b}|R_n(x)|\leq \frac{M_{n+1}}{(n+1)!}\cdot\frac{(b-a)^{n+1}}{2^{n+1}}\cdot$$

$$\max_{-1\leq t\leq 1}\left|\frac{1}{2^n}T_{n+1}(t)\right|=$$

$$\frac{M_{n+1}}{(n+1)!}\cdot\frac{(b-a)^{n+1}}{2^{2n+1}} \qquad ⑳$$

例 6 用多项式插值余项极小化方法求函数 $f(x)=\mathrm{e}^{-x}$ 在 $[0,1]$ 上的近似最佳一致逼近多项式,要求误差不超过 $\frac{1}{2}\times 10^{-3}$.

解 由 $a=0,b=1$ 及式⑳可得

$$\max_{0\leq x\leq 1}|R_n(x)|=\frac{M_{n+1}}{(n+1)!}\cdot\frac{1}{2^{2n+1}}$$

而

$$M_{n+1}=\max_{0\leq x\leq 1}|f^{(n+1)}(x)|=\max_{0\leq x\leq 1}|(-1)^{n+1}\mathrm{e}^{-x}|=1$$

当 $n=3$ 时有

$$\max_{0\leq x\leq 1}|R_3(x)|\leq\frac{1}{2^7\times 4!}=0.000\,325\,5<\frac{1}{2}\times 10^{-3}$$

因此取插值节点为

$$x_k=\frac{1}{2}\left(1+\cos\frac{2k+1}{8}\pi\right) \quad (k=0,1,2,3)$$

即 $x_0=0.961\,939\,8, x_1=0.691\,341\,7, x_2=0.308\,658\,3,$
$x_3=0.038\,060\,23$,得到三次拉格朗日插值多项式为

$$L_3(x)=0.999\,77-0.992\,90x+$$
$$0.463\,23x^2-0.102\,40x^3$$

它可作为 $f(x)=\mathrm{e}^{-x}$ 在 $[0,1]$ 上的近似最佳一致逼近多项式(误差不超过 $\frac{1}{2}\times 10^{-3}$).

第 2 章　什么是逼近

利用 Tschebyscheff 多项式还可以降低逼近多项式的次数,这种方法称为缩减幂级数方法. 我们在一个区间上用多项式逼近函数 $f(x)$,在一定误差范围内逼近的多项式次数越低越好.

由于 $\{x^k\}$ 也可用 Tschebyscheff 多项式 $\{T_n(x)\}$ 表示,如

$$1 = T_0, x = T_1(x), x^2 = \frac{1}{2}(T_0 + T_2), \cdots \quad ㉑$$

若把一般 n 次多项式 $p_n(x) = a_0 + a_1 x + \cdots + a_n x^n$ 中的所有 $x^k (k = 0, 1, 2, \cdots, n)$ 用上述式㉑中的 Tschebyscheff 多项式去代替,即 $p_n(x)$ 可写成

$$p_n(x) = b_0 + b_1 T_1(x) + b_2 T_2(x) + \cdots + b_n T_n(x)$$

在满足一定误差要求下,可用 Tschebyscheff 多项式的极值性质,把 $p_n(x)$ 的高次幂项缩减下来,使它成为 m ($m \leqslant n-1$) 次多项式. 这个 m 次多项式可以作为在已给精度要求下 $f(x)$ 在区间 $[-1, 1]$ 上的近似最佳一致逼近多项式.

例 7　设在区间 $[-1, 1]$ 上找一个低次多项式 $p(x)$ 来代替函数 $f(x) = e^x$,使误差 $\varepsilon < 10^{-2}$.

解　因为

$$f(x) = e^x = 1 + x + \frac{1}{2!}x^2 + \frac{1}{3!}x^3 + \frac{1}{4!}x^4 + \frac{1}{5!}x^5 + \cdots$$

若取前六项之和

$$p_5(x) = 1 + x + \frac{1}{2!}x^2 + \frac{1}{3!}x^3 + \frac{1}{4!}x^4 + \frac{1}{5!}x^5$$

作为 $f(x)$ 的近似,则其截断误差为

$$|R_5(x)| = \left| \frac{1}{6!} e^x \cdot x^6 \right| \leqslant \frac{1}{6!} e = \frac{1}{720} e < 0.003\,775 < 0.01$$

Tschebyscheff 逼近定理

即在满足误差要求下 $p_5(x)$ 可作为近似最佳一致逼近多项式. 利用 $\{x^k\}$ 与 $\{T_n(x)\}$ 之间的关系，$p_5(x)$ 可表示为

$$p_5(x) = \frac{81}{64}T_0 + \frac{217}{192}T_1 + \frac{13}{48}T_2 + \frac{17}{384}T_3 + \frac{1}{194}T_4 + \frac{1}{1\,920}T_5$$

由此式可见，T_k 的下标 k 越大，T_k 的系数越小. 由于当 $x \in [-1,1]$ 时，$|T_k(x)| \leq 1$ ($k=0,1,2,\cdots$)，故可以略去高次项 $T_k(x)$ 以降低逼近多项式的次数. 若用

$$p_3(x) = \frac{81}{64}T_0 + \frac{217}{192}T_1 + \frac{13}{48}T_2 + \frac{17}{384}T_3$$

来近似代替 e^x，与 $p_5(x)$ 相比所增加的误差为 $\frac{1}{192} + \frac{1}{1\,920} < 0.005\,737$，故用 $p_3(x)$ 代替 e^x 总的误差不超过 $0.003\,775 + 0.005\,737 = 0.009\,512 < 0.01$，也满足要求. 从而根据 $\{T_k(x)\}$ 和 $\{x^k\}$ 之间的关系可知

$$p_3(x) = \frac{1}{384}(382 + 383x + 208x^2 + 68x^3)$$

即用 x 的一个三次多项式来代替 e^x 比原来用 x 的五次多项式来近似代替 e^x 从次数上来说降低了二次，计算量大为减少. 这种方法称为缩减幂级数方法.

Tschebyscheff 多项式的一些重要性质：

（i）它满足二阶常微分方程

$$(1-x^2)y'' - xy' + n^2 y = 0 \qquad ㉒$$

（ii）$\qquad T_n(1) = T_n'(1) = 1 \qquad ㉓$

$$T_n^{(k)}(1) = \frac{n^2 \cdot (n^2-1^2) \cdot (n^2-2^2) \cdot \cdots \cdot (n^2-(k-1)^2)}{(2k-1)!!}$$

$$(k=1,2,\cdots,n) \qquad ㉔$$

事实上，$T_n(1) = \cos(n\arccos 1) = 1$，得到

$$\lim_{x\to 1}(Q_n'(x))^2 = \lim_{x\to 1}\frac{n^2(E_{n-1}-Q_n^2(x))}{(1-x)(1-x)} =$$

$$\lim_{x\to 1}\frac{-2Q_n(x)Q_n'(x)}{-2x} =$$

$$Q_n(1)Q_n'(1)$$

由此得到 $Q_n'(1)=Q_n(1)$，因而有

$$T_n'(1)=T_n(1)=1$$

(iii) 对所有的 $k=0,1,\cdots,n$，有

$$|T_n^{(k)}(x)|\leqslant T_n^{(k)}(1)\quad (x\in[-1,1])\qquad \text{㉕}$$

事实上，由 $T_n(x)=\cos nt, t=\arccos x$，得到

$$T_n'(x)=n\frac{\sin nt}{\sqrt{1-x^2}}=n\frac{\sin nt}{\sin t}=$$

$$2n[\cos(n-1)t+\cos(n-3)t+\cdots]$$

容易证明，将上式不断微商，就有

$$T_n^{(k)}(x)=\sum_{j=1}^{n-k}\lambda_j(k)\cos jt\quad(\lambda_j(k)\geqslant 0)$$

由此得到

$$|T_n^{(k)}(x)|\leqslant \sum_{j=1}^{n-k}\lambda_j(k)=T_n^{(k)}(1)\quad(x\in[-1,1])$$

特别的，令 $k=1$，从式㉔得到

$$|T_n'(x)|\leqslant T_n'(1)=n^2\quad(x\in[-1,1])\qquad \text{㉖}$$

(iv) 对所有的 $n=2,3,\cdots$，有

$$T_n(x)=2xT_{n-1}(x)-T_{n-2}(x)\qquad \text{㉗}$$

事实上，由三角恒等式

$$\cos nt+\cos(n-2)t=2\cos t\cos(n-1)t$$

令 $t=\arccos x$ 代入就得式㉗.

由于 $T_0(x)=1, T_1(x)=x$，由此从递推公式㉗可以依次地写出所有的 Tschebyscheff 多项式，如

Tschebyscheff 逼近定理

$$T_2(x) = 2x^2 - 1$$
$$T_3(x) = 4x^3 - 3x$$
$$T_4(x) = 8x^4 - 8x^2 + 1$$
$$T_5(x) = 16x^5 - 20x^3 + 5x$$
$$T_6(x) = 32x^6 - 48x^4 + 18x^2 - 1$$
$$T_7(x) = 64x^7 - 112x^5 + 56x^3 - 7x$$

由公式㉗还容易看出,当 n 是奇数时,$T_n(x)$ 是奇次多项式;当 n 是偶数时,$T_n(x)$ 是偶次多项式.

Tschebyscheff 多项式

第 3 章

在区间 $[a,b]$ 上与零有最小偏差的

$$p_n(x) = x^n + a_1 x^{n-1} + \cdots + a_n$$

型多项式称为 Tschebyscheff 多项式. 我们可以给出它们的解析表示以及证明它们的唯一性. 就是说,我们来证明:n 次代数多项式

$$T_n(x) = \frac{1}{2^{n-1}} \cos n\arccos x \qquad ㉘$$

即为在区间 $[-1,1]$ 上的 Tschebyscheff 多项式. 至于 $T_n(x)$ 为 n 次代数多项式显然,接下来要证明,这个多项式的最高次项系数为 1(么首的),并且它在区间 $[-1,1]$ 上与零有最小偏差. 首先不去证明第一个事实而是建

Tschebyscheff 逼近定理

立第二个. 为此,我们指出多项式 $T_n(x)$ 的某些性质. 首先指出,这些多项式的范数在区间 $[-1,1]$ 上等于 $\frac{1}{2^{n-1}}$,这是由于 $\|\cos n\arccos x\| = 1$. 现在我们来求多项式在区间上达到其范数的那些点,并且考察多项式在这些点的符号. 于是,我们要求出满足

$$\cos n\arccos x = 1 \text{ 或 } \cos n\arccos x = -1$$

的点. 解第一个方程,得到

$$n\arccos x = 2k\pi,\ \arccos x = \frac{2k\pi}{n},\ x_k = \cos\frac{2k\pi}{n}$$

我们设定 $0 \leqslant \arccos x \leqslant \pi$,因而数 k 不能取任意值,而必须仅限于取值 $k = 0, 1, 2, \cdots, \left[\frac{n}{2}\right]$ ($[x]$ 表 x 的整部). 解第二个方程,得到

$$x_k = \cos\frac{2k+1}{n}\pi \quad (k = 0, 1, 2, \cdots, \left[\frac{n-1}{2}\right])$$

把求得的一切值 x 写成递减的顺序

$$x_0 = \cos 0 = 1,\ x_1 = \cos\frac{\pi}{n}$$

$$x_2 = \cos\frac{2\pi}{n},\ x_3 = \cos\frac{3\pi}{n}$$

$$\vdots$$

㉙

我们求得的一切点 x_k 共有多少? 它们的个数等于

$$1 + \left[\frac{n}{2}\right] + 1 + \left[\frac{n-1}{2}\right] = 2 + \left[\frac{2n-1}{2}\right] = 2 + n - 1 = n + 1$$

用这些点可以把区间 $[-1,1]$ 分成 n 个部分,使得在每一区间的端点上,多项式 $T_n(x)$ 按绝对值而言取得了范数而且符号相异

第 3 章　Tschebyscheff 多项式

$$T_n(x_0) = T_n(1) = \frac{1}{2^{n-1}}$$

$$T_n(x_1) = -\frac{1}{2^{n-1}}$$

$$T_n(x_2) = \frac{1}{2^{n-1}}$$

$$T_n(x_3) = -\frac{1}{2^{n-1}}$$

$$\vdots$$

$$T_n(x_k) = \frac{(-1)^k}{2^{n-1}}$$

$$\vdots$$

$$T_n(x_n) = -\frac{(-1)^n}{2^{n-1}}$$

$$\vdots$$

当以后证明 Tschebyscheff 多项式为在一切么首多项式中于区间 $[-1,1]$ 上与零有最小偏差时,我们将用到这个重要性质.

预先证明关于连续函数零点的一个引理.

引理 1　若在实轴上的连续函数 $\lambda(x)$ 满足

$$(-1)^k \lambda(x_k) \geqslant 0 \quad (x_0 < x_1 < x_2 < \cdots < x_n)$$

则函数 $\lambda(x)$ 在区间 $[x_0, x_n]$ 上的零点个数不小于 n,这里对每个二重零点,当经过它时函数不改变符号,均作两个计算.

证明　我们将称数 $d_{i,k} = k - i, k > i$ 为区间 $[x_i, x_k]$ 的"长". 如果我们建立函数 $\lambda(x)$ 在任何区间 $[x_i, x_k]$ 上的零点个数不小于其"长",则引理将得证.

我们应用数学归纳法来证明. 首先指明引理当区

Tschebyscheff 逼近定理

间 $[x_i, x_k]$ 的"长"为 1 亦即 $k=i+1$ 时为真. 于此情形据引理条件,或者数 $\lambda(x_i)$ 与 $\lambda(x_{i+1})$ 异号从而连续函数 $\lambda(x)$ 在区间内部变为零,或者这两数中之一为零. 不论在哪一种情形 $\lambda(x)$ 在区间 $[x_i, x_{i+1}]$ 上的零点个数都不小于这区间的"长". 现在假定函数 $\lambda(x)$ 在每一"长"为 m 的区间上零点个数不小于 m,$m \leqslant n$,而往证明,在区间 $[x_i, x_{i+n+1}]$ 上函数有不小于 $n+1$ 个零点. 为此考察区间 $[x_i, x_k]$ 与 $[x_k, x_{i+n+1}]$($i<k<i+n+1$),它们之中的第一个的"长"都不超过 n. 据假定函数在这两个区间中第一个上的零点个数不小于 $k-i$,而在第二个上不小于 $i+n+1-k$. 若在这些区间的唯一公共点 x_k 处函数 $\lambda(x)$ 不为零,则在区间 $[x_i, x_{i+n+1}]$ 上函数的零点个数将不小于

$$k-i+i+n+1-k=n+1$$

亦即引理已获证. 现在考虑当 $\lambda(x_k)=0$,$k=i+1$,$i+2,\cdots,i+n$ 的情形. 此时我们已得知函数 $\lambda(x)$ 在区间 $[x_i, x_{i+n+1}]$ 上有 n 个零点. 若函数在这个区间还有异于所列举的零点,则引理获证. 于是,接下来我们要考察当 $\lambda(x)$ 没有其他零点的情形. 特别的,$\lambda(x_i) \neq 0$,$\lambda(x_{i+n+1}) \neq 0$. 由于当经过简单零点函数的符号要变化,所以当经过 n 个简单零点时函数的符号要变化 n 次. 若函数 $\lambda(x)$ 的零点 x_k 为单重的,则数

$$\lambda(x_i), (-1)^n \lambda(x_{i+n+1}) \qquad \circledast$$

的符号相同. 但按引理条件

$$\frac{(-1)^i \lambda(x_i)}{(-1)^{i+n+1} \lambda(x_{i+n+1})} = \frac{-\lambda(x_i)}{(-1)^n \lambda(x_{i+n+1})} > 0$$

亦即式 $\circledast$ 中数的符号相反. 由我们所得的矛盾便指明,

第 3 章　Tschebyscheff 多项式

在函数 $\lambda(x)$ 的零点 x_k 之中有多重的，因而这个函数在区间 $[x_i, x_{i+n+1}]$ 上零点的总数不小于 $n+1$. 这就完全证明了引理 1.

我们转向多项式㉘为在一切么首 n 次多项式中在区间 $[-1,1]$ 与零有最小偏差的证明. 为此目的，假定不然，亦即假定存在另一么首 n 次多项式 $p_n(x)$，其范数在 $[-1,1]$ 上不超过 $\dfrac{1}{2^{n-1}}$，那么我们来考察函数 $\lambda(x) = T_n(x) - p_n(x)$. 这个函数为次数不高于 $n-1$ 的多项式，这是由于多项式 $T_n(x)$ 与 $p_n(x)$ 的最高项系数相同. 在点集㉙中每一点函数 $\lambda(x)$ 满足上述引理的条件. 其实

$$(-1)^k \lambda(x_k) = T_n(x_k)(-1)^k - p_n(x_k)(-1)^k \geq$$
$$\dfrac{1}{2^{n-1}} - \|p_n(x)\| \geq 0$$

这是由于 $\|p_n(x)\| \leq \dfrac{1}{2^{n-1}}$. 根据引理在区间 $[-1,1] = [x_0, x_n]$ 上可知函数 $\lambda(x)$ 的零点个数不少于 n 并且由于它为次数不高于 $n-1$ 的多项式，所以 $\lambda(x) \equiv 0$，亦即 $p_n(x) \equiv T_n(x)$.

接下来还要证明 $T_n(x)$ 为么首的. 这个问题很简单. 记住棣美弗公式，得到

$$\cos n\theta = \dfrac{1}{2}(\cos n\theta + i\sin n\theta) + \dfrac{1}{2}(\cos n\theta - i\sin n\theta) =$$
$$\dfrac{1}{2}(\cos \theta + i\sin \theta)^n + \dfrac{1}{2}(\cos \theta - i\sin \theta)^n$$

令 $\theta = \arccos x$，并利用等式

$$\cos \arccos x = x,\ \sin \arccos x = \sqrt{1-x^2}$$

Tschebyscheff 逼近定理

得到

$$T_n(x) = \frac{1}{2^n}\{(x+i\sqrt{1-x^2})^n + (x-i\sqrt{1-x^2})^n\} \quad ⑳$$

把这个等式除以 x^n 并令 x 趋于 ∞,得到多项式 $T_n(x)$ 的最高次项系数为

$$\lim_{x\to\infty}\frac{T_n(x)}{x^n} = \frac{1}{2^n}\lim_{x\to\infty}\left\{\left(1+i\sqrt{\frac{1}{x^2}-1}\right)^n + \left(1-i\sqrt{\frac{1}{x^2}-1}\right)^n\right\} = \frac{2^n}{2^n} = 1$$

于是,我们不仅证明了与零有最小偏差的多项式存在,而且建立了它的唯一性,并且尤其重要的是,指出了这样的多项式的解析表示,以便于把它们写成通常的代数形式且有助于进而研究它们的深刻性质. 对于由 Tschebyscheff 所研究过的与其他问题相关联的那些性质,我们将不加讨论而只提出其中对于内插理论有用的一个性质.

所指的性质是在区间 $[a,b]$ 上与零有最小偏差的 Tschebyscheff 多项式的范数的数量. 当 $a=-1, b=1$ 时,我们已证明了多项式 $T_n(x)$ 的范数为 $\frac{1}{2^{n-1}}$. 为了要求出它在任意区间 $[a,b]$ 上的范数,必须采用把区间 $a \leqslant y \leqslant b$ 映射到区间 $-1 \leqslant x \leqslant 1$ 的线性变换 $x = \frac{2}{b-a}y - \frac{a+b}{b-a}$.

此时我们得到多项式

$$p(y) = T_n\left(\frac{2}{b-a}y - \frac{a+b}{b-a}\right) = \left(\frac{2}{b-a}y - \frac{a+b}{b-a}\right)^n + \cdots$$

它的最高次项系数非 1 而为 $\frac{2^n}{(b-a)^n}$. 把 $p(y)$ 用这个数

第3章 Tschebyscheff 多项式

来除,我们得到区间 $[a,b]$ 上的 Tschebyscheff 多项式

$$\widetilde{T}_n(y) = \frac{(b-a)^n}{2^n} T_n\left(\frac{2}{b-a}y - \frac{a+b}{b-a}\right)$$

它的最高次项系数已为 1 了. 易见,它的范数等于

$$\|\widetilde{T}_n(y)\| = \frac{(b-a)^n}{2^n} \|T_n(x)\| = \frac{(b-a)^n}{2^{2n-1}} \quad ㉛$$

寻找 Tschebyscheff 多项式的零点并无困难. 限于区间 $[-1,1]$ 的情形,我们由方程

$$\cos n\arccos x = 0$$

求出 Tschebyscheff 多项式的零点为

$$x_k = \cos\theta_k \quad \left(\theta_k = \frac{2k+1}{2n}\pi;\ k = 0,1,\cdots,n-1\right) \quad ㉜$$

着重指出,Tschebyscheff 多项式的零点位于那样的区间上,那里多项式有定义,并且它们两两互异. 正因 Tschebyscheff 多项式零点分布的这一特性才使我们把它作为这样的内插多项式来考虑,即取内插基点阵的第 n 列为多项式 $T_{n+1}(x)$ 的根. 相应的内插多项式将用记号 $L_{T_{n+1}}(f;x)$ 来表示.

当借用多项式 $L_{T_{n+1}}(f;x)$ 以求逼近论的其他重要结果之前,我们来解关于在区间 $[0,1]$ 上逼近函数 $f(x) = e^x$ 精确到 $\dfrac{1}{5\cdot 10^9}$ 的多项式的次数问题.

从解多项式 $L_{T_{n+1}}(f;x)$ 开始,这时指的是应用在区间 $[0,1]$ 上,$a=0, b=1$ 的相应的 Tschebyscheff 多项式.

利用等式㉛,得到

$$\|L_{T_{n+1}}(f;x) - f(x)\| \leqslant \frac{1}{2^{2n-1}} \cdot \frac{\|f^{(n+1)}(x)\|}{(n+1)!} \quad ㉝$$

Tschebyscheff 逼近定理

由于 $f(x)=e^x, f^{(n+1)}(x)=e^x$，所以在区间 $[0,1]$ 上 $\|f^{(n+1)}(x)\|\leqslant e$，因而

$$\|L_{T_{n+1}}(f;x)-f(x)\|\leqslant\frac{2e}{4^n(n+1)!} \qquad ㉞$$

有趣的是把这个结果与所述函数按 x 的乘幂展开的泰勒公式的余项相比较. 由于对泰勒公式

$$R_n(x)=\frac{f^{(n+1)}(\xi)}{(n+1)!}x^{n+1}$$

所以当 $x=1$ 时得到

$$R_n(1)=\frac{e^\xi}{(n+1)!}>\frac{1}{(n+1)!} \qquad ㉟$$

比较不等式㉞与㉟表明：函数 $f(x)=e^x$ 用多项式 $L_{T_{n+1}}(f;x)$ 逼近的阶较同一函数用泰勒级数的部分和所给的要好 4^n 倍. 这一情况对于某些其他的函数也成立，这由比较泰勒级数与多项式 $L_{T_{n+1}}(f;x)$ 的余项所推出. 诚然，如果把泰勒级数取为按区间中点为中心展开，则它的部分和所逼近的阶要低 2^n 倍，那么在多项式 $L_{T_{n+1}}(f;x)$ 与泰勒级数部分和所逼近的阶之间的差异已不能用 4^n 倍而用 2^n 倍来描述.

现在，设 $n=9$，得到

$$\frac{2e}{4^9\cdot 10!}<\frac{e}{46\cdot 10^{10}}<\frac{1}{16\cdot 10^{10}}$$

从而由式㉞推得

$$\|L_{T_{10}}(f;x)-f(x)\|<\frac{1}{16\cdot 10^{10}}$$

这样，当 $n=9$ 时，多项式 $L_{T_{n+1}}(f;x)$ 达到了所要求的逼近估值.

我们来求泰勒级数

第3章 Tschebyscheff 多项式

$$e^x = 1 + x + \frac{x^2}{2!} + \frac{x^3}{3!} + \cdots$$

的部分和而对函数给出同一逼近的阶的次数.

由于

$$R_n(1) = \frac{f^{(n+1)}(\xi)}{(n+1)!} \cdot 1^{n+1} = \frac{e^\xi}{(n+1)!} > \frac{1}{(n+1)!}$$

不等式

$$(n+1)! > 5 \cdot 10^9$$

当 $n = 12$ 时首次达到.

设 $A_n(f;x)$ 为最快收敛于函数的线性正多项式算子,由于这函数的二阶导数为正,所以它用算子 $A_n(f;x)$ 逼近的阶为 $\frac{1}{n^2}$. 为了实现我们所要的逼近函数 e^x 的阶,所以必须取 $n \approx 70\,000$.

最后,利用瓦隆诺夫斯卡雅定理容易求得,多项式 $B_n(f;x)$ 当 $n = 10^9$ 时实现了这样的逼近的阶.

在下列表中引进类似于上述关于与函数的偏差不超过 $\frac{1}{5 \cdot 10^9}$ 的多项式的次数. 表1~表3的第一行,是作者借用关于用多项式逼近解析函数的阶的重要的伯恩斯坦定理而得到的.

表1 $f(x) = e^x, 0 \leqslant x \leqslant 1$

多项式的名称	次 数
$T_n(f;x); L_{T_{n+1}}(f;x)$	9
泰勒级数部分和	12
$A_n(f;x)$	约 70 000
$B_n(f;x)$	约 1 000 000 000

Tschebyscheff 逼近定理

表2 $f(x)=\dfrac{2}{2-x}, 0\leqslant x\leqslant 1$

多项式的名称	次 数
$T_n(f;x)$;$L_{T_{n+1}}(f;x)$	约 12
泰勒级数部分和	32
$A_n(f;x)$	约 70 000
$B_n(f;x)$	约 1 000 000 000

表3 $f(x)=\dfrac{1}{1+x^2}, 0\leqslant x\leqslant 1$

多项式的名称	次 数
$T_n(f;x)$;$L_{T_{n+1}}(f;x)$	约 16
泰勒级数部分和	不存在(级数当 $x=1$ 时发散)
$A_n(f;x)$	约 70 000
$B_n(f;x)$	约 1 000 000 000

多项式动力学和 Fermat 小定理的一个证明

第 4 章

4.1 引 言

我们回忆一下,Fermat 小定理证实了 $p\mid m^p-m$,其中 m 是一个自然数,p 是素数.

长期以来,这个著名的定理已经有了若干个证明. 一个动力系统的证明可以在[2]中找到. 这个证明基于由 $a_m(x)=\{mx\}$ 定义的算术函数 a_m,这里 $\{x\}=x-[x]$ 表示 x 的分数部分,

---

① 译自:The Amer. Math. Monthly, Vol. 120 (2013), No. 2, p. 171-173, Polynomial Dynamics and a Proof of the Fermat Little Theorem, Vladimir Dragović, figure number 1. Copyright © 2013 the Mathematical Association of America. Reprinted with permission. All rights reserved. 美国数学协会授予译文出版许可. 作者的邮箱地址是 vladad@ mi. sanu. ac. rs.

Tschebyscheff 逼近定理

$[x]$ 表示最大整数函数,或 x 的整数部分.在[2]中,提到了[1]中一个更早的动力学证明.在[3]中还提出了另外的动力学证明(参阅下面的"结论"小节).

这里,基于 Tschebyscheff 多项式的多项式动力学,我们提出 Fermat 小定理的动力学证明.证明中只用到正弦函数和余弦函数的一个非常基本的加法定理.

4.2 Tschebyscheff 多项式

作为第一类 Tschebyscheff 多项式(T_n)的一个基础关系式,我们取文献[4]中第 1 章

$$T_n(\cos\theta)=\cos(n\theta) \quad (n\in \mathbf{N}_0) \quad (*)$$

对于 $\theta\in[0,\pi]$,通过公式 $T_n(x)=\cos(n\arccos x)$,这个关系式定义了一个 $x\in[-1,1]$ 的 n 次多项式.这些多项式是由 Tschebyscheff 于 1854 年引进的.从关系式($*$)可以直接得到众所周知的性质:

(A) 复合性质

$$T_n \circ T_m = T_{mn} \quad (**)$$

(B) 端点性质

$$T_n(1)=1, T_n(-1)=(-1)^n \quad (***)$$

(C) 递推关系

$$T_0(x)=1, T_1(x)=x$$

以及

$$T_{n+1}(x)=2xT_n(x)-T_{n-1}(x) \quad (n\geqslant 1)$$

从最后的关系式,我们可以容易地算出前面几个 Tschebyscheff 多项式

$$T_2(x)=2x^2-1$$

第4章 多项式动力学和 Fermat 小定理的一个证明

$$T_3(x) = 4x^3 - 3x$$
$$T_4(x) = 8x^4 - 8x^2 + 1$$

和

$$T_5(x) = 16x^5 - 20x^3 + 5x$$

我们把这些多项式画在图 1 中.（原图为彩色. 请读者按 T_j 的表达式确定其对应的曲线. ——编者）

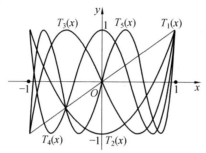

图 1 Tschebyscheff 多项式

现在,我们来计算作为从区间 $[-1,1]$ 到其自身映射的 T_n 的不动点. 它们相应于方程 $T_n(\cos\theta) = \cos n\theta$ 对于 $\theta \in [0,\pi]$ 的解. 我们需要下面的引理.

引理 2 对于数 $\theta \in [0,\pi]$, 下面一些陈述是等价的:

(1) $T_n(\cos\theta) = \cos n\theta$.

(2) $\sin\left(\dfrac{n-1}{2}\theta\right)\sin\left(\dfrac{n+1}{2}\theta\right) = 0.$

(3) $\dfrac{n-1}{2}\theta = l\pi$ 或 $\dfrac{n+1}{2}\theta = k\pi, l, k \in \mathbf{N}^*.$

(4) $0 \leqslant \dfrac{2l}{n-1} \leqslant 1$ 或 $0 \leqslant \dfrac{2k}{n+1} \leqslant 1, l, k \in \mathbf{N}^*.$

我们把此引理的证明留给读者. 从引理中的这些

关系式,并反复利用(＊＊＊),我们推导得 Tschebyscheff 多项式的不动点性质:

(D)对于 $n \geq 2$,多项式 T_n 在区间[-1,1]上的不动点总数等于 n.

参看图1,其中画出了 $T_n(n \leq 5)$ 的图像及其不动点.

不动点性质(D)和复合性质(A)正是我们所需要用来推导 Fermat 小定理的性质. 给定一个素数 p,我们考虑 T_m 的 p 阶周期点的数目. 据刚才提到的两个性质,这个数等于 m^p-m. 这样的周期点自然地被分解为它们的轨道,即形如 $x_0, T_m(x_0), T_{m^2}(x_0), \cdots, T_{m^{p-1}}(x_0)$ 的 p 个不同元素的子集. 这样,我们得到结论:p 整除 m^p-m,因而 Fermat 小定理成立. 这个证明只依赖于三角加法公式.

另一方面,利用 Bolzano(波尔查诺)定理(对于多项式的情形,该定理由 Simon Stevin(斯蒂文)1548—1620 得到),从

(B1)$T_n(\cos \theta_j)=(-1)^j \quad (j=0,1,\cdots,n)$

可以确立(D),其中 $\theta_j=j\pi/n$.

作为最后的观察,根据方程(1),由 Tschebyscheff 多项式 T_n 所诱导的动力学共轭于由算术函数 a_n 定义的动力学.

4.3 结 论

(在 Julia 和 Ritt 之后,见[4]中第4章的相当初等的解释)下述事实是众所周知的:只存在另外一个本

第4章 多项式动力学和 Fermat 小定理的一个证明

质上异于 Tschebyscheff 多项式的多项式族满足复合性质(2): $P_n(x) := x^n$.

在[3]中提出了利用这族多项式的 Fermat 小定理的一个证明. 为了确立不动点性质(D),需要复数及其三角形表示的一些知识. 此外,可以用代数学基本定理和单根的检验做出不动点的计数.

例8 已知 $p_1(x) = x^2 - 2$ 及 $p_j(x) = p_1(p_{j-1}(x))$, $j = 2, 3, \cdots$. 求证:对任何正整数 n,方程 $p_n(x) = x$ 所有的根都是实的,且各不相同.(第18届IMO试题)

证明 首先,因为 $p_n(x) = x$ 是 2^n 次方程,故至多有 2^n 个根. 其次,当 $x > 2$ 时

$$p_1(x) > x$$
$$p_2(x) = p_1(p_1(x_1)) > p_1(x) > x$$

由归纳法的假定可知

$$p_n(x) > x > 2$$

同样,当 $x < -2$ 时

$$p_1(x) > 2 > x$$
$$\vdots$$
$$p_n(x) > 2 > x$$

所以 $p_n(x)$ 的实根都在闭区间 $[-2, 2]$ 内.

假定根有如下形式 $x = 2\cos t$,则

$$p_1(2\cos t) = 4\cos^2 t - 2 = 2\cos 2t$$
$$p_2(2\cos t) = p_1(2\cos 2t) = 2\cos 4t$$

一般的,有 $p_n(2\cos t) = 2\cos(2^n t)$,但 $2\cos(2^n t) = 2\cos t$ 有 2^n 个不同的解

$$t = \frac{2^n m\pi}{2^n - 1}, \quad t = \frac{2^n m\pi}{2^n + 1} \quad (m = 0, 1, \cdots, 2^{n-1} - 1)$$

它们都是实数,且互不相同.

实际上 $p_n(x)$ 就是 Tschebyscheff 多项式,而例 8 则是要证 $p_n(x)$ 的不动点各不相同.

注 "莫斯科数学比赛大会"的 139 题为:

(1) 证明:表达式 $T_n(x) = \frac{1}{2^n}[(x+\sqrt{x^2-1})^n + (x-\sqrt{x^2-1})^n]$ 在打开括号合并同类项以后是最高次项系数为 1 的 n 次多项式(n 次 Tschebyscheff 多项式).

(2) $T_n(x)$ 有 n 个实根.

(3) 多项式 $T_n(x)$ 的所有根包含于 -1 和 1 之间.

(4) 恒等式 $T_n(x) - xT_{n-1}(x) + \frac{1}{4}T_{n-2}(x) \equiv 0$ 成立.

(5) 在 $T_n(x)$ 的两个相邻根之间有 $T_{n-1}(x)$ 的一个根.

参考文献

[1] HAUSNER M. Applications of a simple counting technique[J/OL] Amer. Math. Monthly,1983,90: 127-129[1985-04-06]. http:∥dx.dio.org/10.2307/2975816.

[2] IGA K. A dynamical system proof of Fermat's little theorem,[J/OL]. Mathematics Magazine, 2003, 76:48-51[2004-05-26]. http:∥dx.dio.org/10.2307/3219132.

[3] LEVINE L. Fermats little theorem:A proof by function iteration [J/OL]. Mathematics Magazine,

第4章 多项式动力学和 Fermat 小定理的一个证明

1999,72:308-309[2000-02-15]. http://dx.dio.org/10.2307/2691226.

[4] RIVLIN T J. Chebyshev Polynomials[M]. New York:Wiley,1990.

最佳逼近多项式的特征

第5章

我们来看一道1983年的全国高中联赛试题:

例9 函数 $F(x)=|\cos^2 x+2\sin x\cos x-\sin^2 x+Ax+B|$ 在 $0\leqslant x\leqslant \frac{3}{2}\pi$ 上的最大值 M 与参数 A,B 有关. 问: A,B 取什么值时 M 为最小? 证明你的结论.

解 首先 $F(x)=|\sqrt{2}\sin(2x+\frac{\pi}{4})+Ax+B|$,当 $A=B=0$ 时,有
$$F(x)=f(x)=\sqrt{2}|\sin(2x+\frac{\pi}{4})|\leqslant \sqrt{2}$$
又在区间 $[0,\frac{3\pi}{2}]$ 上有 $x_1=\frac{\pi}{8}$, $x_2=\frac{5\pi}{8}$, $x_3=\frac{9\pi}{8}$ 使 $f(x)_{\max}=\sqrt{2}$,我们猜想

第5章 最佳逼近多项式的特征

$M_{\max} = \sqrt{2}$.

下面证明对任何 A, B 不同时为 0 时,有

$$\max_{0 \le x \le \frac{3\pi}{2}} F(x) > \max_{0 \le x \le \frac{3\pi}{2}} f(x) = \sqrt{2}$$

反证法:假若存在 $\max\limits_{0 \le x \le \frac{3\pi}{2}} F(x) \le \sqrt{2}$,且 A, B 不同时为 0,则应有

$$F(\frac{\pi}{8}) \le \sqrt{2}, \ F(\frac{5\pi}{8}) \le \sqrt{2}, \ F(\frac{9\pi}{8}) \le \sqrt{2}$$

即

$$\begin{cases} |\sqrt{2} + \frac{\pi}{8}A + B| \le \sqrt{2} \\ |-\sqrt{2} + \frac{5\pi}{8}A + B| \le \sqrt{2} \\ |\sqrt{2} + \frac{9\pi}{8}A + B| \le \sqrt{2} \end{cases} \Rightarrow \begin{cases} -2\sqrt{2} \le \frac{\pi}{8}A + B \le 0 \\ 2\sqrt{2} \ge \frac{5\pi}{8}A + B \ge 0 \\ -2\sqrt{2} \le \frac{9\pi}{8}A + B \le 0 \end{cases}$$

由 $\begin{cases} \frac{\pi}{8}A + B \le 0 \\ \frac{5\pi}{8}A + B \ge 0 \end{cases} \Rightarrow A \ge 0$

又由 $\begin{cases} \frac{5\pi}{8}A + B \ge 0 \\ \frac{9\pi}{8}A + B \le 0 \end{cases} \Rightarrow A \le 0$

所以 $A = 0$. 但当 $A = 0, B \ne 0$ 时,有

$$\max_{0 \le x \le \frac{3\pi}{2}} F(x) = \max_{0 \le x \le \frac{3\pi}{2}} |\sqrt{2}\sin(2x + \frac{\pi}{4}) + B| = \sqrt{2} + |B| > \sqrt{2}$$

这与假设矛盾,故猜想正确 $M_{\min} = \sqrt{2}$.

下面讲 $C[a,b]$ 中最佳逼近多项式的特征性质.

$C[a,b]$ 中最佳逼近多项式的特征性质与 $L^p[a,b]$ 中的性质有很大的不同. 我们有

定理 2 设 $\{\varphi_j(x)\}$ $(1\leqslant j\leqslant n)$ 在 $C[a,b]$ 上满足哈尔(Haar)条件. 对任何函数 $f(x)\in C[a,b]$, 若 $P^*(x)=\sum_{j=1}^n a_j^*\varphi_j(x)$ 是 $f(x)$ 在 $C[a,b]$ 上的最佳逼近多项式, 则函数 $f(x)-P^*(x)$ 在 $[a,b]$ 上至少在 $n+1$ 个点所构成的点集 X 上以正负相间的符号取到 $|f(x)-P^*(x)|$ 的最大值, 即

$$E_n=\max_{a\leqslant x\leqslant b}|f(x)-P^*(x)|$$

(今后称 X 上的点为 Tschebyscheff 交错点集)

证明 为了简单起见, 只就 $\varphi_j(x)=x^j$ $(0\leqslant j\leqslant n-1)$ 的情况进行证明. 一般情况下的证明是类似的.

不妨假设 $E_n>0$, 否则 $f(x)=P^*(x)$, 这是显然的.

利用连续函数的性质, $|f(x)-P^*(x)|$ 在 $[a,b]$ 上达到最大值 E_n, 因此必存在点 $x^*\in[a,b]$, 使 $f(x^*)-P^*(x^*)=\pm E_n$. 因此集合 X 是非空的.

用反证法. 设集合 X 至多包含 m 个定理中所指出的交错点 $x_1,x_2,\cdots,x_m$ $(1\leqslant m\leqslant n)$ 满足

$$f(x_i)-P^*(x_i)=\pm(-1)^i E_n$$
$$a\leqslant x_1<x_2<\cdots<x_m\leqslant b$$

利用连续函数的中值定理, 至少存在 $m-1$ 个点 a_i $(2\leqslant m)$ 满足 i

$$a\leqslant x_1<a_2<x_2<\cdots<a_m<x_m\leqslant b$$
$$f(a_i)-P^*(a_i)=0 \quad (i=2,3,\cdots,m)$$

第5章 最佳逼近多项式的特征

不仅如此,由于 $x_1, x_2, \cdots, x_m$ 是 Tschebyscheff 交错点,我们还可以认为在 m 个区间:$[a, a_2], [a_2, a_3], \cdots, [a_m, b]$ 中交错地满足下列等式中的一个

$$-E_n + \mu \leqslant f(x) - P^*(x) \leqslant E_n \quad \text{㊱}$$

或

$$-E_n \leqslant f(x) - P^*(x) \leqslant E_n - \mu \quad \text{㊲}$$

其中 μ 为某个正数(图1).

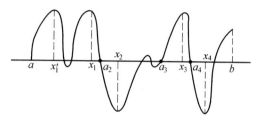

图1 $m=4$ 的情况

现构造 $m-1$ 次多项式($m-1 \leqslant n-1$),有

$$P_{m-1}(x) = \delta \prod_{i=2}^{m}(x - a_i)$$

选择 δ 使其满足条件

$$\|P_{n-1}(x)\|_{C[a,b]} \leqslant \frac{\mu}{2} \quad \text{㊳}$$

$$\text{sign}(P_{m-1}(x_1)) = \text{sign}(f(x_1) - P^*(x_1)) \quad \text{㊴}$$

不妨认为 $f(x_1) - P^*(x_1) = E_n$,则由式㊴知

$$\text{sign}(P_{m-1}(x_1)) = \text{sign}(f(x_1) - P^*(x_1)) = 1$$

利用 $f(x_i) - P^*(x_i) = -(-1)^i E_n$ 及 $P_{m-1}(x)$ 的构造可知,多项式 $P_{m-1}(x)$ 在区间 $[a, a_2], [a_2, a_3], \cdots, [a_m, b]$ 上取值的符号与 $f(x) - P^*(x)$ 在相应点 x_i 上取值的符号相同.

令 $Q_{n-1}(x) = P^*(x) - P_{m-1}(x)$,它也是一个次数不

大于 n 的多项式.

因为 $f(x_1)-P^*(x_1)=E_n$, $f(a_2)-P^*(a_2)=0$, 且由式㊴知,在 $[a,a_2]$ 上 $P_{m-1}(x)\geq 0$, 因此在 $[a,a_2]$ 上有

$$f(x)-Q_{n-1}(x)=f(x)-P^*(x)-P_{m-1}(x)<E_n$$

再利用 $f(x)-P^*(x)$ 满足式㊱, $P_{n-1}(x)$ 满足式㊳,因此在 $[a,a_2]$ 上,有

$$f(x)-Q_{n-1}(x)=f(x)-P^*(x)-P_{m-1}(x)\geq$$
$$-E_n+\mu-\frac{\mu}{2}=-E_n+\frac{\mu}{2}$$

这样一来,在 $[a,a_2]$ 上就有

$$|f(x)-Q_{n-1}(x)|<E_n \qquad ㊵$$

同样,利用 $f(x_2)-P^*(x_2)=-E_n$, $f(a_3)-P^*(a_3)=0$, 在 $[a_2,a_3]$ 上 $P_{m-1}(x)\leq 0$ 以及 $f(x)-P^*(x)$ 满足式㊲, $P_{m-1}(x)$ 满足式㊳,因此在 $[a_2,a_3]$ 上不等式㊵也成立.

依次地可以证明,在 $[a_3,a_4],\cdots,[a_m,b]$ 上不等式㊵全部成立,即在整个区间 $[a,b]$ 上不等式㊵成立. 这样就违反了 E_n 是最佳逼近值的假设,因此得到矛盾. 集合 X 就至少包含有 $n+1$ 个 Tschebyscheff 交错点. 证毕.

现在我们再给出一个 $E_n(f)$ 下界的估计式,这个估计式不仅在理论上,而且在实际上都很有用.

定理 3(Vallée-Poussin) 设 $\{\varphi_j(x)\}$ $(1\leq j\leq n)$ 在 $C[a,b]$ 上满足哈尔条件,$f(x)\in C[a,b]$. 若存在多项式 $Q(x)=\sum_{j=1}^{n}a_j\varphi_j(x)$, 使得 $f(x)-Q(x)$ 在 $[a,b]$ 上至少有 $n+1$ 个点 $x_0,x_1,\cdots,x_n$ 以正负相间的符号取到值

第5章 最佳逼近多项式的特征

$A_0, A_1, \cdots, A_n$,则

$$E_n(f) \geq \lambda = \min_{0 \leq i \leq n} |A_i| \qquad ㊶$$

证明 用反证法. 设 $E_n(f) < \lambda$. 此外,设 $P^*(x)$ 是 $f(x)$ 在 $C[a,b]$ 上的最佳逼近多项式. 我们有

$$P^*(x) - Q(x) = f(x) - Q(x) - (f(x) - P^*(x)) \qquad ㊷$$

由 $|f(x) - P^*(x)| \leq E_n(f)$,及 $E_n(f) < \lambda$,从式㊷可知,多项式 $P^*(x) - Q(x)$ 在 $x = x_i (0 \leq i \leq n)$ 处取值的符号完全由 $f(x_i) - Q(x_i)$ 决定. 因此,根据定理的条件 $f^*(x) - Q(x)$ 在 $x = x_i (0 \leq i \leq n)$ 上交错地取正负值. 根据连续函数的中值定理,在 $[a,b]$ 上多项式 $P^*(x) - Q(x)$ 至少有 n 个零点. 因为函数系 $\{\varphi_j(x)\}(1 \leq j \leq n)$ 在 $C[a,b]$ 上满足哈尔条件,因此必须有 $P^*(x) - Q(x) = 0$,即 $Q(x) = P^*(x)$. 这样就有

$$E_n(f) = \max_{a \leq x \leq b} |f(x) - Q(x)| \geq \max_{0 \leq i \leq n} |A_i| \geq \lambda$$

而这与 $E_n(f) < \lambda$ 是矛盾的. 证毕.

Tschebyscheff 多项式的三角形式在几何中的应用

第 6 章

例 10 单位圆被点 A_0, A_1, A_2, A_3, A_4 分成 5 个等弧. 证明:弦 A_0A_1 和 A_0A_2 之长满足等式

$$(A_0A_1 \cdot A_0A_2)^2 = 5$$

(1899 年匈牙利数学奥林匹克试题)

证明 这个结论可推广到一般情况即:

如果 n 等于任何一个素数 p 或这个素数的乘幂,那么内接于单位圆的不同正 n 边形的边长乘积的平方等于 p.

本题不过是 $p=5$ 时的特例.

第6章 Tschebyscheff 多项式的三角形式在几何中的应用

边 A_0A_1 和 A_0A_2 的长度可以用 $\cos\varphi$ 表示. 这里 φ 是这样的角：$5\varphi = 90°\cdot k, k \in \mathbf{N}, k = 2m+1$，即 $\cos 5\varphi = 0$. 不失一般性，可以认为所研究的仅仅是包含在 $0°\sim 180°$ 范围内的角. 事实上，对于任何超过 $180°$ 的角，总可以在上面所限定的范围内找到另一个角，使得两个角的余弦相等.

我们注意到

$$\cos\frac{1}{5}90° = \cos 18° = \sin 72° = \frac{1}{2}A_0A_2$$

$$\cos\frac{3}{5}90° = \cos 54° = \sin 36° = \frac{1}{2}A_0A_1$$

$$\cos\frac{5}{5}90° = \cos 90° = 0$$

$$\cos\frac{7}{5}90° = \cos 126° = -\sin 36° = -\frac{1}{2}A_0A_1$$

$$\cos\frac{9}{5}90° = \cos 162° = -\sin 18° = -\frac{1}{2}A_0A_2$$

在三角方程 $\cos 5\varphi = 0$ 中作替换 $x = \cos\varphi$，我们可以得到能够求出所有上面的余弦值的方程.

为此可通过 $x = \cos\varphi$ 来表示 $\cos n\varphi$ 和 $\dfrac{\sin(n+1)\varphi}{\sin\varphi}$ 的递推关系式. 设

$$T_n(x) = \cos nx,\ U_n(x) = \frac{\sin(n+1)\varphi}{\sin\varphi}\quad (n=0,1,\cdots)$$

（用 $\dfrac{\sin(n+1)\varphi}{\sin\varphi}$ 来代替 $\sin(n+1)\varphi$ 是为了避免出现二次根式）

当 $n=0$ 时

$$\cos 0 = T_0(x) = 1,\ \frac{\sin\varphi}{\cos\varphi} = U_0(x) = 1$$

65

当 $n=1$ 时

$$\cos\varphi = T_1(x) = x, \quad \frac{\sin 2\varphi}{\sin\varphi} = 2\cos\varphi = U_1(x) = 2x$$

为了得到递推式，我们要利用公式

$$\cos(n+1)\varphi = \cos\varphi\cos n\varphi - \sin\varphi\sin n\varphi =$$
$$\cos\varphi\cos n\varphi - (1-\cos^2\varphi)\frac{\sin n\varphi}{\sin\varphi}$$

即

$$T_{n+1}(x) = xT_n(x) - (1-x^2)U_{n-1}(x) \qquad ㊸$$

$$\frac{\sin(n+2)\varphi}{\sin\varphi} = \frac{1}{\sin\varphi}[\sin\varphi\cos(n+1)\varphi + \cos\varphi\sin(n+1)\varphi]$$

即

$$U_{n+1}(x) = T_{n+1}(x) + xU_n(x) \qquad ㊹$$

利用式㊸,㊹可计算出

$$T_5(x) = 16x^5 - 20x^3 + 5x$$

方程 $T_5(x) = 0$ 的根是 $0, \pm\frac{1}{2}A_0A_1, \pm\frac{1}{2}A_0A_2$. 由韦达定理立即可得

$$\frac{1}{16}(A_0A_1 \cdot A_0A_2)^2 = \frac{5}{16}$$

即

$$(A_0A_1 \cdot A_0A_2)^2 = 5$$

周持中利用 Tschebyscheff 多项式研究了一类三角求值问题.

6.1 第一型 Tschebyscheff 多项式

假如我们要求 $\cos\frac{\pi}{16}, \cos\frac{3\pi}{16}, \cdots, \cos\frac{15\pi}{16}$ 这 8 个三角函数之和，每两个之积之和，……，每 7 个之积之

第6章 Tschebyscheff 多项式的三角形式在几何中的应用

和,以及这 8 个数之积各等于多少,我们自然地会这样设想:若能作出一个八次多项式,使其根恰为 $\cos\dfrac{\pi}{8}$, $\cos\dfrac{3\pi}{8},\cdots,\cos\dfrac{5\pi}{8}$,那么由韦达定理立即得到所需的结果. 这就为我们引出了一个普遍性的问题:求作一个 n 次多项式,使其根恰为 $x_k=\cos\dfrac{2k-1}{2n}\pi\,(k=1,\cdots,n)$.

记 $\theta_k=\dfrac{2k-1}{2n}\pi$,则 $\cos n\theta_k=\cos\dfrac{2k-1}{2}\pi=0$. 此说明 θ_k 为 $\cos n\theta=0$ 之根. 若 $\cos n\theta$ 可展为 $\cos\theta$ 的多项式,则 x_k 就是此多项式的根了. 由棣美弗公式,我们有

$$\cos n\theta=\dfrac{1}{2}[\,(\cos n\theta+\mathrm{i}\sin n\theta)+(\cos n\theta-\mathrm{i}\sin n\theta)\,]=$$
$$\dfrac{1}{2}[\,(\cos\theta+\mathrm{i}\sin\theta)^n+(\cos\theta-\mathrm{i}\sin\theta)^n\,]$$

用二项式定理展开并化简得

$$\cos n\theta=\sum_{k=0}^{[\frac{n}{2}]}(-1)^k\binom{n}{2k}\sin^{2k}\theta\cos^{n-2k}\theta$$

令 $x=\cos\theta,\theta\in[0,\pi]$,得一个 n 次多项式
$T_n(x)=\cos n(\arccos x)=$

$$\sum_{k=0}^{[\frac{n}{2}]}(-1)^k\binom{n}{2k}(1-x^2)^k x^{n-2k}\quad(-1\leqslant x\leqslant 1)\quad ㊺$$

我们称之为第一型 Tschebyscheff 多项式. 它具有如下性质

性质1 $T_n(x)$ 的 n 个根为

$$x_k=\cos\dfrac{2k-1}{2n}\pi\quad(k=1,\cdots,n)$$

性质 2 设 $T_n(x) = \sum_{i=0}^{n} a_i x^{n-i}$,则 $a_{2t+1} = 0$,而

$$a_{2t} = (-1)^t \sum_{k=0}^{[\frac{n}{2}]} \binom{n}{2k} \binom{k}{k-t} \quad (t = 0, 1, \cdots, [\frac{n}{2}])$$

㊻

证明 由式 ㊺

$$T_n(x) = \sum_{k=0}^{[\frac{n}{2}]} (-1)^k \binom{n}{2k} \sum_{j=0}^{k} (-1)^j \binom{k}{j} x^{n-2k+2j}$$

令 $n - 2k + 2j = n - i$,得 $j = k - \frac{i}{2}$. 当 $i = 2t, j = k - t$, 所以

$$a_{2t} = \sum_{k=0}^{[\frac{n}{2}]} (-1)^k \binom{n}{2k} (-1)^{k-t} \binom{k}{k-t}$$

由此即证.

推论 1 由以上性质,有
$$a_0 = 2^{n-1} \qquad ㊼$$

证明 由式 ㊻ 有 $a_0 = \sum_{k=0}^{[\frac{n}{2}]} \binom{n}{2k} = \frac{1}{2} \sum_{i=0}^{n} \binom{n}{i} = 2^{n-1}$.

推论 2 当 $2 \nmid n$ 时,$a_n = 0$,当 $2 \mid n$ 时
$$a_n = (-1)^{\frac{n}{2}} \qquad ㊽$$

证明 当 $2 \mid n$ 时,在式 ㊻ 中令 $t = \frac{n}{2}$. 因为 $k \geq t$,故必有 $k = \frac{n}{2}$. 由此即证.

我们以 $\sigma_i(x_1, x_2, \cdots, x_n)$ 表 $x_1, x_2, \cdots, x_n$ 每 i 个积

第6章 Tschebyscheff 多项式的三角形式在几何中的应用

之和,$i = 1,\cdots,n$,则由韦达定理有 $\sigma_i(\cos\frac{\pi}{2n},\cos\frac{3\pi}{2n},\cdots,\cos\frac{2n-1}{2n}\pi) = (-1)^i\frac{a_i}{a_0}$,故由式㊻,㊼得

推论 3 $\sigma_{2t+1}(\cos\frac{\pi}{2n},\cos\frac{3\pi}{2n},\cdots,\cos\frac{2n-1}{2n}\pi) = 0$,

而

$$\sigma_{2t}(\cos\frac{\pi}{2n},\cos\frac{3\pi}{2n},\cdots,\cos\frac{2n-1}{2n}\pi) = $$

$$\frac{(-1)^t}{2^{n-1}}\sum_{k=0}^{[\frac{n}{2}]}\binom{n}{2k}\binom{k}{k-t} \qquad ㊽$$

特别的,结合推论2可得

推论 4 当 $2\nmid n$ 时,$\prod_{k=1}^{n}\cos\frac{2k-1}{2n}\pi = 0$;当 $2\mid n$ 时,

$\prod_{k=1}^{n}\cos\frac{2k-1}{2n}\pi = (-1)^{\frac{n}{2}}\cdot\frac{1}{2^{n-1}}$.

由于利用式㊻计算 $T_n(x)$ 之系数的过程较复杂,我们常可利用如下的递推公式

性质 3 $T_n(x)$ 适合 $T_0(x) = 1$,$T_1(x) = x$ 及

$$T_{n+1}(x) = 2xT_n(x) - T_{n-1}(x) \quad (n \geqslant 1) \qquad ㊿$$

证明 $T_0(x) = \cos(0\cdot\arccos x) = 1$,$T_1(x) = \cos(1\cdot\arccos x) = x$. 当 $n \geqslant 1$ 时,由 $\cos(n+1)\theta + \cos(n-1)\theta = 2\cos n\theta\cos\theta$,以 $\theta = \arccos x$ 代入即得式㊿,利用式㊿,我们可算得

$$T_2(x) = 2x^2 - 1$$
$$T_3(x) = 4x^3 - 3x$$
$$T_4(x) = 8x^4 - 8x^2 + 1$$
$$T_5(x) = 16x^5 - 20x^3 + 5x$$

$$T_6(x)=32x^6-48x^4+18x^2-1$$
$$\vdots \qquad (51)$$

由以上讨论可知,要求某些三角函数的对称式的值,关键在于作出以这些三角函数为根的多项式.下面我们将把重点放在后一问题上.

6.2 第二型 Tschebyscheff 多项式

假如要求以 $x_k=\cos\frac{k\pi}{n+1}(k=1,\cdots,n)$ 为根的 n 次多项式.令 $\theta_k=\frac{k\pi}{n+1}$,则 $\sin(n+1)\theta_k=0$,由此我们可考虑是否能把 $\sin(n+1)\theta$ 展开为 $\cos\theta$ 的多项式.因为

$$\begin{aligned}\sin(n+1)\theta=&\left(\frac{1}{2\mathrm{i}}\right)[(\cos\theta+\mathrm{i}\sin\theta)^{n+1}-\\&(\cos\theta-\mathrm{i}\sin\theta)^{n+1}]=\\&\sin\theta\sum_{k=0}^{[\frac{n}{2}]}\binom{n+1}{2k+1}\cdot(\mathrm{i}\sin\theta)^{2k}(\cos\theta)^{n-2k}\end{aligned}$$

可知 $\frac{\sin(n+1)\theta}{\sin\theta}$ 符合要求,即

$$\begin{aligned}\frac{\sin(n+1)\theta}{\sin\theta}=&\sum_{k=0}^{[\frac{n}{2}]}(-1)^k\cdot\binom{n+1}{2k+1}\cdot\\&(1-\cos^2\theta)^k(\cos\theta)^{n-2k}\end{aligned}$$

令 $x=\cos\theta,\theta\in(0,\pi)$,得一个 n 次多项式

$$U_n(x)=\frac{\sin[(n+1)\arccos x]}{\sin(\arccos x)}=$$

第6章 Tschebyscheff 多项式的三角形式在几何中的应用

$$\sum_{k=0}^{[\frac{n}{2}]} (-1)^k \binom{n+1}{2k+1} (1-x^2)^k x^{n-2k}$$
$$(-1 \leq x \leq 1) \quad \text{⑤}$$

我们称之为第二型 Tschebyscheff 多项式,其中 $x = \pm 1$ 是用极限方法补充定义的. 仿前,$U_n(x)$ 也有如下一些性质.

性质 4 $U_n(x)$ 的 n 个根为

$$x_k = \cos\frac{k\pi}{n+1} \quad (k=1,\cdots,n)$$

性质 5 设 $U_n(x) = \sum_{i=0}^{n} b_i x^{n-i}$,则 $b_{2t+1} = 0$,而

$$b_{2t} = (-1)^t \sum_{k=0}^{[\frac{n}{2}]} \binom{n+1}{2k+1} \cdot \binom{k}{k-t} \quad (t=0,1,\cdots,[\frac{n}{2}])$$
$$\text{⑤}$$

推论 1 $b_0 = 2^n$.

推论 2 当 $2 \nmid n$ 时,$b_n = 0$;$2 \mid n$ 时

$$b_n = (-1)^{\frac{n}{2}} \quad \text{⑤}$$

推论 3 $\sigma_{2t+1}(\cos\frac{\pi}{n+1}, \cos\frac{2\pi}{n+1}, \cdots, \cos\frac{n\pi}{n+1}) = 0$,

而

$$\sigma_{2t}(\cos\frac{\pi}{n+1}, \cos\frac{2\pi}{n+1}, \cdots, \cos\frac{n\pi}{n+1}) =$$
$$\frac{(-1)^t}{2^n} \cdot \sum_{k=0}^{[\frac{n}{2}]} \binom{n+1}{2k+1}\binom{k}{k-t}$$
$$\text{⑤}$$

推论 4 $\prod_{k=1}^{n} \cos\frac{k\pi}{n+1}$,当 $2 \nmid n$ 时值为 0;当 $2 \mid n$ 时,值为

$$(-1)^{\frac{n}{2}}\cdot(\frac{1}{2^n}) \tag{56}$$

性质 6 $U_n(x)$适合 $U_0(x)=1, U_1(x)=2x$ 及

$$U_{n+1}(x)=2xU_n(x)-U_{n-1}(x) \quad (n\geqslant 1) \tag{57}$$

注 $U_0(x)=\frac{\sin(\arccos x)}{\sin(\arccos x)}=1$, $U_1(x)=\frac{\sin(2\arccos x)}{\sin(\arccos x)}=2\cos(\arccos x)=2x$, 当 $n\geqslant 1$ 时, 由 $\sin(n+2)\theta+\sin n\theta=2\sin(n+1)\theta\cos\theta$ 两边同除以 $\sin\theta$ 并令 $\theta=\arccos x$ 即得式(57).

利用式(57)可算得

$$U_2(x)=4x^2-1$$
$$U_3(x)=8x^3-4x$$
$$U_4(x)=16x^4-12x^2+1$$
$$U_5(x)=32x^5-32x^3+6x$$
$$U_6(x)=64x^6-80x^4+24x^2-1$$
$$\vdots \tag{58}$$

以 Tschebyscheff 多项式为基础, 可构造许多以其他三角函数为根的多项式.

例 11 求以 $\pm\sin\frac{k\pi}{2n+1}(k=1,\cdots,n)$ 为根的 $2n$ 次多项式.

解 由式(52), 在 $U_{2n}(x)$ 中以 $y^2=1-x^2$ 代入得所求为

$$V_{2n}(y)=\sum_{k=0}^{n}(-1)^k\binom{2n+1}{2k+1}\cdot y^{2k}(1-y^2)^{n-k} \tag{59}$$

在求三角函数对称式之值中的应用:

利用前面导出的各多项式及韦达定理, 可直接求

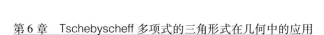

第 6 章　Tschebyscheff 多项式的三角形式在几何中的应用

出相应的三角函数对称式之值. 在此不一一列举. 这里只举若干变化的例子.

例 12　求 $\sin^2 \dfrac{\pi}{7} + \sin^2 \dfrac{2\pi}{7} + \sin^2 \dfrac{3\pi}{7}$ 之值.

解法 1　由以上知 $V_6(y)$ 有根 $\pm \sin \dfrac{\pi}{7}$, $\pm \sin \dfrac{2\pi}{7}$, $\pm \sin \dfrac{3\pi}{7}$. 在式㊺中令 $x^2 = 1 - y^2$, 得 $V_6(y) = -64y^6 + 112y^4 - 56y^2 + 7$. 令 $z = y^2$, 得 $G_3(z) = -64z^3 + 112z^2 - 56z + 7$ 有根为 $\sin^2 \dfrac{\pi}{7}, \sin^2 \dfrac{2\pi}{7}, \sin^2 \dfrac{3\pi}{7}$. 故由韦达定理得 $\sin^2 \dfrac{\pi}{7} + \sin^2 \dfrac{2\pi}{7} + \sin^2 \dfrac{3\pi}{7} = \dfrac{112}{64} = \dfrac{7}{4}$.

解法 2　$\sin^2 \dfrac{\pi}{7} + \sin^2 \dfrac{2\pi}{7} + \sin^2 \dfrac{3\pi}{7} = \dfrac{1}{2}(3 - \cos \dfrac{2\pi}{7} - \cos \dfrac{4\pi}{7} - \cos \dfrac{6\pi}{7}) = \dfrac{3}{2} - \dfrac{1}{2}(-\cos \dfrac{\pi}{7} + \cos \dfrac{2\pi}{7} - \cos \dfrac{3\pi}{7})$. 又 $-\cos \dfrac{\pi}{7}, \cos \dfrac{2\pi}{7}, -\cos \dfrac{3\pi}{7}$ 为 $F_3(x) = U_3(x) + U_2(x) = 8x^3 + 4x^2 - 4x - 1$ 之根, 所以 $-\cos \dfrac{\pi}{7} + \cos \dfrac{2\pi}{7} - \cos \dfrac{3\pi}{7} = -\dfrac{4}{8} = -\dfrac{1}{2}$, 由此可得所求之同一结果.

例 13　求 $\cos^5 \dfrac{2\pi}{7} + \cos^5 \dfrac{4\pi}{7} + \cos^5 \dfrac{6\pi}{7}$ 之值.

解　记 $x_k = \cos \dfrac{2k\pi}{7} (k = 1, 2, 3)$, $S_r = x_1^r + x_2^r + x_3^r$. 由上例知, $F_3(x_k) = 0$, 可得 $x_k^3 = -\dfrac{1}{2} x_k^2 + \dfrac{1}{2} x_k + \dfrac{1}{8}$, 两

边同乘 x_k^r 得 $x_k^{r+3} = -\frac{1}{2}x_k^{r+2} + \frac{1}{2}x_k^{r+1} + \frac{1}{8}x_k^r$,两边对 $k=1,2,3$ 求和得 $S_{r+3} = -\frac{1}{2}S_{r+2} + \frac{1}{2}S_{r+1} + \frac{1}{8}S_r$. 因为 $S_0 = 3$, $S_1 = -\frac{1}{2}$, $S_2 = \frac{1}{2}(3+\cos\frac{4\pi}{7}+\cos\frac{8\pi}{7}+\cos\frac{12\pi}{7}) = \frac{1}{2}(3+\cos\frac{2\pi}{7}+\cos\frac{4\pi}{7}+\cos\frac{6\pi}{7}) = \frac{1}{2}\times(3-\frac{1}{2}) = \frac{5}{4}$,所以反复应用上述递推公式得 $S_3 = -\frac{1}{2}\times\frac{5}{4}+\frac{1}{2}\times(-\frac{1}{2})+\frac{1}{8}\times 3 = -\frac{1}{2}$,同理 $S_4 = \frac{13}{16}$, $S_5 = -\frac{1}{2}$,此即所求.

例 14 设 $1 \leqslant k \leqslant m-1$,求 $A_m = \prod_{r=1}^{m-1}(\cos\frac{k\pi}{m} - \cos\frac{r\pi}{m})$ 之值($r \neq k$).

解 因为 $\cos\frac{r\pi}{m}(r=1,\cdots,m-1)$ 为 $U_{m-1}(x)$ 之根,所以有分解式 $U_{m-1}(x) = 2^{m-1}\prod_{r=1}^{m-1}(x-\cos\frac{r\pi}{m})$. 于是

$$A_m = \frac{1}{2^{m-1}}\lim_{x\to\cos(\frac{k\pi}{m})}\frac{U_{m-1}(x)}{x-\cos(\frac{k\pi}{m})} = \frac{1}{2^{m-1}}U'_{m-1}(\cos\frac{k\pi}{m}).$$

另一方面由 $U_n(x)$ 之推导过程可知

$$U_{m-1}(x) = \frac{1}{2\mathrm{i}\sqrt{1-x^2}} \cdot [(x+\mathrm{i}\sqrt{1-x^2})^m - (x-\mathrm{i}\sqrt{1-x^2})^m]$$

由此求导得

第 6 章 Tschebyscheff 多项式的三角形式在几何中的应用

$$U'_{m-1}\left(\cos\frac{r\pi}{m}\right) = (-1)^{k+1} m \left(\sin\frac{k\pi}{m}\right)^{-2}$$

故所求为

$$A_m = (-1)^{k+1} \frac{m}{2^{m-1}} \csc^2 \frac{k\pi}{m}$$

Tschebyscheff 多项式的三角形式不等式

第 7 章

例 15 求证：对于任意 $a \in (-\infty, +\infty)$，有
$$|2\cos^2\alpha + a\cos\alpha + b| \leq 1$$
当且仅当
$$2\cos^2\alpha + a\cos\alpha + b = \cos 2\alpha$$
（第五届全俄中学生数学竞赛）

证明 充分性：当 $2\cos^2\alpha + a\cos\alpha + b = \cos 2\alpha$ 时，显然有
$$|2\cos^2\alpha + a\cos\alpha + b| \leq 1$$

必要性：若
$$|2\cos^2\alpha + a\cos\alpha + b| \leq 1$$
则当 $b > -1$ 时，有

第7章 Tschebyscheff 多项式的三角形式不等式

$$2\cos^2\alpha + a\cos\alpha + b = 2\cos^2\alpha - 1 + a\cos\alpha + b + 1$$

由假设知,$b+1>0$,取 α 值使 $a\cos\alpha = |a|$,则有

$$|2\cos^2\alpha + a\cos\alpha + b| = 1 + |a| + b + 1 > 1$$

矛盾.

又当 $b<-1$ 时,若令 $\cos\alpha = 0$,则有

$$|2\cos^2\alpha + a\cos\alpha + b| = |b| > 1$$

矛盾. 所以,$b = -1$.

当 $b = -1$ 时,取 α 使 $a\cos\alpha = |a|$,则

$$|2\cos^2\alpha + a\cos\alpha - 1| = 1 + |a| > 1$$

矛盾.

综上所述,必有 $b = -1, a = 0$,即有

$$2\cos^2\alpha + a\cos\alpha + b = 2\cos^2\alpha - 1 = \cos 2\alpha$$

江苏江阴要塞中学的李尧明和江苏江阴青阳中学的李尧亮在 1987 年第四期《中等数学》中以"一道竞赛题的启示及推广"为题,对上述命题作了一些变换,得到了许多结论.

令 $\cos\alpha = x$,则有:对于任意 $x \in [-1,1]$,$|2x^2 + ax + b| \leq 1$,当且仅当 $a = 0, b = -1$.

令 $\cos\alpha = \frac{1}{2}x$,则有:对于任意 $x \in [-2,2]$,$|2x^2 + ax + b| \leq 4$,当且仅当 $a = 0, b = -4$.

令 $\cos\alpha = \frac{1}{3}(2x-1)$,则有:对于任意 $x \in [-1, 2]$,$|8x^2 + ax + b| \leq 9$,当且仅当 $a = -8, b = -7$.

令 $\cos\alpha = \frac{1}{n-m}(2x-m-n)$,$m<n$,则有:对于任意 $x \in [m,n]$,$|8x^2 + ax + b| \leq (n-m)^2$,当且仅当 $a = -8(m+n), b = 2(m+n)^2 - (n-m)^2$.

Tschebyscheff 逼近定理

因 $|\cos 3\alpha| \leqslant 1, \alpha \in (-\infty, +\infty)$，于是由例 15 我们又可猜测到以下命题.

命题 1 对于任意 $\alpha \in (-\infty, +\infty)$，有 $|4\cos^3\alpha + a\cos^2\alpha + b\cos\alpha + c| \leqslant 1$，当且仅当 $4\cos^3\alpha + a\cos^2\alpha + b\cos\alpha + c = \cos 3\alpha$.

更一般的，$|\cos n\alpha| \leqslant 1$，同时，应用棣美弗公式可以证明 $\cos n\alpha$ 能够展开为关于 $\cos \alpha$ 的 n 次实系数多项式，且首项系数为 2^{n-1}，于是根据命题我们自然会猜想出以下命题.

命题 2 若 $f_n(\cos \alpha)$ 是关于 $\cos \alpha$ 的 n 次实系数多项式，且首项系数（最高次项）为 2^{n-1}，对于任意 $\alpha \in (-\infty, +\infty)$，有 $|f_n(\cos \alpha)| \leqslant 1$，当且仅当 $f_n(\cos \alpha) = \cos n\alpha$（$n$ 为自然数）.

证明 对于 $n=1$，我们容易证明命题是成立的. 对于 $n=2$，我们在前面已给出了证明. 由于 $|\cos n\alpha| \leqslant 1$，所以，我们只要证明当 $n \geqslant 3$ 时，上述命题的必要性成立.

设 $f_n(\cos \alpha) = \cos n\alpha + g(\cos \alpha)$，其中 $g(\cos \alpha)$ 是次数低于 n 次的实系数多项式. 于是我们只要证明 $g(\cos \alpha) = 0, \alpha \in (-\infty, +\infty)$.

下面，我们对 $g(\cos \alpha)$ 的项数采用第二数学归纳法来加以证明：

当 $g(\cos \alpha)$ 仅有一项时，可设 $g(\cos \alpha) = a\cos^k\alpha$，其中 $0 \leqslant k \leqslant n-1$.

若 $a > 0$，则取 $\alpha = 0$，便有 $|f_n(\cos \alpha)| = |\cos n\alpha + a\cos^k\alpha| = 1 + a > 1$，矛盾.

若 $a < 0$，则取 $\alpha = \dfrac{\pi}{n} \in \left(0, \dfrac{\pi}{2}\right)$（因为 $n \geqslant 3$），有

第7章 Tschebyscheff 多项式的三角形式不等式

$|f_n(\cos\alpha)| = |\cos n\alpha + a\cos^k\alpha| = 1-a > 1$,矛盾. 由此可见,$a=0$,即 $g(\cos\alpha)=0$.

假设当 $g(\cos\alpha)$ 不超过 $n-1$ 项时,由 $|\cos n\alpha + g(\cos\alpha)| \leq 1$, $\alpha \in (-\infty, +\infty)$,可以导出 $g(\cos\alpha)=0$.

当 $g(\cos\alpha)$ 有 n 项时,令 $g(\cos\alpha) = g_1(\cos\alpha) + g_2(\cos\alpha)$,其中 $g_1(\cos\alpha)$ 为 g 中所有关于 $\cos\alpha$ 是奇数次项的和,$g_2(\cos\alpha)$ 是 g 中关于 $\cos\alpha$ 是偶数次项的和再加上常数项. 显然,g_1 和 g_2 的项数都不超过 $n-1(n \geq 3)$.

由于 $|\cos n\alpha + g(\cos\alpha)| \leq 1$ 对于任意 $\alpha \in (-\infty, +\infty)$ 均能成立,故 $|\cos n(\pi-\alpha) + g[\cos(\pi-\alpha)]| \leq 1$ 对于任意 $\alpha \in (-\infty, +\infty)$ 亦能成立.

当 n 为奇数时,$\cos n(\pi-\alpha) = -\cos n\alpha$,所以
$|\cos n\alpha + g(\cos\alpha)| =$

$\dfrac{1}{2}|\cos n\alpha + g(\cos\alpha) + \cos n\alpha - g(-\cos\alpha)| \leq$

$\dfrac{1}{2}\{|\cos n\alpha + g(\cos\alpha)| + |\cos n\alpha - g(-\cos\alpha)|\} =$

$\dfrac{1}{2}\{|\cos n\alpha + g(\cos\alpha)| + |\cos n(\pi-\alpha) + g[\cos(\pi-\alpha)]|\} \leq \dfrac{1}{2}(1+1) = 1$

由归纳假设可知:$g_1(\cos\alpha) = 0$. 将 $g_1(\cos\alpha) = 0$ 代入 $|\cos n\alpha + g(\cos\alpha)| \leq 1$,得 $|\cos n\alpha + g_2(\cos\alpha)| \leq 1$. 从而有 $g_2(\cos\alpha) = 0$. 所以 $g(\cos\alpha) = 0$.

当 n 为偶数时,$\cos n(\pi-\alpha) = \cos n\alpha$. 所以

Tschebyscheff 逼近定理

$$|\cos n\alpha + g_2(\cos\alpha)| \leq \frac{1}{2}\{|\cos n\alpha + g(\cos\alpha)| +$$
$$|\cos n\alpha + g(-\cos\alpha)|\} =$$
$$\frac{1}{2}\{|\cos n\alpha + g(\cos\alpha)| +$$
$$|\cos n(\pi-\alpha) +$$
$$g[\cos(\pi-\alpha)]|\} \leq$$
$$\frac{1}{2}(1+1) = 1$$

故由归纳假设可知:$g_2(\cos\alpha) = 0$. 将 $g_2(\cos\alpha) = 0$ 代入 $|\cos n\alpha + g(\cos\alpha)| \leq 1$,得 $|\cos n\alpha + g_1(\cos\alpha)| \leq 1$,故有 $g_1(\cos\alpha) = 0$,所以,$g(\cos\alpha) = 0$.

由归纳法原理知上述命题成立.

Tschebyscheff 多项式的拉格朗日形式

例 16 在 $[-1,1]$ 内给定 n 个不同点 $x_1, x_2, \cdots, x_n (n \geq 2)$. 记 $t_k (1 \leq k \leq n)$ 为点 x_k 到其他所有点 $x_i (i \neq k)$ 的距离的乘积,求证

$$\sum_{k=1}^{n} \frac{1}{t_k} \geq 2^{n-2}$$

(选自《美国数学月刊》1954 年第 54 页,1994 年国家数学集训队第三次测验题)

证明 引进 Tschebyscheff 多项式和拉格朗日多项式:

我们知道:

(1) Tschebyscheff 多项式 $T_{n-1}(x)$

的首项是 $2^{n-2}x^{n-1}$.

(2)当 $-1\leqslant x\leqslant 1$ 时, $|T_n(x)|\leqslant 1$.

我们对 Tschebyscheff 多项式 $T_{n-1}(x)$ 使用拉格朗日插值公式,有

$$T_{n-1}(x)=$$

$$T_{n-1}(x_1)\frac{(x-x_2)\cdot(x-x_3)\cdot\cdots\cdot(x-x_n)}{(x_1-x_2)\cdot(x_1-x_3)\cdot\cdots\cdot(x_1-x_n)}+$$

$$T_{n-1}(x_2)\frac{(x-x_1)\cdot(x-x_3)\cdot\cdots\cdot(x-x_n)}{(x_2-x_1)\cdot(x_2-x_3)\cdot\cdots\cdot(x_2-x_n)}+\cdots+$$

$$T_{n-1}(x_n)\frac{(x-x_1)\cdot(x-x_2)\cdot\cdots\cdot(x-x_{n-1})}{(x_n-x_1)\cdot(x_n-x_2)\cdot\cdots\cdot(x_n-x_{n-1})} \quad ⑥⓪$$

这里 $x_1,x_2,\cdots,x_n$ 是题中 $[-1,1]$ 内给定的 n 个不同点,比较式⑥⓪两端首项系数,有

$$2^{n-2}=\frac{T_{n-1}(x_1)}{(x_1-x_2)\cdot(x_1-x_3)\cdot\cdots\cdot(x_1-x_n)}+$$

$$\frac{T_{n-1}(x_2)}{(x_2-x_1)\cdot(x_2-x_3)\cdot\cdots\cdot(x_2-x_n)}+\cdots+$$

$$\frac{T_{n-1}(x_n)}{(x_n-x_1)\cdot(x_n-x_2)\cdot\cdots\cdot(x_n-x_{n-1})} \quad ⑥①$$

由所给条件

$$t_1=|(x_1-x_2)\cdot(x_1-x_3)\cdot\cdots\cdot(x_1-x_n)|$$

$$t_2=|(x_2-x_1)\cdot(x_2-x_3)\cdot\cdots\cdot(x_2-x_n)|$$

$$\vdots$$

$$t_n=|(x_n-x_1)\cdot(x_n-x_2)\cdot\cdots\cdot(x_n-x_{n-1})|$$

对式⑥①两端取绝对值得

$$2^{n-2}\leqslant\frac{|T_{n-1}(x_1)|}{|(x_1-x_2)\cdot(x_1-x_3)\cdot\cdots\cdot(x_1-x_n)|}+$$

$$\frac{|T_{n-1}(x_2)|}{|(x_2-x_1)\cdot(x_2-x_3)\cdot\cdots\cdot(x_2-x_n)|}+\cdots+$$

第8章 Tschebyscheff 多项式的拉格朗日形式

$$\frac{|T_{n-1}(x_n)|}{|(x_n-x_1)\cdot(x_n-x_2)\cdot\cdots\cdot(x_n-x_{n-1})|} \leqslant \frac{1}{t_1}+\frac{1}{t_2}+\cdots+\frac{1}{t_n}.$$

证毕.

例 17 令 $\cos\theta = x$,表达式

$$T_n(x) = \cos n\theta = \cos(n\arccos x)$$

是 x 的 n 阶多项式,称为 Tschebyscheff 多项式. $T_n(x)$ 的首项系数等于 2^{n-1}. 这类多项式的前五个是

$$T_1(x)=x,\ T_2(x)=2x^2-1,\ T_3(x)=4x^3-3x$$
$$T_4(x)=8x^4-8x^2+1,\ T_5(x)=16x^5-20x^3+5x$$

T_n 的根都是位于区间 $(-1,1)$ 内的不同的实数,即

$$\cos\frac{(2k-1)\pi}{2n} \quad (k=1,2,\cdots,n)$$

设 $x_1,x_2,\cdots,x_n$ 是任意 n 个不同的实数或复数,且 $f(x)=a_0(x-x_1)\cdot(x-x_2)\cdot\cdots\cdot(x-x_n)$, $a_0\neq 0$. 令

$$f_k(x)=\frac{1}{f'(x_k)}\cdot\frac{f(x)}{x-x_k}=$$

$$\frac{(x-x_1)\cdot\cdots\cdot(x-x_{k-1})\cdot(x-x_{k+1})\cdot\cdots\cdot(x-x_n)}{(x_k-x_1)\cdot\cdots\cdot(x_k-x_{k-1})\cdot(x_k-x_{k+1})\cdot\cdots\cdot(x_k-x_n)}$$

每一个 $n-1$ 次多项式 p 都可用它在点 $x_1,x_2,\cdots,x_n$ 处的值表示为 $p(x)=p(x_1)f_1(x)+p(x_2)f_2(x)+\cdots+p(x_n)f_n(x)$,这就是拉格朗日插值公式. 多项式 f_k 叫作基本插值公式. 证明下述结果:设

$$x_k=\cos\frac{(2k-1)\pi}{2n} \quad (k=1,2,\cdots,n)$$

是 Tschebyscheff 多项式 T_n 的根. 若 Q 是次数小于或等于 $n-1$ 的多项式,则

Tschebyscheff 逼近定理

$$Q(x) = \frac{1}{n} \sum_{k=1}^{n} (-1)^{k-1} \sqrt{1-x_k^2} Q(x_k) \frac{T_n(x)}{x-x_k}$$

证明 因为 $T_n(x) = \cos(n\arccos x)$，我们得

$$T_n'(x) = \frac{n}{\sqrt{1-x^2}} \sin(n\arccos x)$$

从 $\arccos x_k = \frac{(2k-1)\pi}{2n}$ 得，$\sin(n\arccos x_k) = \sin\frac{(2k-1)\pi}{2} = (-1)^{k-1}$，所以

$$T_n'(x_k) = \frac{(-1)^{k-1} n}{\sqrt{1-x_k^2}}$$

为了证明上述等式，我们注意到等式两边是次数不大于 $n-1$ 的多项式，所以，只要证明等式两边对 n 个 x_k 的值相同即可. 当 $x \to x_k$ 时

$$\frac{T_n(x)}{x-x_k} \to T_n'(x_k) = \frac{n(-1)^{k-1}}{\sqrt{1-x_k^2}}$$

而当 $x = x_k$ 时右边的每一项除第 k 项外都为零，因为对 $i = 1, 2, \cdots, n$ 及 $k \neq i$，当 x_i 是 T_n 的根时，有

$$\frac{T_n(x_i)}{x_i - x_k} = 0$$

因此，右边的表达式关于 a 有拉格朗日插值多项式.

记 $T_n(x) = \cos\theta(x = \cos\theta)$ 为 Tschebyscheff 多项式，$x_k = \cos\theta_k = \cos\frac{2k-1}{2n}\pi \, (k=1,\cdots,n)$ 是它的 n 个零点. 以 $(1-x^2)T_n(x)$ 的零点为节点的拉格朗日插值多项式有如下形式

$$L_n(f,x) = L_n(f,\theta) = \cos^2\frac{\theta}{2} \cos n\theta f(1) +$$

84

第 8 章 Tschebyscheff 多项式的拉格朗日形式

$$(-1)^n \sin^2 \frac{\theta}{2} \cos n\theta f(-1) +$$

$$\sum_{k=1}^{n} f(\cos \theta_k) \frac{\sin^2 \theta}{\sin \theta_k} l_k(\theta) \quad (x = \cos \theta)$$

这里 $l_k(\theta) = \dfrac{(-1)^{k+1} \sin \theta_k \cos n\theta}{n(\cos \theta - \cos \theta_k)}$ $(k=1,2,\cdots,n)$.

再谈最佳逼近多项式

第 9 章

第 39 届(1978 年 12 月 2 日)美国大学生数学竞赛的试题 B5 为:

例 18 求出一个最大的数 A,使得存在一个实系数多项式
$$p(x) = Ax^4 + Bx^3 + Cx^2 + Dx + E$$
当 $-1 \leq x \leq 1$ 时满足 $0 \leq p(x) \leq 1$.

解 如果知道下述结论:

在满足条件 $-1 \leq f(x) \leq 1$,$-1 \leq x \leq 1$ 的所有四次多项式 $f(x)$ 中 Tschebyscheff 多项式 $C_4(x) = 8x^4 - 8x^2 + 1 = \cos(4\arccos x)$ 具有最大的首项系数,则取 $p(x) = [C_4(x)+1]/2$(这样可以保证将 $p(x)$ 限制在 $[0,1]$ 内),就知 A 的最大值为 4.

第9章 再谈最佳逼近多项式

如果不用这个结论可按下法来求.

先令 $Q(x) = [p(x) + p(-x)]/2$，则原来的条件可改写为

$$0 \leq Q(x) = Ax^4 + Cx^2 + E \leq 1 \quad (-1 \leq x \leq 1)$$

令 $x^2 = y$，则化为

$$0 \leq R(y) = Ay^2 + Cy + E \leq 1 \quad (0 \leq y \leq 1)$$

令 $y = \dfrac{z+1}{2}$，$S(z) = R(\dfrac{z+1}{2})$，则为

$$0 \leq S(z) = \dfrac{A}{4}z^2 + Fz + G \quad (-1 \leq z \leq 1)$$

对于 $T(z) = \dfrac{S(z) + S(-z)}{2}$，得到

$$0 \leq \dfrac{A}{4}z^2 + G \leq 1 \quad (-1 \leq z \leq 1)$$

再令 $z^2 = w$，最后得到

$$0 \leq \dfrac{A}{4}w + G \leq 1 \quad (0 \leq w \leq 1)$$

显然，当 $G = 0$ 时，即当

$$T(z) = z^2, \ R(y) = (2y-1)^2, \ Q(x) = 4x^4 - 4x^2 + 1$$

时，得出 A 的最大值为 4.

下面讲最佳逼近多项式的存在性：

对于定义在 $[a, b]$ 上的连续函数 $f(x)$ 与 $\varphi(x)$，通常称 $\max\limits_{a \leq x \leq b} |f(x) - \varphi(x)|$ 为 $f(x)$ 与 $\varphi(x)$ 的偏差. 如果 $\tilde{x} \in [a, b]$ 使得 $|f(\tilde{x}) - \varphi(\tilde{x})| = \max\limits_{a \leq x \leq b} |f(x) - \varphi(x)|$，则 $\tilde{x}$ 称为近似函数 $\varphi(x)$ 的偏差点. 特别的，若有

$$\varphi(\tilde{x}) - f(\tilde{x}) = \max\limits_{a \leq x \leq b} |f(x) - \varphi(x)|$$

则称 $\tilde{x}$ 为 $\varphi(x)$ 的正偏差点；若有

$$\varphi(\tilde{x}) - f(\tilde{x}) = -\max\limits_{a \leq x \leq b} |f(x) - \varphi(x)|$$

则称 $\tilde{x}$ 为 $\varphi(x)$ 的负偏差点. 由于假设 $f(x)$ 与 $\varphi(x)$ 在 $[a,b]$ 上连续,故偏差点总是存在的,但正、负偏差点不一定同时存在.

关于最佳逼近多项式 $p_n^*(x)$ 的存在性,有如下定理.

定理4(波莱尔存在性定理) 对任意给定的 $[a,b]$ 上连续的函数 $f(x)$,总存在 $p_n^*(x) \in P_n(x)$,使得

$$\max_{a \le x \le b} |f(x) - p_n^*(x)| \le \min_{p_n(x) \in P_n(x)} \{\max_{a \le x \le b} |f(x) - p_n(x)|\}$$

成立.

定理5 若 $p_n^*(x) \in P_n(x)$ 是 $f(x)$ 在区间 $[a,b]$ 上的最佳逼近多项式,则 $p_n^*(x)$ 一定同时存在正、负偏差点.

证明 不妨设 $p_n^*(x)$ 与 $f(x)$ 不存在负偏差点而仅存在正偏差点. 设 $\tilde{x}$ 是其中一个正偏差点,则有

$$p_n^*(\tilde{x}) - f(\tilde{x}) = \max_{a \le x \le b} |f(x) - p_n^*(x)| \triangleq u$$

由于 $p_n^*(x) - f(x)$ 是 $[a,b]$ 上的连续函数,则必存在最大值 M 和最小值 m,有

$$-\mu < m \le p_n^*(x) - f(x) \le M = u$$

令

$$s = \frac{m+\mu}{2} > 0$$

且

$$-\frac{\mu-m}{2} = m-s \le p_n^*(x) - s - f(x) \le \mu - s = \frac{\mu-m}{2}$$

$$\max_{a \le x \le b} |(p_n^*(x) - s) - f(x)| = \frac{\mu-m}{2} = \mu - s < \mu =$$

$$\max_{a \le x \le b} |f(x) - p_n^*(x)|$$

此式与 $p_n^*(x)$ 是函数 $f(x)$ 的最佳逼近多项式相矛盾,

第9章 再谈最佳逼近多项式

故 $p_n^*(x)$ 同时存在正负偏差点.

定理 6(Tschebyscheff 定理) $p_n^*(x)$ 是 $f(x)$ 在 $[a,b]$ 上的 n 次最佳逼近多项式的充要条件是在区间 $[a,b]$ 上 $p_n^*(x)$ 至少具有 $n+2$ 个依次轮流为正负的偏差点 $x_i(i=1,2,\cdots,n+2)$,即

$$a \leqslant x_1 < x_2 < \cdots < x_{n+2} \leqslant b$$

证明 充分性:假设 $p_n^*(x)$ 不是 $f(x)$ 的 n 次最佳逼近多项式,由定理 4 知, $f(x)$ 的最佳逼近多项式是存在的,记为 $p(x)$. 于是有

$$\max_{a \leqslant x \leqslant b} |p(x)-f(x)| < \max_{a \leqslant x \leqslant b} |p_n^*(x)-f(x)| \quad ⑥②$$

由于 $a \leqslant x_1 < x_2 < \cdots < x_{n+2} \leqslant b$ 是近似函数 $p_n^*(x)$ 的 $n+2$ 个依次轮流为正负的偏差点,从而有

$$|p_n^*(x_i)-f(x_i)| = \max_{a \leqslant x \leqslant b} |p_n^*(x)-f(x)|$$
$$(i=1,2,\cdots,n+2) \quad ⑥③$$

对于不超过 n 次的多项式 $p_n^*(x)-p(x) = (p_n^*(x)-f(x))-(p(x)-f(x))$,由式 ⑥②,⑥③ 可知 $p_n^*(x_i)-p(x_i)$ 和 $p_n^*(x_i)-f(x_i)$ 的符号相同 ($1 \leqslant i \leqslant n+2$). 而 $\{p_n^*(x_i)-f(x_i)\}(i=1,2,\cdots,n+2)$ 的符号交错,从而 $\{p_n^*(x_i)-p(x_i)\}(i=1,2,\cdots,n+2)$ 的符号也交错. 在区间 $[x_i,x_{i+1}](i=1,2,\cdots,n+1)$ 上对函数 $p_n^*(x)-p(x)$ 应用连续函数的零点定理有: $p_n^*(x)-p(x)$ 在 (x_i,x_{i+1}) 上至少有一个零点,从而在 $[a,b]$ 内至少有 $n+1$ 个零点. 这与 $p_n^*(x)-p(x)$ 为不超过 n 次的多项式矛盾,故充分性正确.

必要性:由定理 5 知,最佳逼近多项式 $p_n^*(x)$ 在区间 $[a,b]$ 上一定同时存在正负偏差点. 现假设它仅有 $m(m<n+2)$ 个依次轮流为正负的偏差点,因此存在 m

个没有公共内部的子区间

$$I_1=[a,\xi_1],I_2=[\xi_1,\xi_2],\cdots,I_m=[\xi_{m-1},b]$$

使得在每个子区间上所包含的偏差点，或者全是正偏差点，或者全是负偏差点. 也即有 $\xi_i(i=1,2,\cdots,m-1)$ 都不是交错点. 由于 $p_n^*(x)$ 有 m 个交错偏差点，故这 m 个子区间的任何相邻两个子区间所包含的偏差点类型必然是相反的. 记含正偏差点的子区间个数为 m_1，含负偏差点的子区间个数为 $m_2=m-m_1$.

我们仅对 $I_1,I_3,\cdots,I_{2m_1-1}$ 含正偏差点，其他子区间含负偏差点的情形进行讨论，另一相反情形完全可类似讨论.

因为 $p_n^*(x)-f(x)$ 是连续函数，故在仅含正偏差点的子区间 I_k 上有最大值 $\widetilde{M}_k$ 和最小值 $\widetilde{m}_k$，且 $M_k=\max\limits_{a\leqslant x\leqslant b}|p_n^*(x)-f(x)|$，$m_k>-\max\limits_{a\leqslant x\leqslant b}|p_n^*(x)-f(x)|$ $(k=1,3,5,\cdots,2m_1-1)$. 因而对 $x\in\Delta_1\xlongequal{\triangle}I_1\cup I_3\cup\cdots\cup I_{2m_1-1}$ 有不等式

$$-\max_{a\leqslant x\leqslant b}|p_n^*(x)-f(x)|<\widetilde{m}\leqslant p_n^*(x)-f(x)\leqslant \max_{a\leqslant x\leqslant b}|p_n^*(x)-f(x)|\xlongequal{\triangle}\mu \quad (64)$$

其中 $\widetilde{m}=\min\{m_1,m_3,m_5,\cdots,m_{2m_1-1}\}$. 令 $s_1=\dfrac{(\widetilde{m}+\mu)}{2}$，有

$$-\mu+s_1<p_n^*(x)-f(x)\leqslant\mu \quad (65)$$

同样，对于 $x\in\Delta_2\xlongequal{\triangle}I_2\cup I_4\cup\cdots\cup I_{2m_2}$，有

$$-\mu\leqslant p_n^*(x)-f(x)\leqslant\mu-s_2 \quad (66)$$

第9章 再谈最佳逼近多项式

其中

$$s_2 = \frac{(\mu - \widetilde{M})}{2}$$

且 $\widetilde{M} = \max\{M_2, M_4, \cdots, M_{2m_2}\}$,而 $M_2, M_4, \cdots, M_{2m_2}$ 是连续函数 $p_n^*(x) - f(x)$ 在含负偏差点子区间 $I_2, I_4, \cdots, I_{2m_2}$ 上的相应最大值.

令 $s = \min\{s_1, s_2\}$,由式⑥⑤,⑥⑥可得对任何 $x \in \Delta_1$,有

$$-\mu + s < p_n^*(x) - f(x) < \mu \qquad ⑥⑦$$

对任何 $x \in \Delta_2$,有

$$-\mu \leqslant p_n^*(x) - f(x) \leqslant \mu - s \qquad ⑥⑧$$

构造 $m-1$ 次多项式 $\Phi(x) = (x - \xi_1) \cdot (x - \xi_2) \cdot \cdots \cdot (x - \xi_{m-1})$,它是区间 $[a, b]$ 上的连续函数,故存在绝对值充分小的非零实数 β,使得对于任何 $x \in [a, b]$ 都有

$$|\beta \Phi(x)| < s \qquad ⑥⑨$$

由于多项式 $\Phi(x)$ 在这 m 个子区间上的函数值符号是交错的,可选择实数 β 的符号,使得当 $x \in \Delta_1$ 时,有 $\beta \Phi(x) \geqslant 0$;而当 $x \in \Delta_2$ 时,有 $\beta \Phi(x) \leqslant 0$.

由式⑥⑦,⑥⑧和式⑥⑨可知,当 $x \in \Delta_1$ 时,有

$$-\mu + \beta \Phi(x) < p_n^*(x) - f(x) \leqslant \mu$$

当 $x \in \Delta_2$ 时,有

$$-\mu \leqslant p_n^*(x) - f(x) < \mu + \beta \Phi(x) \qquad ⑦⓪$$

于是,有

$$\left| p_n^*(x) - f(x) - \frac{\beta \Phi(x)}{2} \right| \leqslant \mu - \frac{\beta \Phi(x)}{2} \quad (x \in \Delta_1)$$

$$\left| p_n^*(x) - f(x) - \frac{\beta \Phi(x)}{2} \right| \leqslant \mu + \frac{\beta \Phi(x)}{2} \quad (x \in \Delta_2) \qquad ⑦①$$

Tschebyscheff 逼近定理

从而对任何 $x \in [a,b]$，有

$$\left| p_n^*(x) - \frac{\beta \Phi(x)}{2} - f(x) \right| \leq \mu - \frac{\beta \Phi(x)}{2} \quad ⑦2$$

当 $x \neq \xi_i (i = 1, 2, \cdots, m-1)$ 时，有

$$\left| \left(p_n^*(x) - \frac{\beta \Phi(x)}{2} \right) + f(x) \right| < \mu \quad ⑦3$$

当 $x = \xi_i (i = 1, 2, \cdots, m-1)$ 时，因 ξ_i 不是 $p_n^*(x)$ 的交错点，同样得到不等式⑦3. 故总有

$$\max_{a \leq x \leq b} \left| \left(p_n^*(x) - \frac{\beta \Phi(x)}{2} \right) - f(x) \right| < \mu$$

由 $m < n+2$ 知 $\Phi(x)$ 是不超过 n 次的多项式，即有 $p_n^*(x) - \frac{\beta \Phi(x)}{2}$ 是不超过 n 次的多项式. 上述不等式与 $p_n^*(x)$ 是 $f(x)$ 在 $[a,b]$ 上的最佳逼近多项式相矛盾，故假设 $m < n+2$ 不成立，从而必要性成立.

定理 7(唯一性) 若函数 $f(x)$ 在 $[a,b]$ 上是连续函数，则 $f(x)$ 的最佳逼近多项式 $p_n^*(x) \in P_n(x)$ 是唯一的.

证明 若 $p(x), q(x)$ 都是 $f(x)$ 在区间 $[a,b]$ 上且是 $P_n(x)$ 中的最佳逼近多项式，则当任何 $x \in [a,b]$ 都有

$$-\varepsilon \leq p(x) - f(x) \leq \varepsilon$$
$$-\varepsilon \leq q(x) - f(x) \leq \varepsilon$$

即有

$$-\varepsilon \leq \frac{p(x) + q(x)}{2} - f(x) \leq \varepsilon$$

因此 $r(x) \stackrel{\Delta}{=\!=} \frac{1}{2}(p(x) + q(x))$ 也是函数 $f(x)$ 在 $P_n(x)$ 中的最佳逼近多项式.

第9章 再谈最佳逼近多项式

由定理 3 知 $r(x)$ 存在 $n+2$ 个依次轮流为正负的偏差点 $x_i(i=1,2,\cdots,n+2)$,满足

$$\varepsilon = |r(x_i)-f(x_i)| = |\frac{1}{2}(p(x_i)+q(x_i))-f(x_i)| \leq$$

$$\frac{1}{2}|p(x_i)-f(x_i)|+\frac{1}{2}|q(x_i)-f(x_i)| \leq \varepsilon$$

于是对于 $1 \leq i \leq n+2$,有

$$|p(x_i)-f(x_i)| = \varepsilon, \quad |q(x_i)-f(x_i)| = \varepsilon$$

这就是说 $x_i(i=1,2,\cdots,n+2)$ 也是 $p(x)$ 和 $q(x)$ 关于 $f(x)$ 的偏差点.

又由

$$|r(x_i)-f(x_i)| = \left|\frac{p(x_i)-f(x_i)}{2}+\frac{q(x_i)-f(x_i)}{2}\right| = \varepsilon$$

可知 $p(x_i)-f(x_i)$ 与 $q(x_i)-f(x_i)$ 同号,从而有

$$p(x_i)-f(x_i) = q(x_i)-f(x_i) \quad (i=1,2,\cdots,n+2)$$

即

$$[p(x)-q(x)]_{x=x_i} = 0 \quad (i=1,2,\cdots,n+2)$$

而 $p(x),q(x)$ 均是不超过 n 次的多项式,故 $p(x) = q(x)$,即最佳逼近多项式唯一.

最小偏差多项式

第 10 章

例 19 设 M 是所有形如
$$p(x) = ax^3 + bx^2 + cx + d \quad (a, b, c, d \in \mathbf{R})$$
且当 $x \in [-1, 1]$ 时满足 $|p(x)| < 1$ 的多项式的集合. 证明：必存在某个数 k，使得对所有 $p(x) \in M$，都有 $|a| \le k$，并求最小的 k. (捷克 1974 年数学奥林匹克试题)

证明 因为 Tschebyscheff 多项式 $p_0(x) = 4x^3 - 3x$ 满足 $p_0(-1) = 1$，$p_0(1) = 1$，并且在它的极值点上有
$$p\left(-\frac{1}{2}\right) = 1, \quad p\left(\frac{1}{2}\right) = -1$$
所以 $p_0(x) \in M$. 现在证明,对任意 $p(x) \in M$,有 $|a| \le 4$. 否则,设存在多项式

第10章 最小偏差多项式

$$p(x) = ax^3 + bx^2 + cx + d \in M$$

且 $|a| > 4$，则非零多项式

$$Q(x) = p_0(x) - \frac{4}{a} p(x)$$

的次数不超过 2，由于当 $|x| \leqslant 1$ 时，$\left|\frac{4}{a} p(x)\right| < 1$，所以

$$Q(-1) < 0,\ Q(-\frac{1}{2}) > 0,\ Q(\frac{1}{2}) < 0,\ Q(1) > 0$$

故多项式 $Q(x)$ 至少有 3 个根，矛盾. 于是所求的 k 等于 4.

最小偏差等于零的代数多项式——Tschebyscheff 多项式

考虑 $f(x) = x^n$，n 是自然数，我们将研究它在 $C[-1,1]$ 上被次数不大于 $n-1$ 的代数多项式来逼近时的最佳逼近多项式及最佳逼近值

$$E_{n-1}(x^n) = \inf_{-1 \leqslant x \leqslant 1} \max |P_n(x)| = \inf_{\{a_j\}} \max_{-1 \leqslant x \leqslant 1} \left| x^n - \sum_{j=0}^{n-1} a_j x^j \right|$$

显然，$\{x^j\}$ $(0 \leqslant j \leqslant n-1)$ 在 $C[-1,1]$ 上满足哈尔条件，因此设 $P_{n-1}^*(x) = \sum_{j=0}^{n-1} a_j^* x^j$ 是最佳逼近多项式，则多项式 $Q_n(x) = x^n - P_{n-1}^*(x)$ 在 $[-1,1]$ 上至少有 $n+1$ 个点以正负相间的符号取得值 $E_{n-1}(x^n)$. 现在我们来研究 n 次多项式 $Q_n(x)$ 的性质：

（1）$Q_n(x)$ 在 $(-1,1)$ 内部有且只有 $n-1$ 个点使其在这些点上以正负相间的符号取到值 $E_{n-1}(x^n) = \max_{-1 \leqslant x \leqslant 1} |Q_n(x)|$，且在这些点上 $Q_n'(x) = 0$.

事实上，在 $(-1,1)$ 内这种点至少有 $n-1$ 个. 另一

Tschebyscheff 逼近定理

方面,若在 $(-1,1)$ 上有至少 n 个这样的点,则由于这样点是极值点,因此在 $(-1,1)$ 上就至少有 n 个点使得 $Q'_n(x) = 0$. 因为 $Q'_{n-1}(x)$ 是一个 $n-1$ 次多项式,因此就得到了矛盾.

今后将这些点记作 $x_i, -1 < x_i < 1, 1 \leq i \leq n-1$. 显然有 $Q_n(x_i) = \pm(-1)^i E_{n-1}(x^n)$, $Q'_n(x_i) = 0, 1 \leq i \leq n-1$.

(2) $Q_n(1) = \pm E_{n-1}(x^n)$, $Q_n(-1) = \pm E_{n-1}(x^n)$.

事实上, $Q_n(x)$ 在 $[-1,1]$ 上至少有 $n+1$ 个点取到 $\pm E_{n-1}(x^{n-1})$, 而根据(1), 在 $(-1,1)$ 上只可能有 $n-1$ 个点取到 $\pm E_{n-1}(x^{n-1})$, 因此还有两个点必在区间的端点上.

(3) $Q_n(x)$ 满足方程

$$(1-x)(1+x)(Q'_n(x))^2 = n^2([E_{n-1}(x^n)]^2 - Q_n^2(x)) \quad ⑭$$

事实上, 式⑭左边是一个 $2n$ 次多项式, 它以 x_i ($1 \leq i \leq n-1$) 为重根, 以 $x = \pm 1$ 为单根; 而式⑭右边也是一个 $2n$ 次多项式, 它以 $x = \pm 1$ 及 $x = x_i (1 \leq i \leq n-1)$ 为根. 由于 $\dfrac{\mathrm{d}}{\mathrm{d}x}([E_{n-1}(x^n)]^2 - Q_n^2(x)) = -2Q_n(x) \cdot Q'_n(x)$ 也以 $x = x_i (1 \leq i \leq n-1)$ 为根. 因此两边的根完全相同(包括根的重数). 因此两边只相差一个常数倍数, 但比较其最高项系数, 可以看出此常数为 1. 因此式⑭成立.

从式⑭得到

$$\frac{\mathrm{d}(\pm Q_n(x))}{\sqrt{E_{n-1}^2 - Q_n^2(x)}} = \frac{n\mathrm{d}x}{\sqrt{1-x^2}}, \quad E_{n-1} = E_{n-1}(x^n)$$

即

$$\arccos\left(\frac{\pm Q_n(x)}{E_{n-1}}\right) = n \arccos x + c$$

第10章 最小偏差多项式

由此得到 $Q_n(x) = \pm E_{n-1}\arccos(n\arccos x + c)$

其中 c 为某个常数，由于 $Q_n(1) = \pm E_n$，代入上式后，可得 $c = 0$ 或 π. 因而就有

$$Q_n(x) = \pm E_{n-1}\cos(n\arccos x) \qquad ⑦⑤$$

由于 $Q_n(x)$ 是一个首项系数为 1 的 n 次多项式，由式⑦⑤就可以求出其前面应该取正号. 为此，令 $t = \arccos^{-1} x$，则

$$\cos(n\arccos x) = \cos nt = \frac{e^{int} + e^{-int}}{2} =$$

$$\frac{(\cos t + i\sin t)^n + (\cos t - i\sin t)^n}{2} =$$

$$\frac{1}{2}[(x + \sqrt{x^2-1})^n + (x - \sqrt{x^2-1})^n] =$$

$$\frac{1}{2}\sum_{k=0}^{n}\binom{0}{k}((x^k\sqrt{x^2-1})^{n-k} +$$

$$(-1)^{n-k}x^k(\sqrt{x^2-1})^{n-k})$$

是一个 n 次多项式. 为了求其最高项系数，求

$$\lim_{x\to\infty}\frac{\arccos(n\arccos x)}{x^n} =$$

$$\lim_{x\to\infty}\frac{\frac{1}{2}[(x+\sqrt{x^2-1})^n + (x-\sqrt{x^2-1})^n]}{x^n} = 2^{n-1} \qquad ⑦⑥$$

比较式⑦⑤与式⑦⑥，就得到 E_{n-1} 前的符号取正值，且

$$E_{n-1} = \frac{1}{2^{n-1}}, \quad Q_n(x) = \frac{1}{2^{n-1}}\arccos(n\arccos x) \qquad ⑦⑦$$

由式⑦⑦所确定的 n 次多项式称为 Tschebyscheff 多项式，其首项系数为 1，有时也用 $T_n(x) = \cos(n\arccos x)$ 表示，其首项系数为 2^{n-1}.

高次 Tschebyscheff 逼近

第 11 章

11.1 一道集训队试题

在 2001 年,中国国家集训队考试试题中有如下题目:

例 20 记 $F = \max\limits_{1 \leqslant x \leqslant 3} |x^3 - ax^2 - bx - c|$,当 a,b,c 取遍所有实数时,求 F 的最小值.

对这道试题,湖南师范大学的叶军教授给出了一个很直接的解法.

这个问题实际上是求解下面的极值问题:

设 $f(x) = x^3 + px^2 + qx + r$. 其中 $p, q, r \in \mathbf{R}$. 求 $\min\limits_{p,q,r \in \mathbf{R}} \max\limits_{x \in [-1,1]} |f(x)|$.

事实上

第 11 章 高次 Tschebyscheff 逼近

$$f(1)-f(-1)=2+2q,\ f(\tfrac{1}{2})-f(-\tfrac{1}{2})=\tfrac{1}{4}+q$$

消去 q 得

$$\tfrac{3}{4}=\tfrac{1}{2}f(1)-\tfrac{1}{2}f(-1)+f(-\tfrac{1}{2})-f(\tfrac{1}{2})$$

所以

$$\tfrac{3}{4}\leqslant(\tfrac{1}{2}+\tfrac{1}{2}+1+1)\max\{|f(1)|\cdot|f(-1)|\cdot|f(-\tfrac{1}{2})|\cdot|f(\tfrac{1}{2})|\}\leqslant 3\max_{-1\leqslant x\leqslant 1}|f(x)| \qquad ⑱$$

所以

$$\max_{-1\leqslant x\leqslant 1}|f(x)|\geqslant\tfrac{1}{4} \qquad ⑲$$

式 ⑱ 成立当且仅当

$$\begin{cases} \arg f(1)=\arg(-f(-1))=\arg f(-\tfrac{1}{2})= \\ \qquad \arg(-f(\tfrac{1}{2}))=0 \\ |f(1)|=|f(-1)|=|f(\tfrac{1}{2})|=|f(-\tfrac{1}{2})|=\tfrac{1}{4} \end{cases} \Leftrightarrow$$

$$\begin{cases} f(1)=\tfrac{1}{4} \\ f(-1)=-\tfrac{1}{4} \\ f(-\tfrac{1}{2})=\tfrac{1}{4} \\ f(\tfrac{1}{2})=-\tfrac{1}{4} \end{cases} \Leftrightarrow \begin{cases} 1+p+q+r=\tfrac{1}{4} \\ -1+p-q+r=-\tfrac{1}{4} \\ -\tfrac{1}{8}+\tfrac{1}{4}p-\tfrac{1}{2}q+r=\tfrac{1}{4} \\ \tfrac{1}{8}+\tfrac{1}{4}p+\tfrac{1}{2}q+r=-\tfrac{1}{4} \end{cases} \Leftrightarrow$$

$$\begin{cases} p=r=0 \\ q=-\tfrac{3}{4} \end{cases}$$

Tschebyscheff 逼近定理

即当且仅当 $f(x)=x^3-\dfrac{3}{4}x$ 时,式⑱取等号,以下只需证当 $f(x)=x^3-\dfrac{3}{4}x$ 时

$$\max_{-1\leqslant x\leqslant 1}|f(x)|=\frac{1}{4}$$

而这由平均值不等式及柯西不等式是容易做到的.

11.2 П.Л. Tschebyscheff 定理

下面我们将建立最佳逼近多项式的某些性质. 设 $f(x)\in C[a,b]$,固定某一 $n\geqslant 0$,并用 $P(x)$ 表示集合 H_n 中 $f(x)$ 的最佳逼近多项式之一①. 就是说

$$\max|P(x)-f(x)|=E_n$$

如果 $E_n=0$,则这就表示函数 $f(x)$ 是一个次数不超过 n 的多项式. 我们把这种显而易见的情况丢开不论而当做是 $E_n>0$.

因为在闭区间 $[a,b]$ 上的连续函数 $|P(x)-f(x)|$ 能达到它的最大值,所以至少可以求出这样的一点 x_0,使得

$$|P(x_0)-f(x_0)|=E_n$$

我们把任何这样的点都称为多项式 $P(x)$ 的 (e) 点. 如果

$$P(x_0)-f(x_0)=E_n$$

---

① 就事实而论,在每个 H_n 中都只有一个最佳逼近多项式. 而由于没有证明这一点(本节中便要做到这一点),所以我们还不得不说成是这种多项式中"之一".

我们便称这种(e)点x_0为$(+)$点,如果
$$P(x_0)-f(x_0)=-E_n$$
就称它为$(-)$点.

定理 8 $(+)$点及$(-)$点都是存在的.

证明 我们假定,例如,多项式$P(x)$没有$(-)$点. 这时对于$[a,b]$中的一切x,都有
$$P(x)-f(x)>-E_n$$

特别的,连续函数$P(x)-f(x)$的最小值也大于$-E_n$. 用$-E_n+2h$来表示这个最小值,其中$h>0$. 这时对于$[a,b]$中的一切x,都有
$$-E_n+2h\leqslant P(x)-f(x)\leqslant E_n$$
从而
$$-E_n+h\leqslant [P(x)-h]-f(x)\leqslant E_n-h$$
于是
$$|[P(x)-h]-f(x)|\leqslant E_n-h$$

但这就表示多项式$P(x)-h$与$f(x)$的偏差小于E_n,显然这是与E_n的定义相违背的.

已证明的定理在几何上是十分明显的.

其实,我们试考察曲线
$$y=f(x)+E_n,\ y=f(x)-E_n \qquad ⑳$$

多项式$P(x)$的圆形($a\leqslant x\leqslant b$时)位于曲线⑳之间的带形中. 已证的定理就意味着,这个圆形应当与式⑳的上、下两条曲线至少各相切一次,这是十分明显的,因为,若曲线$y=P(x)$一次也不与下边这条曲线$y=f(x)-E_n$(缺乏$(-)$点)相切,则我们可以把它稍微向下移动一下,就得到一条位于曲线$y=f(x)$附近的一个较狭的带形内的曲线. 前面提到的证明,实际说来,就是

Tschebyscheff 逼近定理

把这段讨论赋予精确的形式而已.

然而,正如 Tschebyscheff 所指出过的,曲线 $y=P(x)$ 与边界曲线⑧的交点却非常之多. 确切地说,我们有

定理 9 在闭区间 $[a,b]$ 上存在着由 $n+2$ 个点所构成的点组

$$x_1<x_2<\cdots<x_{n+2}$$

它们交错为(+)点及(-)点.

这样的点组,我们把它们称为 Tschebyscheff 交错组.

证明 我们用点组

$$u_0=a<u_1<u_2<\cdots<u_m=b$$

把 $[a,b]$ 分成这样小的闭区间 $[u_k,u_{k+1}]$,使得在其中的每一个上,连续函数 $P(x)-f(x)$ 的振幅都小于 $\dfrac{E_n}{2}$.

若闭区间 $[u_k,u_{k+1}]$ 至少含有一个(e)点,则称之为(e)区间. 容易看出,差 $P(x)-f(x)$ 在(e)区间上不能为零,因而必保持常号. 所以我们可以把(e)区间的集合分成两类:一类是(+)区间,在其上差 $P(x)-f(x)$ 是正的;而另一类是(-)区间,在其上差是负的.

其次,我们把所有的(e)区间依从左到右的顺序一一编号

$$d_1,d_2,d_3,\cdots,d_N \qquad ⑧$$

为了确定起见,假定 d_1 为(+)区间.

我们把序列⑧依下列格式分成若干组

$$\begin{cases} d_1,d_2,\cdots,d_{k_1},[(+)\text{区间}] \\ d_{k_1+1},d_{k_1+2},\cdots,d_{k_2},[(-)\text{区间}] \\ \qquad\qquad\vdots \\ d_{k_{m-1}+1},d_{k_{m-1}+2},\cdots,d_{k_m},[(-1)^{m-1}\text{区间}] \end{cases} \qquad ⑧$$

上面的每一组中,都至少含有一个闭区间,并且在第一组的每个闭区间中至少要有一个(+)点,第二组的每个闭区间中至少要有一个(-)点,如此,等等. 因此,要证明本定理,只需证明

$$m \geqslant n+2 \qquad \text{⑧}$$

即可(前面的定理只保证了 $m \geqslant 2$). 今假定

$$m < n+2 \qquad \text{⑭}$$

由于差 $P(x)-f(x)$ 在闭区间 d_{k_1} 及 d_{k_1+1} 上的符号相异,闭区间 d_{k_1} 的右端点就不会与闭区间 d_{k_1+1} 的左端点重合. 所以我们可以在闭区间 d_{k_1} 之右与 d_{k_1+1} 之左选取一点 z_1,用符号来表示可以写成

$$d_{k_1} < z_1 < d_{k_1+1}$$

与此完全相仿,我们可以取这样的一些点 z_2, $z_3, \cdots, z_{m-1}$,而使

$$d_{k_2} < z_2 < d_{k_2+1}$$
$$\vdots$$
$$d_{k_{m-1}} < z_{m-1} < d_{k_{m-1}+1}$$

完成上述步骤之后,我们令

$$\rho(x) = (z_1-x) \cdot (z_2-x) \cdot \cdots \cdot (z_{m-1}-x)$$

根据我们的假定⑭,有 $m-1 \leqslant n$,所以多项式 $\rho(x)$ 属于 H_n,而且除了诸点 z_i 之外,多项式 $\rho(x)$ 没有其他的极点,因之它在诸闭区间 d_k 上都不能为零,因而在这些闭区间上保持常号. 不难看出,在式⑫的第一组中的每个闭区间上,多项式 $\rho(x)$ 是正的(因为每个因子 z_i-x 都是正的). 在式⑫的第二组中的每个闭区间上,多项式 $\rho(x)$ 是负的(因为它含有一个负因子 z_1-x),这样继续讨论下去,我们可以断定,在式⑪的所有(e)区间上

Tschebyscheff 逼近定理

多项式 $\rho(x)$ 与 $P(x)-f(x)$ 的符号相同.

设 $[u_i, u_{i+1}]$ 是原分划中的任一闭区间,若它不是 (e) 区间,则量

$$\max_{u_i \leqslant x \leqslant u_{i+1}} |P(x)-f(x)| \qquad \text{⑧⑤}$$

将严格小于 E_n,因此,如果用 E^* 表示式⑧⑤中的最大数,就得到

$$E^* < E_n$$

令

$$R = \max_{a \leqslant x \leqslant b} |\rho(x)|$$

并取这样小的正数 λ,使其满足①

$$\lambda R < E_n - E^*, \quad \lambda R < \frac{1}{2} E_n \qquad \text{⑧⑥}$$

如果令

$$Q(x) = P(x) - \lambda \rho(x)$$

则可证明多项式 $Q(x)$(显然属于 H_n)与 $f(x)$ 的偏差将小于 E_n. 由于这是不可能的,因而我们就得到了所期望的矛盾. 于是最后便归结到要证明不等式

$$\Delta(Q) < E_n \qquad \text{⑧⑦}$$

---

① 容易看出,$E_n - E^* < \frac{1}{2} E_n$,所以不等式⑧⑥中的第二个可以从第一个推出来. 事实上,如果 u_p 是闭区间 d_{k_1} 的右端点,则 $P(u_p) - f(u_p) > \frac{1}{2} E_n$(这可以根据 d_{k_1} 含有 $(+)$ 点以及 $P(x) - f(x)$ 在 d_{k_1} 上的振幅小于 $\frac{1}{2} E_n$ 来推出). 另一方面,u_p 是闭区间 $[u_p, u_{p+1}]$ 的左端点,这个闭区间紧靠着 d_{k_1} 之右,它已经不是 (e) 区间,所以 $|P(u_p) - f(u_p)| \leqslant E^*$. 从而知 $E^* > \frac{1}{2} E_n$.

第 11 章 高次 Tschebyscheff 逼近

设 $[u_i, u_{i+1}]$ 为原分划中的闭区间而不是 (e) 区间,并设 $x \in [u_i, u_{i+1}]$,那么

$$|Q(x)-f(x)| \leq |P(x)-f(x)|+\lambda|\rho(x)| \leq$$
$$E^* + \lambda R < E_n$$

现在假定 x 属于 (e) 区间 d_k 中的某一个. 这时,两个数 $P(x)-f(x)$ 及 $\lambda\rho(x)$ 便有着相同的符号. 同时 $|P(x)-f(x)| > \lambda|\rho(x)|$,因为

$$|P(x)-f(x)| > \frac{1}{2}E_n, \ \lambda|\rho(x)| < \frac{1}{2}E_n$$

从而

$$|Q(x)-f(x)| = |P(x)-f(x)-\lambda\rho(x)| =$$
$$|P(x)-f(x)| - \lambda|\rho(x)|$$

所以

$$|Q(x)-f(x)| \leq E_n - \lambda|\rho(x)| < E_n$$

这是因为在 (e) 区间上 $\rho(x) \neq 0$ 的缘故.

于是,对于 $[a,b]$ 中的任何 x 都有

$$|Q(x)-f(x)| < E_n$$

因此便推得式⑧⑦,定理获证.

我们指出,作多项式 $Q(x)$ 时可以不依赖于式⑧④的假定,并且对于它来说,不等式⑧⑦仍然是正确的,可是在 $m \geq n+2$ 时我们却得不到矛盾,因为对于这样的 m,多项式 $Q(x)$ 便不属于 H_n 了.

尽管这个定理的证明相当复杂,但其基本观点却与前述定理相似. Tschebyscheff 曾试图证明:当缺乏 $n+2$ 项的交错组时,$P(x)$ 与 $f(x)$ 的偏差在从 $P(x)$ 中减去用适当的方法选出的多项式 $\rho(x)$ 时是可以减小的. 因为要达到这个目的,就必须在所有 (e) 点处减小差 $P(x)-f(x)$ 的绝对值,所以在这些点处,$\rho(x)$ 应当与

Tschebyscheff 逼近定理

所述的差同号. 如果这个差的变号小于 $n+2$ 次,则次数不超过 n 的多项式 $\rho(x)$ 便会满足我们所提出的要求. 这时在其他点处可能发生多项式 $P(x)-\rho(x)$ 与 $f(x)$ 的偏差大于 E_n 的危险,把 $\rho(x)$ 乘以一个足够小的正因子 λ 很容易就可免除这样的危险. 于是推得,在缺乏交错组时,多项式 $P(x)$ 便不可能是最佳逼近多项式.

从已证明的定理便直接推得最佳逼近多项式的唯一性.

定理 10 在 H_n 中只存在着一个最小偏差多项式.

证明 假定在 H_n 中有两个最小偏差多项式 $P(x)$ 及 $Q(x)$,则对于 $[a,b]$ 中的一切 x,都有

$$-E_n \leq P(x)-f(x) \leq E_n,\ -E_n \leq Q(x)-f(x) \leq E_n$$

把这两个不等式相加后除以 2 所得结果便是

$$-E_n \leq \frac{P(x)+Q(x)}{2}-f(x) \leq E_n$$

这就表明,所得的半和

$$R(x)=\frac{P(x)+Q(x)}{2}$$

也是 $f(x)$ 的一个最小偏差多项式. 因而这时对于 $R(x)$ 便存在着 Tschebyscheff 交错点组

$$x_1<x_2<\cdots<x_{n+2}$$

设 x_k 是 $R(x)$ 的一个 $(+)$ 点,这就表示

$$\frac{P(x_k)-f(x_k)}{2}+\frac{Q(x_k)-f(x_k)}{2}=E_n$$

但

$$Q(x_k)-f(x_k) \leq E_n$$

$$|P(x_i)-f(x_i)|\leqslant E_n<A$$

从而便显然可知,两个差 $Q(x_i)-P(x_i)$ 与 $Q(x_i)-f(x_i)$ 的符号相同,因而差 $Q(x_i)-P(x_i)$ 在经过每一点 x_i 到下一点 x_{i+1} 时也变号,所以在区间 (x_1,x_2), (x_2,x_3), $\cdots$, (x_{n+1},x_{n+2}) 中的每一个内都有差 $Q(x)-P(x)$ 的根. 这样一来,由于有 $n+1$ 个根,于是这个差 $Q(x)-P(x)$(因为是一个次数小于或等于 n 的多项式)就应当恒等于零了. 然而这是不可能的,因为

$$\Delta(Q)=A>E_n=\Delta(P)$$

即多项式 $Q(x)$ 与 $P(x)$ 并不恒等. 所得的矛盾使我们断定 $A=E_n$ 为真. 而在这种情况下,便得到

$$\Delta(Q)=E_n$$

因而 $Q(x)$ 是最小偏差多项式.

依同样的推理方式,我们还可以建立一个估计 E_n 下界的定理.

定理 11 设对于函数 $f(x)\in C[a,b]$ 能够求出这样的多项式 $Q(x)\in H_n$ 以及这样的点组

$$x_1<x_2<\cdots<x_{n+2}$$

使得每次经过点 x_i 到 x_{i+1} 时,差 $Q(x_i)-f(x_i)$ 都变号. 如果用 A 表示诸数

$$|Q(x_i)-f(x_i)|\quad(i=1,2,\cdots,n+2)$$

中的最小者,则

$$A\leqslant E_n$$

证明 事实上,若假定 $A>E_n$,那么,逐字重复前一段的讨论,便仍然会得出矛盾来.

包瑞尔定理及 Tschebyscheff 定理确定了任一连续函数的最小偏差

Tschebyscheff 逼近定理

$$\frac{P(x_k)-f(x_k)}{2}+\frac{E_n}{2}\geqslant E_n$$

因而

$$P(x_k)-f(x_k)\geqslant E_n \qquad ⑧⑨$$

因为差 $P(x)-f(x)$ 不超过 E_n,所以在式⑧⑨中有等号,换句话说,x_k 是 $P(x)$ 的一个(+)点. 同理可知它也是 $Q(x)$ 的一个(+)点. 这样一来

$$P(x_k)-f(x_k)=E_n=Q(x_k)-f(x_k)$$

因而 $P(x_k)=Q(x_k)$. 仿此可确定 $P(x)$ 与 $Q(x)$ 在交错组⑧⑧中的(-)点处也都相等. 我们看到,次数都不高于 n 的多项式 $P(x)$ 及 $Q(x)$ 应当在式⑧⑧的 $n+2$ 个点处相等. 这只有当它们恒等时才是可能的.

不难证明,Tschebyscheff 交错点组的存在便是最佳逼近多项式的特征.

定理 12 设 $f(x)\in C[a,b]$,且 $Q(x)$ 为 H_n 中的任一多项式. 令

$$A=\max|Q(x)-f(x)|$$

如果在闭区间 $[a,b]$ 上存在着这样的点组

$$x_1<x_2<\cdots<x_{n+2} \qquad ⑨⓪$$

使得

$$|Q(x_i)-f(x_i)|=A \quad (i=1,2,\cdots,n+2) \qquad ⑨①$$

并且在经过每一点 x_i 到下一点 x_{i+1} 时,差 $Q(x_i)-f(x_i)$ 变号,那么,$A=E_n$,并且 $Q(x)$ 便是 $f(x)$ 的最佳逼近多项式.

证明 因为 $A=\Delta(Q)$,所以 $A\geqslant E_n$. 我们来证明 $A=E_n$. 如果不是这样,则

$$A>E_n \qquad ⑨②$$

用 $P(x)$ 表示 $f(x)$ 的最佳逼近多项式. 那么

$$Q(x_i) - P(x_i) = \{Q(x_i) - f(x_i)\} - \{P(x_i) - f(x_i)\}$$

差多项式的存在性及唯一性,可是却没有给出实际上用以求出这种多项式的方法. 后面这个问题是极其困难的,直到如今对于一般的情形都还没有解决. 下面我们所要讲的是 $n=0$ 及 $n=1$ 这两种最简单的情形.

在 $n=0$ 时解答极为简单. 那就是,如果用 m 及 M 分别表示定义在 $[a,b]$ 上的连续函数 $f(x)$ 的最小值及最大值,则一切常量中与 $f(x)$ 有最小偏差的量为

$$P = \frac{m+M}{2}$$

这在几何上是非常明显的,因为 $\Delta(P) = \frac{M-m}{2}$,而上下移动直线 $y=P$ 时,我们都显然会把偏差增大.

正式的证明也并不怎么困难. 事实上,如果 x_1 及 x_2 是这样的点,使得

$$f(x_1) = M, \ f(x_2) = m$$

则

$$P - f(x_1) = -\frac{M-m}{2}, \ P - f(x_2) = \frac{M-m}{2}$$

又因 $\Delta(P) = \frac{M-m}{2}$,所以点 x_1 及 x_2 就构成 Tschebyscheff 交错点组,而它的存在便表征出 P 是一个最佳逼近多项式.

在 $n=1$ 时问题也很简单,如果假定 $f(x)$ 是二次可微的并且 $f''(x)$ 不变号,例如

$$f''(x) > 0 \tag{93}$$

事实上,设 $P(x) = Ax + B$ 是最佳逼近多项式. 在含有 3 个点的 Tschebyscheff 交错组 $x_1 < x_2 < x_3$ 中,中间的点 x_2

Tschebyscheff 逼近定理

必位于 $[a,b]$ 的内部,即它是差 $f(x)-P(x)$ 的极值点①. 所以 $f'(x_2)-P'(x_2)=0$,从而
$$A = f'(x_2)$$

根据式㉓,导数 $f'(x)$ 为严格的递增函数,因而它只能取值 A 一次. 这就证明,差 $f(x)-P(x)$ 在 $[a,b]$ 内部除了点 x_2 以外不再有其他的极值点,因而交错点组中的其余两点 x_1 及 x_3 便与闭区间 $[a,b]$ 的端点重合
$$x_1 = a,\ x_3 = b$$

为了简单起见,把点 x_2 记作 c. 于是所引起的讨论就证明了
$$f(a)-P(a)=f(b)-P(b)=-\{f(c)-P(c)\}$$
或者,写得更详细一些
$$f(a)-Aa-B=f(b)-Ab-B=Ac+B-f(c)$$
由第一个等式便得出
$$A = \frac{f(b)-f(a)}{b-a} \qquad ㉔$$
而这时也容易求出 B
$$B = \frac{f(a)+f(c)}{2} - \frac{f(b)-f(a)}{b-a} \cdot \frac{a+c}{2}$$
这两个等式便把问题解决了,因为这时点 c 可由下面的方程求出来
$$f'(c) = A = \frac{f(b)-f(a)}{b-a} \qquad ㉕$$

所得到的解答具有很简单的几何意义. 就是,等式㉔指出了直线 $y=P(x)$ 平行于联结 $M(a, f(a))$ 及

---

① 由条件㉓显然可知,这便是差 $f(x)-P(x)$ 的极小值点,亦即,x_2 为 $P(x)$ 的一个 $(+)$ 点.

$N(b,f(b))$ 二点的弦 MN(图 2). 其次,把这条直线的方程改写成

$$y - \frac{f(a)+f(c)}{2} = A\left(x - \frac{a+c}{2}\right)$$

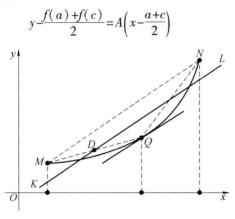

图 2

的形式以后,我们便看出,它通过联结 M 与 $Q(c,f(c))$ 二点的弦 MQ 的中点 D.

这样一来,我们便得出下面的法则:要作出函数 $f(x)$ 的一次最佳逼近多项式,我们可以如下进行:

(1) 作弦 MN.

(2) 在 $\widehat{MN}$ 上求出一点 Q,使在这点的切线与弦 MN 平行.

(3) 联结 Q 及 M,Q 及 N,并引出 $\triangle MQN$ 的中间线 KL. 直线 KL 就是所求多项式的图形.

我们来考虑,例如,函数 $y=\sqrt{x}$ 在闭区间 $[0,1]$ 上的一个最小偏差多项式.

这里弦 MN 的斜率等于 1,方程㉕就成为 $\dfrac{1}{2\sqrt{c}}=1$,

Tschebyscheff 逼近定理

从而 $c = \dfrac{1}{4}$,因而点 Q 及 D 便分别为 $Q\left(\dfrac{1}{4}, \dfrac{1}{2}\right)$ 及 $D\left(\dfrac{1}{8}, \dfrac{1}{4}\right)$,所以直线 KL 的方程便是

$$y - \dfrac{1}{4} = x - \dfrac{1}{8}$$

因而所求的多项式就是 $x + \dfrac{1}{8}$.

现在来考虑这样一个问题:在 H_{n-1} 中求一多项式 $P(x)$,使它在闭区间 $[-1,1]$ 上与函数 $f(x) = x^n$ 的偏差为最小.

设所求的多项式为

$$P(x) = ax^{n-1} + bx^{n-2} + \cdots + r$$

并且令

$$\widetilde{R}(x) = x^n - (ax^{n-1} + bx^{n-2} + \cdots + r)① \qquad ⑯$$

则所考虑的问题便归结为如何去选择系数 $a, b, \cdots, r$ 使得量

$$M = \max_{-1 \leqslant x \leqslant 1} |\widetilde{R}(x)|$$

尽可能的取最小值. 因为我们可以把任一个最高次项系数为 1 的多项式改写成式⑯的形式,而量 M 不是别的,它正是这个多项式与零的偏差. 于是我们便看出,提出的问题完全相当于这样的问题:从最高次项系数为 1 的一切多项式中,去求出在闭区间 $[-1,1]$ 上与零有最小偏差的多项式.

------

① 记号 $\widetilde{R}(x)$(或 $\widetilde{P}(x)$ 等)用以表示(根据 B·Л·巩恰罗夫的记法)多项式 $\widetilde{R}(x)$ 的最高次项系数为 1.

112

第 11 章 高次 Tschebyscheff 逼近

为了解决这个非常重要的问题,我们需要一些辅助命题

引理 2 恒等式

$$\cos n\theta = 2^{n-1}\cos^n\theta + \sum_{k=0}^{n-1}\lambda_k^{(n)}\cos^k\theta \quad (n=1,2,3,\cdots)$$

是成立的,其中 $\lambda_0^{(n)},\lambda_1^{(n)},\cdots,\lambda_{n-1}^{(n)}$ 为某些常数.

证明 在 $n=1$ 时引理是显而易见的. 今假定它一直到某一个 n(包括 n 在内)都成立.

由于 $\cos\alpha + \cos\beta = 2\cos\dfrac{\alpha-\beta}{2}\cos\dfrac{\alpha+\beta}{2}$

所以 $\cos(n+1)\theta + \cos(n-1)\theta = 2\cos\theta\cos n\theta$

从而

$$\cos(n+1)\theta = 2\cos\theta\left(2^{n-1}\cos^n\theta + \sum_{k=0}^{n-1}\lambda_k^{(n)}\cos^k\theta\right) - \sum_{k=0}^{n-1}\mu_k\cos^k\theta$$

于是 $\cos(n+1)\theta = 2^n\cos^{n+1}\theta + \sum_{k=0}^{n}\nu_k\cos^k\theta$

因而引理获证.

于特例,对于 $-1\leqslant x\leqslant 1$,令

$$\theta = \arccos x$$

便得①

推论 1 当 $-1\leqslant x\leqslant 1$ 时,恒等式

$$\cos(n\arccos x) = 2^{n-1}x^n + \sum_{k=0}^{n-1}\lambda_k^{(n)}x^k \quad (n\geqslant 1) \qquad ⑨7$$

成立.

---

① 我们记住,记号 $\arccos x$ 表示(唯一地)这样的角 θ,其余弦等于 x 并且满足不等式 $0\leqslant\theta\leqslant\pi$.

Tschebyscheff 逼近定理

定义 1 称多项式①

$$T_n(x) = \cos(n\arccos x) \qquad ⑨⑧$$

为 Tschebyscheff 多项式.

等式⑨⑧只在 $-1 \leq x \leq 1$ 时规定了这个多项式,然而,与每个多项式一样,$T_n(x)$ 对于一切实数 x(甚至复数)都是有定义的;恒等式⑨⑦的右方便对于不限定在 $[-1,1]$ 上的 x 给出了它的表示法. 我们指出

$$T_0(x) = 1$$

对于 $x \in [-1,1]$ 来计算 $T_n(x)$ 的值时,宜分成两步进行:

(1)在 $[0, \pi]$ 中求出这样的角 θ 使得

$$\cos \theta = x$$

(2)计算

$$T_n(x) = \cos n\theta$$

Tschebyscheff 多项式给出了前面提出的问题的答案. 就是

定理 13 在最高次项系数为 1 的一切多项式中,在闭区间 $[-1,1]$ 上与零有最小偏差的多项式为

$$\widetilde{T}_n(x) = \frac{1}{2^{n-1}} T_n(x) = \frac{1}{2^{n-1}} \cos(n\arccos x)$$

证明 我们已经确定,要求出所需的多项式

$$\widetilde{R}(x) = x^n - (ax^{n-1} + bx^{n-2} + \cdots + r)$$

就必须求出这样的多项式

$$P(x) = ax^{n-1} + bx^{n-2} + \cdots + r$$

使其成为函数 $f(x) = x^n$ 在闭区间 $[-1,1]$ 上的最佳逼

---

① 这样的多项式是 П. Л. Tschebyscheff[2] 于 1857 年提出的.

第 11 章 高次 Tschebyscheff 逼近

近多项式. 要使多项式 $P(x)$ 满足这种要求, 只需存在 $n-1$ 项①的 Tschebyscheff 交错点组, 亦即这样的一组点

$$x_1 < x_2 < \cdots < x_{n+1} \quad (-1 \leqslant x_k \leqslant 1)$$

使得多项式 $\widetilde{R}(x)$ 在其中每一点上都达到它的最大绝对值, 并且在每经过点 x_k 到 x_{k+1} 时 $\widetilde{R}(x)$ 都变号.

我们来验证, 多项式 $\widetilde{T}_n(x)$ 便具有这样的交错组. 其实, 如果令

$$\theta_0 = 0, \theta_1 = \frac{\pi}{n}, \cdots, \theta_k = \frac{k\pi}{n}, \cdots, \theta_n = \pi$$

则
$$\cos n\theta_k = (-1)^k$$

所以对于

$$x_k = \cos \theta_k \quad (k=0,1,\cdots,n) \qquad ⑨$$

将有
$$\widetilde{T}_n(x_k) = \frac{(-1)^k}{2^{n-1}}$$

另一方面, 由于

$$\max_{-1 \leqslant x \leqslant 1} |\widetilde{T}_n(x)| = \frac{1}{2^{n-1}} \qquad ⑩$$

因而点⑨便构成了所需的交错组.

附注 Tschebyscheff 大致是这样求得多项式的: 设 $\widetilde{T}(x)$ 为所求的多项式, 其最高次项系数为 1, 并且在闭区间 $[-1,1]$ 上与零有着最小偏差, 把这个偏差亦即在 $-1 \leqslant x \leqslant 1$ 时的 $\max|\widetilde{T}(x)|$ 记作 M. 像已经说过的那

---

① 在这里我们要注意, 具有最小偏差(对 x^n 而言)的多项式不在 H_n 内而是在 H_{n-1} 内, 由于这个缘故, 交错组中的点数便减少一个.

Tschebyscheff 逼近定理

样,多项式 $\widetilde{T}(x)$ 应当在闭区间 $[-1,1]$ 上的 $n+1$ 个点处达到了值 $\pm M$. 对那些位于 $[-1,1]$ 内部的点,必然要有 $\widetilde{T}'(x)=0$. 但是 $\widetilde{T}'(x)$ 是一个 $n-1$ 次多项式,这就表明 $\widetilde{T}'(x)$ 不能有多于 $n-1$ 个根. 由此可知,有 $n-1$ 个偏差点都落在闭区间 $[-1,1]$ 的内部,而其余两个点位于闭区间的两端. 于是,多项式 $M^2 - \widetilde{T}^2(x)$ 及 $(1-x^2)\widetilde{T}'^2(x)$ 有相同的根,并且位于 $(-1,1)$ 内部的这些根都是上面两个多项式的二重根. 由此可知这两个多项式只能相差一个常数因子. 比较它们的最高次项系数,便得到

$$M^2 - \widetilde{T}^2(x) = \frac{(1-x^2)\widetilde{T}'^2(x)}{n^2} \qquad ⑩$$

这个关系式本身具有独立意义. 从式⑩可知

$$\sqrt{M^2 - \widetilde{T}^2(x)} = \pm \frac{1}{n}\sqrt{1-x^2}\,\widetilde{T}'(x)$$

导数 $\widetilde{T}'(x)$ 在经过每个交错点时都变号,假设在区间 (α,β) 上它是正的,则在这个区间上便有

$$\frac{\widetilde{T}'(x)}{\sqrt{M^2-\widetilde{T}^2(x)}} = \frac{n}{\sqrt{1-x^2}}$$

两边取积分,便得

$$\arccos \frac{\widetilde{T}(x)}{M} = n\arccos x + C$$

就是说

$$\widetilde{T}(x) = M\cos[n\arccos x + C]$$

或

$$\widetilde{T}(x) = M[\cos C \cos(n\arccos x) - \sin C \sin(n\arccos x)]$$

由于 $\widetilde{T}(x)$ 是一个多项式,就应当有 $\sin C = 0$,从而 $\cos C = \pm 1$. 但 $\cos(n\arccos x)$ 的最高次项系数是 2^{n-1},这就表明 $\cos C = 1$,而且 $M = 2^{-n+1}$. 从而

$$\widetilde{T}(x) = \frac{1}{2^{n-1}} \cos(n\arccos x)$$

这个等式只是在区间 (α, β) 内建立的,但因这个等式的两方都是多项式,所以它便处处都成立.

回到已证明的定理上,与等式⑩比较,便得

推论 2 对于每一个最高次系数为 1 的 n 次多项式 $\widetilde{P}_n(x)$,不等式

$$\max_{-1 \leqslant x \leqslant 1} |\widetilde{P}_n(x)| \geqslant \frac{1}{2^{n-1}}$$

都成立.

如果 n 次多项式 $P_n(x)$ 的最高次项系数不为 1 而为 A_0,则

$$\max_{-1 \leqslant x \leqslant 1} |P_n(x)| \geqslant \frac{A_0}{2^{n-1}}$$

注意到这一点后,我们来考虑 n 次多项式

$$P_n(x) = c_0 x^n + c_1 x^{n-1} + \cdots + c_n$$

令

$$x = \frac{a+b}{2} + \frac{b-a}{2} y$$

我们便看出,$P_n(x)$ 可以当作是变数 y 的多项式,且其最高次项系数是

$$c_0 \left(\frac{b-a}{2}\right)^n$$

于是

Tschebyscheff 逼近定理

$$\max_{a \leq x \leq b} |P_n(x)| = \max_{-1 \leq y \leq 1} |P_n(x)| \geq c_0 \frac{(b-a)^n}{2^{2n-1}}$$

Tschebyscheff 多项式的进一步性质

$$T_n(x) = \cos(n \arccos x)$$

除了与零有最小偏差的多项式外,还具有很多其他的重要性质. 这一节中我们将要谈到其中的一些.

首先,在公式

$$\cos n\theta = 2\cos\theta \cos(n-1)\theta - \cos(n-2)\theta \quad (n=2,3,\cdots)$$

中施行代换 $\theta = \arccos x$,便可得到联系三个 Tschebyscheff 多项式的递推关系

$$T_n(x) = 2x T_{n-1}(x) - T_{n-2}(x) \quad (n=2,3,\cdots) \quad ⑩$$

知道了 $\quad T_0(x) = 1, T_1(x) = x$

我们从等式⑩便可以相继求得

$$T_2(x) = 2x^2 - 1$$
$$T_3(x) = 4x^3 - 3x$$
$$T_4(x) = 8x^4 - 8x^2 + 1$$
$$T_5(x) = 16x^5 - 20x^3 + 5x$$
$$T_6(x) = 32x^6 - 48x^4 + 18x^2 - 1$$
$$\vdots$$

这可以无限制地继续排列下去.

寻求多项式 $T_n(x)$ 的另一方法是以所谓"母函数"为基础的. 为了阐述它,我们需要下面的公式

$$\sum_{n=0}^{+\infty} t^n \cos n\theta = \frac{1 - t\cos\theta}{1 - 2t\cos\theta + t^2} \quad (-1 < t < 1) \quad ⑩$$

上式左方显然是一个绝对收敛的级数.

为了证明这个公式,我们指出它的左方是几何级数

第 11 章　高次 Tschebyscheff 逼近

$$\sum_{n=0}^{+\infty} t^n e^{n\theta i} = \frac{1}{1-te^{\theta i}}$$

之和的实部.

可是
$$\frac{1}{1-te^{\theta i}} = \frac{1}{1-t\cos\theta - it\sin\theta} = \frac{1-t\cos\theta + it\sin\theta}{(1-t\cos\theta)^2 + t^2\sin^2\theta}$$

因为这个分式的实部正好与式⑩的右方相同.

如果在式⑩中令 $\theta = \arccos x$,则得

$$\frac{1-tx}{1-2tx+t^2} = \sum_{n=0}^{+\infty} t^n T_n(x) \quad (-1<t<1) \qquad ⑩$$

因之一系列的 Tschebyscheff 多项式就成为函数

$$T(t,x) = \frac{1-tx}{1-2tx+t^2}$$

的展开式中 t 的不同乘幂的底系数. 我们称函数 $T(t,x)$ 为 Tschebyscheff 多项式的母函数. 由于 $T(t,x)$ 依 t 的乘幂的展式在形式上可以用 $1-2tx+t^2$ 除 $1-tx$ 得到,所以我们又得到一个实际上计算多项式 $T_n(x)$ 的新方法

$$\begin{array}{r}
1-tx \\
\underline{1-2tx+t^2} \\
tx-t^2
\end{array} \bigg| \begin{array}{l}
1-2tx+t^2 \\
\overline{1+tx+t^2(2x^2-1)+t^3(4x^3-3x)+\cdots}
\end{array}$$

$$\begin{array}{r}
tx - 2t^2x^2 + t^3x \\
\hline
t^2(2x^2-1) - t^3x
\end{array}$$

$$\begin{array}{r}
t^2(2x^2-1) - 2t^3(2x^3-x) + t^4(2x^2-1) \\
\hline
t^3(4x^3-3x) - t^4(2x^2-1)
\end{array}$$

$$\vdots$$

商

$$1+tx+t^2(2x^2-1)+t^3(4x^3-3x)+\cdots$$

的系数实际上都与 $T_0(x), T_1(x), T_2(x), T_3(x), \cdots$ 相

Tschebyscheff 逼近定理

合.

要写出多项式 $T_n(x)$ 的显式并不困难.

为了这一目的,我们先来证明其本身具有独立意义的定理

定理 14 多项式 $y = T_n(x)$ 满足微分方程
$$(1-x^2)y'' - xy' + n^2 y = 0 \qquad ⑮$$

证明 事实上,设 $\quad y = \cos(n\arccos x)$

则 $\quad y' = \dfrac{n}{\sqrt{1-x^2}} \sin(n\arccos x)$

因而 $\quad \sqrt{1-x^2}\, y' = n\sin(n\arccos x)$

微分这个恒等式,便得

$$\sqrt{1-x^2}\, y'' - \dfrac{xy'}{\sqrt{1-x^2}} = -\dfrac{n^2}{\sqrt{1-x^2}} \cos(n\arccos x)$$

从而便推得①式⑮. 此外,还有更简单的方法能够获得式⑮,就是对于满足 Tschebyscheff 多项式的关系式

$$M^2 - y^2 = \dfrac{(1-x^2)y'^2}{n^2}$$

取微分.

现在转到对我们具有意义的求 $T_n(x)$ 的显式的问题上来. 令

$$T_n(x) = \sum_{k=0}^{n} a_k x^{n-k}$$

并且将这个表达式代入式⑮中,这就给出了等式

---

① 当然,我们的讨论只证明 $T_n(x)$ 的方程⑮在 $-1 \leq x \leq 1$ 情形下的解. 但由于 $y = T_n(x)$ 是多项式,所以 $(1-x^2)y'' - xy' + n^2 y$ 也是多项式,因之它在 $-1 \leq x \leq 1$ 时为零,从而便推得定理成立的普遍性.

第 11 章 高次 Tschebyscheff 逼近

$$(1-x^2)\sum_{k=0}^{n}(n-k)(n-k-1)a_k x^{n-k-2} -$$

$$x\sum_{k=0}^{n}(n-k)a_k x^{n-k-1} + n^2\sum_{k=0}^{n}a_k x^{n-k} = 0$$

它可以表示成

$$\sum_{k=0}^{n}(n-k)(n-k-1)a_k x^{n-k-2} +$$

$$\sum_{k=0}^{n}[n^2-(n-k)^2]a_k x^{n-k} = 0 \qquad ⑯$$

这里在第一个和中对应于值 $k=n$ 及 $k=n-1$ 的两项消失了. 所以引用新的求和附标 $i=k+2$ 并仍记成 k,则这个和就可表示成

$$\sum_{k=2}^{n}(n-k+2)(n-k+1)a_{k-2} x^{n-k}$$

在第二个和中对应于值 $k=0$ 的项消失了,因之恒等式⑯呈下列形式

$$[n^2-(n-1)^2]a_1 x^{n-1} + \sum_{k=2}^{n}\{(n-k+2)(n-k+1)a_{k-2} +$$

$$[n^2-(n-k)^2]a_k\}x^{n-k} = 0$$

从而便推得 $a_1=0$,并且

$$a_k = -\frac{(n-k+2)(n-k+1)}{k(2n-k)}a_{k-2}$$

从这个关系式立刻推出所有带奇附标的系数都等于零. 把 k 换成 $2k$ 时,便得到

$$a_{2k} = -\frac{(n-2k+2)(n-2k+1)}{4k(n-k)}a_{2k-2}$$

但因 $a_0 = 2^{n-1}$,从而推知

$$a_2 = -\frac{n(n-1)}{1\cdot(n-1)}2^{n-3}$$

Tschebyscheff 逼近定理

$$a_4 = \frac{n(n-1)(n-2)(n-3)}{2!\ (n-1)(n-2)} 2^{n-5}$$

而且一般的

$$a_{2k} = (-1)^k \frac{n \cdot (n-1) \cdot \cdots \cdot (n-2k+1)}{k!\ \cdot (n-1) \cdot \cdots \cdot (n-k)} 2^{n-2k-1}$$

它是很容易用归纳法验证的. 其次, 注意到

$$\frac{n \cdot (n-1) \cdot \cdots \cdot (n-2k+1)}{k!\ \cdot (n-1) \cdot \cdots \cdot (n-k)} =$$

$$\frac{n}{n-k} \cdot \frac{(n-k) \cdot (n-k-1) \cdot \cdots \cdot (n-2k+1)}{k!} =$$

$$\frac{n}{n-k} C_{n-k}^k$$

最后便得

$$T_n(x) = \sum_{k=0}^{\left[\frac{n}{2}\right]} (-1)^k \frac{n}{n-k} C_{n-k}^k 2^{n-2k-1} x^{n-2k}$$

其中记号 $\left[\frac{n}{2}\right]$ 表示 $\frac{n}{2}$ 的整数部分.

还可以写出 Tschebyscheff 多项式的另一个显式. 那就是, 从棣美弗公式

$$\cos n\theta + i\sin n\theta = (\cos \theta + i\sin \theta)^n$$

推得

$$\cos n\theta - i\sin n\theta = (\cos \theta - i\sin \theta)^n$$

从而

$$\cos n\theta = \frac{1}{2}[(\cos \theta + i\sin \theta)^n + (\cos \theta - i\sin \theta)^n]$$

如果在上式中令 $\theta = \arccos x$(算作$|x| \leq 1$), 则便得到

$$T_n(x) = \frac{1}{2}[(x + i\sqrt{1-x^2})^n + (x - i\sqrt{1-x^2})^n]$$

第11章 高次 Tschebyscheff 逼近

我们只在 $|x|\leq 1$ 时建立了这个等式,可是容易看出,它对于所有的实数及复数 x 都能成立. 因为它的右方恰是一个多项式. 要断定这一点,把它写成

$$\frac{1}{2}\sum_{k=0}^{n} C_n^k x^{n-k} i^k (\sqrt{1-x^2})^k (1+(-1)^k)$$

的形式,并且注意,这里只是对应于偶数值 k 的那些加数异于零就可以了.

对于适合 $|x|>1$ 的那些实数 x,宜于把表示 $T_n(x)$ 的最后这个公式做一些改变. 就是,若 k 为偶数,则

$$i^k(\sqrt{1-x^2})^k = (\sqrt{x^2-1})^k$$

这就表明

$$T_n(x) = \frac{1}{2}\sum_{k=0}^{n} C_n^k x^{n-k} (\sqrt{x^2-1})^k (1+(-1)^k)$$

从而

$$T_n(x) = \frac{1}{2}[(x+\sqrt{x^2-1})^n + (x-\sqrt{x^2-1})^n]$$

这个表达式中不包含虚部(在 $|x|>1$ 时). 于特例,从它可以推得 $T_n(x)$ 的下述重要估值.

定理15 若 x 为实数,并且 $|x|>1$,则

$$|T_n(x)| \leq (|x|+\sqrt{x^2+1})^n$$

其实,$x+\sqrt{x^2-1}$ 及 $x-\sqrt{x^2-1}$ 中每一个的绝对值都不大于 $|x|+\sqrt{x^2-1}$.

现在我们来研究多项式 $T_n(x)$ 的根.

由于 $T_n(x) = \cos n\theta$,其中 $\theta = \arccos x$,显然可见,使对应的值 θ 能满足方程 $\cos n\theta = 0$ 的那些值 x 便是 $T_n(x)$ 的根.

这样的值 θ(我们记住 $0 \leq \theta \leq \pi$)显然便是

Tschebyscheff 逼近定理

$$\theta_1 = \frac{\pi}{2n}, \theta_2 = \frac{3\pi}{2n}, \cdots, \theta_k = \frac{(2k-1)\pi}{2n}, \cdots, \theta_n = \frac{(2n-1)\pi}{2n}$$

所以,诸数

$$x_1^{(n)} = \cos\frac{\pi}{2n}, x_2^{(n)} = \cos\frac{3\pi}{2n}, \cdots, x_n^{(n)} = \cos\frac{(2n-1)\pi}{2n}$$

便是多项式 $T_n(x)$ 的所有根,并且其中的每一个都是单根.

这样一来,我们便建立了

定理 16 多项式 $T_n(x)$ 的根可以用公式

$$x_k^{(n)} = \cos\frac{(2k-1)\pi}{2n} \quad (k = 1, 2, \cdots, n)$$

来给出.

它们全都是实的单根,而且位于区间$(-1,1)$的内部.

从所引出的公式便推得了 Tschebyscheff 多项式的相邻两根的"介插"性质.

定理 17 在多项式 $T_n(x)$ 的相邻两根 $x_k^{(n)}$ 及 $x_{k+1}^{(n)}$ 之间,一定有而且只有一个前一多项式 $T_{n-1}(x)$ 的根.

事实上,多项式 $T_{n-1}(x)$ 的第 k 个根为

$$x_k^{(n-1)} = \cos\frac{(2k-1)\pi}{2n-2}$$

我们来确定①

$$x_k^{(n)} > x_k^{(n-1)} > x_{k+1}^{(n)}$$

或者,相当于建立

$$\frac{2k-1}{2n} < \frac{2k-1}{2n-2} < \frac{2k+1}{2n}$$

---

① 因为 $\cos\theta$(在 $0 \leqslant \theta \leqslant \pi$ 时)是减函数,所以根 $x_1^{(n)}, x_2^{(n)}, \cdots, x_n^{(n)}$ 便能排成一个递减的次序.

第 11 章 高次 Tschebyscheff 逼近

这个双边不等式的第一部分是显而易见的,而它的第二部分则与不等式

$$2k < 2n-1$$

等价,由此便推出 $k<n$. 于是不等式⑩便得到证明. 因为在 n 个根 $x_k^{(n)}$ 之间有 $n-1$ 个区间,其中每一个区间都含有多项式 $T_{n-1}(x)$ 的根,于是显然可知在这些区间中都正好各有一个根,因为,这些根的个数也是 $n-1$.

推论 两个相邻的多项式 $T_n(x)$ 及 $T_{n-1}(x)$ 不能有公共根.

我们指出,这个推论可以很简单地由递推式⑩得出. 其实,假定 $T_n(x)$ 及 $T_{n-1}(x)$ 有一公共根 x_0,则从式⑩便推得

$$T_{n-2}(x_0) = 0$$

即 x_0 将成为多项式 $T_{n-2}(x)$ 的根. 可是,重复这种推理方法,我们便可断定 x_0 将成为多项式 $T_{n-3}(x)$ 的根,从而将成为 $T_{n-4}(x)$ 的根,等等. 最后 x_0 就应当成为多项式 $T_0(x) \equiv 1$ 的根,这便显然是荒谬可笑了.

现在我们对于甚大的 n 来考虑多项式 $T_n(x)$ 的根的分布问题. 为了格外清晰起见,我们对这个问题给出它的力学解释. 那就是,我们设想某一个质量为 M 的物质,其质量均匀分布于多项式 $T_n(x)$ 的 n 个根 $x_1^{(n)}$, $x_2^{(n)}, \cdots, x_n^{(n)}$ 上. 于是在其每一个根上就荷载质量 $\frac{M}{n}$. 我们来阐明,在某个闭区间 $[a, b] \subset [-1, 1]$ 上所具有物质的质量 $\overset{b}{\underset{a}{M}}_n$ 是多少. 显然

$$\overset{b}{\underset{a}{M}}_n = \frac{M}{n} \tau_n$$

Tschebyscheff 逼近定理

其中 τ_n 是位于 $[a,b]$ 上的根 $x_k^{(n)}$ 的个数. 换句话说, 我们需要从整数 $1,2,\cdots,n$ 中求出满足不等式

$$a \leqslant \cos\frac{(2k-1)\pi}{2n} \leqslant b$$

的那些 k 的个数.

上面的不等式与不等式

$$\alpha \geqslant \frac{(2k-1)\pi}{2n} \geqslant \beta$$

是等价的, 其中

$$\alpha = \arccos a, \quad \beta = \arccos b$$

而后一不等式可以写成这样的形式

$$\frac{1}{2}\left(1+\frac{2n\beta}{\pi}\right) \leqslant k \leqslant \frac{1}{2}\left(1+\frac{2n\alpha}{\pi}\right)$$

于是, 我们便需要解决这样的问题: 给出了数 P 及 Q, 其中 $P<Q$, 要确定满足不等式

$$P \leqslant k \leqslant Q$$

的那些自然数 k 的个数.

如果

$$i<P \leqslant i+1<i+2<\cdots<i+m \leqslant Q<i+m+1$$

则所求的个数就等于 m. 容易看出

$$m-1 \leqslant Q-P < m+1$$

从而

$$Q-P-1 < m \leqslant Q-P+1$$

于是

$$m = Q-P+\mu \quad (-1<\mu\leqslant 1)$$

在我们的有趣的问题中

$$P=\frac{1}{2}\left(1+\frac{2n\beta}{\pi}\right), \quad Q=\frac{1}{2}\left(1+\frac{2n\alpha}{\pi}\right)$$

第 11 章 高次 Tschebyscheff 逼近

因此

$$\tau_n = n\frac{\alpha-\beta}{\pi}+\mu_n \quad (-1<\mu_n\leqslant 1)$$

从而

$$\overset{b}{\underset{a}{M}}_n = \frac{M}{\pi}(\alpha-\beta)+\frac{M\mu_n}{n} \quad (-1<\mu_n\leqslant 1)$$

或

$$\overset{b}{\underset{a}{M}}_n = \frac{M}{\pi}(\arccos a-\arccos b)+\frac{M\mu_n}{n}$$

这个等式指出,当 n 无限增大时,质量的分布便趋于一种极限的情况,在这时,分布在闭区间 $[a,b]$ 上的质量是

$$\overset{b}{\underset{a}{M}}_\infty = \frac{M}{\pi}(\arccos a-\arccos b)$$

像通常在力学中所讨论的一样,我们力求以其密度来刻画出质量的极限分布状态.

因为位于区间 $[x,x+\Delta x]$ 上的质量等于

$$\overset{x+\Delta x}{\underset{x}{M}}_\infty = \frac{M}{\pi}[\arccos x-\arccos(x+\Delta x)]$$

所以这个区间的平均密度为

$$\frac{1}{\Delta x}\overset{x+\Delta x}{\underset{x}{M}}_\infty = \frac{M}{\pi}\cdot\frac{\arccos x-\arccos(x+\Delta x)}{\Delta x}$$

因此,在点 x 的点密度便为

$$p(x) = \frac{M}{\pi}\lim_{\Delta x\to 0}\frac{\arccos x-\arccos(x+\Delta x)}{\Delta x}$$

其中右方的极限只与函数 $\arccos x$ 在点 x 的导数相差一个符号,于是

$$p(x) = \frac{M}{\pi}\cdot\frac{1}{\sqrt{1-x^2}}$$

Tschebyscheff 逼近定理

这就是在闭区间 $[-1,1]$ 上 Tschebyscheff 多项式 $T_n(x)$ 的根的极限分布密度。当然，这里只是因子 $\dfrac{1}{\sqrt{1-x^2}}$ 起着重要作用，所以，若取原来的质量为 π（这联系着测量质量的单位的选择），我们便得到 $p(x)$ 的更简单的表达式

$$p(x) = \frac{1}{\sqrt{1-x^2}} \qquad ⑱$$

公式⑱可以给出多项式 $T_n(x)$ 落在小区间 $[x, x+\Delta x]$ 上的根的个数的近似表达式，就是说，如果 Δx 很小时，则

$$\overset{x+\Delta x}{\underset{x}{M}}_\infty \approx \frac{\Delta x}{\sqrt{1-x^2}}$$

若数 n 甚大，则上式左方的 $\overset{x+\Delta x}{\underset{x}{M}}_\infty$ 可以写成 $\overset{x+\Delta x}{\underset{x}{M}}_n$，而由于 $\overset{b}{\underset{a}{M}}_n = \dfrac{\pi}{n}\tau_n$，所以对于所求的根的个数有表达式

$$\frac{n\Delta x}{\pi\sqrt{1-x^2}}$$

由此显然可见，当 n 甚大时多项式 $T_n(x)$ 的根都聚集到闭区间 $[-1,1]$ 的两端去了。

在函数⑱与 Tschebyscheff 多项式之间有紧密的关系以及完全另外的性质。为了叙述它，我们引入下面的定义。

定义 2 两个函数 $f(x)$ 及 $g(x)$ 称为对于权函数 $p(x)$ 在区间 $[a,b]$ 上是互相直交的，如果

$$\int_a^b p(x)f(x)g(x)\,\mathrm{d}x = 0$$

128

第 11 章 高次 Tschebyscheff 逼近

定理 18 任何两个相异的 Tschebyscheff 多项式 $T_n(x)$ 及 $T_m(x)$ 对于权函数

$$p(x) = \frac{1}{\sqrt{1-x^2}}$$

在区间 $[-1,1]$ 上都是互相正交的.

证明 要证明这一点只需计算积分

$$I_{nm} = \int_{-1}^{1} \frac{T_n(x) T_m(x)}{\sqrt{1-x^2}} dx$$

为了这一目的,我们引用代换 $x = \cos\theta$. 因为 $T_n(\cos\theta) = \cos n\theta (0 \leq \theta \leq \pi)$,所以

$$I_{nm} = \int_0^\pi \cos n\theta \cos m\theta \, d\theta$$

从而

$$I_{nm} = \frac{1}{2} \int_0^\pi [\cos(n-m)\theta + \cos(n+m)\theta] d\theta = \frac{1}{2} \left[\frac{\sin(n-m)\theta}{n-m} + \frac{\sin(n+m)\theta}{n+m} \right]_0^\pi$$

于是

$$I_{nm} = 0$$

定理 19 设 $P(x)$ 为次数不高于 n 的多项式,而 M 为其在区间 $[-1,1]$ 上的绝对值的最大者. 若 x_0 是一个实数且 $|x_0| > 1$,则

$$|P(x_0)| \leq M |T_n(x_0)|$$

证明 假如定理不真,那么便有①

$$M < \left| \frac{P(x_0)}{T_n(x_0)} \right|$$

---

① 注意,$T_n(x_0) \neq 0$. 因为 $T_n(x)$ 所有的根都位于区间 $(-1,1)$ 内.

Tschebyscheff 逼近定理

我们引入多项式

$$R(x) = \frac{P(x_0)}{T_n(x_0)} T_n(x) - P(x)$$

来讨论. 在点 $y_i = \cos \frac{i\pi}{n}$ ($i=0,1,2,\cdots,n$) 处将有

$$T_n(y_i) = \cos i\pi = (-1)^i$$

这就表明, 在差

$$R(y_i) = (-1)^i \frac{P(x_0)}{T_n(x_0)} - P(y_i)$$

之中, 被减数的绝对值大于减数的绝对值(因为 $-1 \leqslant y_i \leqslant 1$, 所以 $|P(y_i)| \leqslant M$). 所以, 所提的差 $R(x)$ 在经过每一点 y_i 到 y_{i+1} 时都变号. 由此便推得, 在 n 个区间

$$(y_0, y_1), (y_1, y_2), \cdots, (y_{n-1}, y_n)$$

的每一个中都含有多项式 $R(x)$ 的根. 但是, 除此之外, 还有 $R(x_0) = 0$. 这就表明, n 次多项式 $R(x)$ 有了 $n+1$ 个根. 这只有在 $R(x)$ 恒等于零时才有可能, 所以

$$P(x) = \frac{P(x_0)}{T_n(x_0)} T_n(x)$$

在这里令 $x=1$ 并注意 $T_n(1)=1$ (由于 $T_n(1) = \cos(n\arccos 1) = \cos 0$), 便得到

$$P(1) = \frac{P(x_0)}{T_n(x_0)}$$

而这却与式⑩的假定相矛盾.

故定理得证.

比较定理 6 与定理 2, 我们便得到重要的

推论 在定理 6 的记号下

$$|P(x_0)| \leqslant M(|x_0| + \sqrt{x_0^2 - 1})^n \qquad ⑩$$

在固定最高次项系数时, Tschebyscheff 多项式与

第11章 高次 Tschebyscheff 逼近

零有最小偏差. 可以表明, 它的其余的系数与最高次项系数都具有这一性质. 为了建立属于马尔柯夫的这一结果, 我们需要

引理3 函数
$$f(x) = A_1 x^{\lambda_1} + A_2 x^{\lambda_2} + \cdots + A_{m+1} x^{\lambda_{m+1}}$$
不能有多于 m 个正根①, 其中 $\lambda_1 < \lambda_2 < \cdots < \lambda_{m+1}$ 为任意实数.

证明 事实上, 在 $m=1$ 时引理显然成立. 今假定, 它对于某一个值 m 成立, 而对于值 $m+1$ 不成立. 于是我们便发现函数
$$f(x) = A_1 x^{\lambda_1} + A_2 x^{\lambda_2} + \cdots + A_{m+1} x^{\lambda_{m+1}} + A_{m+2} x^{\lambda_{m+2}}$$
的根. 可是, 这时按照罗尔定理, 这个函数的正根将多于 m 个. 然而这与假定矛盾, 由于所说的导数具有
$$B_1 x^{\mu_1} + B_2 x^{\mu_2} + \cdots + B_{m+1} x^{\mu_{m+1}}$$
的形式之故. 引理证毕.

定理 20(A·A·马尔柯夫) 若 n 与 $p \leq n$ 同是奇数或同是偶数, 那么, 在所有次数不高于 n 而 x^p 项的系数为 1 的那些多项式中, 在区间 $[-1,1]$ 上与零有最小偏差的多项式为

$$\frac{T_n(x)}{A_p^{(n)}} \qquad ⑪$$

其中 $T_n(x)$ 为 Tschebyscheff 多项式, 而 $A_p^{(n)}$ 是这个多项式中 x^p 项的系数. 若 n 与 p 的奇偶性不同, 那么, 具有这种性质的多项式便是

$$\frac{T_{n-1}(x)}{A_p^{(n-1)}} \qquad ⑫$$

---

① 我们假设所有的数 A_i 都不为零.

Tschebyscheff 逼近定理

证明 这个定理包含下列四种可能情形①:

(1) p 为偶数,n 为偶数.
(2) p 为奇数,n 为奇数.
(3) p 为偶数,n 为奇数.
(4) p 为奇数,n 为偶数.

我们假定,在情形(1)下求出多项式 $P(x) \in H_n$,它的 x^p 项的系数为 1 且与零的偏差小于多项式⑪与零的偏差,亦即,对于一切 $x \in [-1,1]$,有

$$|P(x)| < \frac{1}{|A_p^{(n)}|}$$

于是多项式 $P(-x)$ 也具有这些同样的性质(因 p 为偶数),这就表明

$$\left|\frac{P(x)+P(-x)}{2}\right| < \frac{1}{|A_p^{(n)}|}$$

注意到这一点后,我们来考虑差

$$R(x) = \frac{T_n(x)}{A_p^{(n)}} - \frac{P(x)+P(-x)}{2}$$

这是一个次数不高于 n 且不含 x^p 项的多项式.

不难看出,$R(x)$ 只由指数 m 为偶数的那些项 x^m 所组成.这样,有

$$R(x) = \sum_{k=0}^{\frac{n}{2}} c_k x^{2k} \quad (c_{\frac{p}{2}} = 0)$$

令

$$Q(y) = \sum_{k=0}^{\frac{n}{2}} c_k y^k$$

---

① 我们这里引用的是 С·Н·伯恩斯坦的证明.

第 11 章 高次 Tschebyscheff 逼近

因为 $Q(y)$ 含有 $\frac{n}{2}$ 个项(或者还要少一些,如果除了 $c_{\frac{p}{2}}$ 之外还有些系数 c_k 为零的话),所以 $Q(y)$ 的正根个数不大于 $\frac{n}{2}-1$.

然而,令

$$y_i = \cos^2 \frac{i\pi}{n}$$

我们便得

$$Q(y_i) = R\left(\cos\frac{i\pi}{n}\right) = \frac{T_n\left(\cos\frac{i\pi}{n}\right)}{A_p^{(n)}} - \frac{P\left(\cos\frac{i\pi}{n}\right)+P\left(-\cos\frac{i\pi}{n}\right)}{2}$$

从而由等式

$$T_n\left(\cos\frac{i\pi}{n}\right) = \cos i\pi = (-1)^i$$

便推得, $Q(y_i)$ 与 $(-1)^i$ 的符号相同. 这就是说在 $\frac{n}{2}$ 个区间

$$(y_{\frac{n}{2}}, y_{\frac{n}{2}-1}), (y_{\frac{n}{2}-1}, y_{\frac{n}{2}-2}), \cdots, (y_2, y_1), (y_1, y_0)$$

的每一个中都含有 $Q(y)$ 的一个根,所得的矛盾就证明了定理的情形(1).

情形(3)可仿此讨论. 只需注意到 $\frac{P(x)+P(-x)}{2} \in H_{n-1}$,那么

$$R(x) = \frac{T_{n-1}(x)}{A_p^{(n-1)}} - \frac{P(x)+P(-x)}{2} = \sum_{k=0}^{\frac{n-1}{2}} c_k x^{2k} \quad (c_{\frac{p}{2}}=0)$$

往后的证明可以像情形(1)一样来完成.

最后,对于(2)及(4)这两种情形,应当考虑 $P(x)-P(-x)$ 而不考虑 $P(x)+P(-x)$. 例如,对于(4)这种情形,我们将有

$$R(x)=\frac{T_{n-1}(x)}{A_p^{(n-1)}}-\frac{P(x)-P(-x)}{2}=\sum_{k=0}^{\frac{n}{2}-1}c_k x^{2k+1}$$

并应引入多项式 $Q(y)=\sum_{k=0}^{\frac{n-1}{2}}c_k y^k$. 往后的证明细节留给读者来完成.

可以证明,多项式⑪(或者,对应的式⑫)是唯一的一个多项式,其中 x^p 项的系数等于1,并且在[-1,1]上它与零的偏差为最小,但是我们不去讲它了. 代替这个我们引出A·A·马尔柯夫定理的若干推论.

推论1 设 $P(x)$ 是一个 n 次多项式,它的 x^p 项 ($p \leqslant n$)的系数等于1,则视 n 与 p 的奇偶性相同或否,而有不等式

$$\max_{-1 \leqslant x \leqslant 1}|P(x)| \geqslant \frac{1}{2^{p-1}} \cdot \frac{p!}{n} \cdot \frac{\left(\frac{n-p}{2}\right)!}{\left(\frac{n+p}{2}-1\right)!}$$

或

$$\max_{-1 \leqslant x \leqslant 1}|P(x)| \geqslant \frac{1}{2^{p-1}} \cdot \frac{p!}{n-1} \cdot \frac{\left(\frac{n-p-1}{2}\right)!}{\left(\frac{n+p-3}{2}\right)!}$$

这个推论由系数 $A_p^{(n)}$ 及 $A_p^{(n-1)}$ 的表达式便立即推得.

显然,若多项式 $P(x)$ 中 x^p 项的系数不是1而是

第 11 章　高次 Tschebyscheff 逼近

c_p,则上述不等式的右方就应当乘上一个因子 $|c_p|$. 从而便推得

推论 2　若 $P(x)$ 是一个 n 次多项式,则它的 x^p 项的系数 c_p 视 $n-p$ 是偶数与否而有估值

$$|c_p| \leqslant 2^{p-1}\frac{n}{p!} \cdot \frac{\left(\frac{n+p}{2}-1\right)!}{\left(\frac{n-p}{2}\right)!} \max_{-1 \leqslant x \leqslant 1} |P(x)|$$

或

$$|c_p| \leqslant 2^{p-1}\frac{n-1}{p!} \cdot \frac{\left(\frac{n+p-3}{2}\right)!}{\left(\frac{n-p-1}{2}\right)!} \max_{-1 \leqslant x \leqslant 1} |P(x)|$$

Tschebyscheff 多项式与不等式

第 12 章

复旦大学黄宣国教授给出了 Tschebyscheff 多项式的一个有趣应用.

例 21 正整数 $n \geqslant 2$, $d_2, d_3, \cdots, d_n$ 是 $n-1$ 个复数. 求证:在 $[0,1]$ 内必有一个实数 x, 使得

$$|x+d_2x^2+d_3x^3+\cdots+d_nx^n| \geqslant \frac{1}{n}\tan\frac{\pi}{4n}$$

证明 记

$$d_j = a_j + \mathrm{i}b_j \quad (j=2,3,\cdots,n) \quad ⑬$$

这里 a_j, b_j 全是实数. 对于任一实数 x

$$|x+d_2x^2+d_3x^3+\cdots+d_nx^n| =$$
$$|(x+a_2x^2+a_3x^3+\cdots+a_nx^n) +$$
$$\mathrm{i}(b_2x^2+b_3x^3+\cdots+b_nx^n)| \geqslant$$
$$|x+a_2x^2+a_3x^3+\cdots+a_nx^n| \quad ⑭$$

如果能证明对于任一 n 次($n \geqslant 2$)实系数多项式
$$f(x) = x + a_2 x^2 + a_3 x^3 + \cdots + a_n x^n \qquad ⑮$$
在 $[0,1]$ 内有一实数 x,使得 $|f(x)| \geqslant \dfrac{1}{n}\tan\dfrac{\pi}{4n}$,则本题结论成立. 下面就对实系数多项式⑮进行研究.

我们知道 Tschebyscheff 多项式
$$T_1(x) = x$$
$$T_2(x) = 2x^2 - 1$$
$$\vdots$$
$$T_n(x) = 2x T_{n-1}(x) - T_{n-2}(x) \quad (n \geqslant 3)$$
$$T_n(\cos\theta) = \cos n\theta \quad (n \in \mathbf{N}) \qquad ⑯$$

引入一个多项式
$$F(x) = (-1)^{n+1}\dfrac{1}{n}\tan\dfrac{\pi}{4n} T_n\left(\left(1+\cos\dfrac{\pi}{2n}\right)x - \cos\dfrac{\pi}{2n}\right) \qquad ⑰$$

由于 $T_n(x)$ 是 x 的 n 次多项式(从式⑯前三个公式立即可以知道),则 $F(x)$ 是 x 的 n 次多项式. 这里限制 $x \in [0,1]$.

$F(x)$ 满足以下两个简单的性质:

(1) $F(0) = 0$.

(2) 当 $x \in [0,1]$ 时,$|F(x)| \leqslant \dfrac{1}{n}\tan\dfrac{\pi}{4n}$ $(n \in \mathbf{N})$.

从式⑰和式⑯的最后一个公式,有
$$F(0) = (-1)^{n+1}\dfrac{1}{n}\tan\dfrac{\pi}{4n} T_n\left(-\cos\dfrac{\pi}{2n}\right) =$$
$$(-1)^{n+1}\dfrac{1}{n}\tan\dfrac{\pi}{4n} T_n\left(\cos\left(\pi - \dfrac{\pi}{2n}\right)\right) =$$
$$(-1)^{n+1}\dfrac{1}{n}\tan\dfrac{\pi}{4n} \cos n\left(\pi - \dfrac{\pi}{2n}\right) = 0 \qquad ⑱$$

（利用 $\cos n\left(\pi - \dfrac{\pi}{2n}\right) = \cos\left(n\pi - \dfrac{\pi}{2}\right) = 0$）于是，性质（1）成立．

当 $0 \leqslant x \leqslant 1$ 时，有

$$\left|\left(1 + \cos \dfrac{\pi}{2n}\right)x - \cos \dfrac{\pi}{2n}\right| = \left|x - (1-x)\cos \dfrac{\pi}{2n}\right| \leqslant$$

$$|x| + \left|(1-x)\cos \dfrac{\pi}{2n}\right| \leqslant$$

$$|x| + |1-x| = x + (1-x) = 1 \qquad ⑲$$

于是，对于固定的 $x \in [0,1]$，必有 $\psi \in [0,\pi]$，使得

$$\cos \psi = \left(1 + \cos \dfrac{\pi}{2n}\right)x - \cos \dfrac{\pi}{2n} \qquad ⑳$$

利用式⑰，⑳和式⑯的最后一个公式，有

$$F(x) = (-1)^{n+1}\dfrac{1}{n}\tan \dfrac{\pi}{4n}T_n(\cos \psi) =$$

$$(-1)^{n+1}\dfrac{1}{n}\tan \dfrac{\pi}{4n}\cos n\psi \qquad ㉑$$

从式㉑立即有

$$|F(x)| \leqslant \dfrac{1}{n}\tan \dfrac{\pi}{4n} \quad (n \in \mathbf{N}) \qquad ㉒$$

有了以上这些预备知识，就可以来证明本题了．用反证法，如果对于 $[0,1]$ 内的任意实数 x，由式⑮定义的实系数多项式 $f(x)$ 满足

$$|f(x)| < \dfrac{1}{n}\tan \dfrac{\pi}{4n} \qquad ㉓$$

我们的方法是研究 $F(x) - f(x)$，这里 $x \in [0,1]$．取

第12章 Tschebyscheff 多项式与不等式

$$x_k = \frac{\cos\frac{\pi}{2n} - \cos\frac{k\pi}{n}}{1+\cos\frac{\pi}{2n}} \quad (k=1,2,\cdots,n) \qquad ⑫$$

明显的,从式⑫,有

$$0 < x_1 < x_2 < \cdots < x_n = 1 \qquad ⑫$$

和

$$x_k\left(1+\cos\frac{\pi}{2n}\right) - \cos\frac{\pi}{2n} = -\cos\frac{k\pi}{n} = \cos\left(\pi - \frac{k\pi}{n}\right) \qquad ⑫$$

于是

$$T_n\left(\left(1+\cos\frac{\pi}{2n}\right)x_k - \cos\frac{\pi}{2n}\right) = T_n\left(\cos\left(\pi-\frac{k\pi}{n}\right)\right) =$$
$$\cos n\left(\pi-\frac{k\pi}{n}\right) =$$
$$\cos(n-k)\pi =$$
$$(-1)^{n-k} = (-1)^{n+k} \qquad ⑫$$

从式⑰和式⑫,有

$$F(x_k) = (-1)^{n+1}\frac{1}{n}\tan\frac{\pi}{4n}(-1)^{n+k} = (-1)^{k+1}\frac{1}{n}\tan\frac{\pi}{4n} \qquad ⑫$$

这里 $k=1,2,\cdots,n$. 在 $[x_k, x_{k+1}]$ ($k=1,2,\cdots,n$) 内考虑函数 $F(x) - f(x)$. 利用式⑫,有

$$F(x_{k+1}) - f(x_{k+1}) = (-1)^{k+2}\frac{1}{n}\tan\frac{\pi}{4n} - f(x_{k+1}) \qquad ⑫$$

$$F(x_k) - f(x_k) = (-1)^{k+1}\frac{1}{n}\tan\frac{\pi}{4n} - f(x_k) \qquad ⑬$$

如果 k 是偶数,从式⑫,⑫,⑫和式⑬,有

$$F(x_{k+1}) - f(x_{k+1}) > 0, \quad F(x_k) - f(x_k) < 0 \qquad ⑬$$

如果 k 是奇数,类似有

Tschebyscheff 逼近定理

$$F(x_{k+1})-f(x_{k+1})<0, \ F(x_k)-f(x_k)>0 \qquad ⑬②$$

由于多项式函数 $F(x)-f(x)$ 在 $[x_k,x_{k+1}]$ 内关于 x 是连续的，那么，在 (x_k,x_{k+1}) 内必有一点 y_k，使得

$$F(y_k)-f(y_k)=0 \quad (k=1,2,\cdots,n-1) \qquad ⑬③$$

利用式⑪⑤和式⑱，又可以知道

$$f(0)=0=F(0) \qquad ⑬④$$

从式⑬④，我们可写

$$F(x)=b_1 x+b_2 x^2+\cdots+b_n x^n \qquad ⑬⑤$$

这里 $b_j(1\leqslant j\leqslant n)$ 全是实数。明显的，我们有

$$b_1=\lim_{x\to 0}\frac{F(x)}{x} \qquad ⑬⑥$$

当 $x\to 0$，从式⑫⓪，有 $\psi\to \pi-\dfrac{\pi}{2n}$，且

$$x=\frac{\cos\psi+\cos\dfrac{\pi}{2n}}{1+\cos\dfrac{\pi}{2n}} \qquad ⑬⑦$$

从式⑫①，⑬⑥和式⑬⑦，有

$$b_1=\lim_{\psi\to\pi-\frac{\pi}{2n}}\frac{(-1)^{n+1}\dfrac{1}{n}\tan\dfrac{\pi}{4n}\cos n\psi\left(1+\cos\dfrac{\pi}{2n}\right)}{\cos\psi+\cos\dfrac{\pi}{2n}} \qquad ⑬⑧$$

令

$$\psi=\pi-\frac{\pi}{2n}+t \qquad ⑬⑨$$

当 $\psi\to\pi-\dfrac{\pi}{2n}$ 时，有 $t\to 0$。另外，我们可以看到

$$\cos\psi+\cos\frac{\pi}{2n}=\cos\frac{\pi}{2n}-\cos\left(\frac{\pi}{2n}-t\right)=$$

140

第12章　Tschebyscheff 多项式与不等式

$$-2\sin\left(\frac{\pi}{2n}-\frac{1}{2}t\right)\sin\frac{1}{2}t \qquad ⑭⓪$$

$$\tan\frac{\pi}{4n}\left(1+\cos\frac{\pi}{2n}\right)=\tan\frac{\pi}{4n}2\cos^2\frac{\pi}{4n}=\sin\frac{\pi}{2n} \qquad ⑭①$$

$$\cos n\psi = \cos n\left(\pi-\frac{\pi}{2n}+t\right)+\cos\left(n\pi-\frac{\pi}{2}+nt\right)=$$
$$(-1)^n\sin nt \qquad ⑭②$$

将式⑭⓪,⑭①和式⑭②代入式⓵⓷⑧,兼顾式⓵⓷⑨,有

$$b_1=\lim_{t\to 0}\frac{(-1)^{n+1}\frac{1}{n}\sin\frac{\pi}{2n}(-1)^n\sin nt}{-2\sin\frac{1}{2}t\sin\left(\frac{\pi}{2n}-\frac{1}{2}t\right)}=$$

$$\frac{1}{2n}\lim_{t\to 0}\frac{\sin nt}{\sin\frac{1}{2}t}=\frac{1}{2n}\lim_{\theta\to 0}\frac{\sin 2n\theta}{\sin\theta} \quad (\text{令}\ t=2\theta) \qquad ⑭③$$

现在我们来计算式⑭③右端的极限值.

当 n 是正整数时,我们知道

$$(\cos\theta+i\sin\theta)^n=\cos n\theta+i\sin n\theta \qquad ⑭④$$

这里 $\theta\in[0,2\pi]$. 用地方展开公式,从式⑭④,有

$$\cos n\theta+i\sin n\theta=\cos^n\theta+C_n^1\cos^{n-1}\theta(i\sin\theta)+$$
$$C_n^2\cos^{n-2}\theta(i\sin\theta)^2+$$
$$C_n^3\cos^{n-3}\theta(i\sin\theta)^3+$$
$$C_n^4\cos^{n-4}\theta(i\sin\theta)^4+$$
$$C_n^5\cos^{n-5}\theta(i\sin\theta)^5+\cdots+$$
$$C_n^{n-1}\cos\theta(i\sin\theta)^{n-1}+(i\sin\theta)^n \qquad ⑭⑤$$

比较式⑭⑤两端实部和虚部,有

$$\cos n\theta=\cos^n\theta-C_n^2\cos^{n-2}\theta\sin^2\theta+C_n^4\cos^{n-4}\theta\sin^4\theta+\cdots+$$

Tschebyscheff 逼近定理

$$\begin{cases} (-1)^{\frac{n}{2}} \sin^n \theta, & \text{当 } n \text{ 为偶数时} \\ (-1)^{\frac{n-1}{2}} C_n^{n-1} \cos \theta \sin^{n-1} \theta, & \text{当 } n \text{ 为奇数时} \end{cases} \qquad ⑯$$

$$\sin n\theta = C_n^1 \cos^{n-1} \theta \sin \theta - C_n^3 \cos^{n-3} \theta \sin^3 \theta +$$
$$C_n^5 \cos^{n-5} \theta \sin^5 \theta + \cdots +$$
$$\begin{cases} (-1)^{\frac{n-2}{2}} C_n^{n-1} \cos \theta \sin^{n-1} \theta, & \text{当 } n \text{ 为偶数时} \\ (-1)^{\frac{n-1}{2}} \sin^n \theta, & \text{当 } n \text{ 为奇数时} \end{cases} \qquad ⑰$$

在式⑰中，用 $2n$ 代替 n，有
$$\sin 2n\theta = 2n\cos^{2n-1} \theta \sin \theta - C_{2n}^3 \cos^{2n-3} \theta \sin^3 \theta + \cdots +$$
$$(-1)^{n-1} C_{2n}^{2n-1} \cos \theta \sin^{2n-1} \theta \qquad ⑱$$

从式⑭和式⑱，有
$$b_1 = 1 \qquad ⑲$$

从式⑮，⑮和式⑲，有
$$F(x) - f(x) =$$
$$x^2 \left[(b_2 - a_2) + (b_3 - a_3)x + \cdots + (b_n - a_n)x^{n-2} \right] \qquad ⑳$$

从式㉕和式⑱，可以知道 $0 < y_1 < y_2 < \cdots < y_{n-1} < 1$，及 $y_1, y_2, \cdots, y_{n-1}$ 是多项式 $(b_2 - a_2) + (b_3 - a_3)x + \cdots + (b_n - a_n)x^{n-2}$（$x$ 的至多 $n-2$ 次多项式）的 $n-1$ 个两两不同实根，因此，对于任意实数 x，有
$$(b_2 - a_2) + (b_3 - a_3)x + \cdots + (b_n - a_n)x^{n-2} = 0 \qquad ㉑$$

从式⑳和式㉑，对于任意实数 x，有
$$F(x) = f(x) \qquad ㉒$$

这与不等式⑬，⑬是矛盾的.

Tschebyscheff 多项式与马尔可夫定理

第 13 章

例 22 设 Q 是次数不大于 $n-1$ 的多项式,且对 $x \in [-1,1]$,有

$$|Q(x)| \geq \frac{1}{\sqrt{1-x^2}}$$

证明:在 $[-1,1]$ 内有估计式 $|Q(x)| \leq n$ 成立.

证明 利用上一章问题的符号表示,若 $-x_1 = x_n \leq x \leq x_1$,则

$$\sqrt{1-x^2} \geq \sqrt{1-x_1^2} = \sin\frac{\pi}{2n} \geq \frac{1}{n}$$

于是断言对 $x_n \leq x \leq x_1$ 成立. 对于 $[-1,1]$ 内其余的点,我们把上一章问题中求出的拉格朗日插值公式应用到多项式 $Q(x)$

$$Q(x) = \frac{1}{n} \sum_{k=1}^{n} (-1)^{k-1} \sqrt{1-x_k^2}\, Q(x_k) \frac{T_n(x)}{x-x_k}$$

因为不论 $x<x_n$ 还是 $x>x_1$，数 $x-x_k$ 都是同号的. 于是

$$|Q(x)| \leq \frac{1}{n} \sum_{k=1}^{n} \left| \frac{T_n(x)}{x-x_k} \right|$$

但是

$$T_n(x) = 2^{n-1} \sum_{k=1}^{n} (x-x_k)$$

所以

$$\frac{T_n'(x)}{T_n(x)} = \sum_{k=1}^{n} \frac{1}{x-x_k}$$

因此

$$|Q(x)| \leq \frac{1}{n} |T_n'(x)|$$

而且，由于 $x = \cos\theta$，有

$$T_n'(x) = \frac{n \sin n\theta}{\sin \theta}$$

利用数学归纳法易验证，当 $-\infty < \theta < +\infty$ 时，有 $|\sin n\theta| \leq n|\sin\theta|$，因此

$$|T_n'(x)| \leq n^2$$

13.1 多项式与三角多项式的导数增长的阶

我们已考虑了一系列正算子(伯恩斯坦多项式、费耶尔算子、用以证明杰克逊定理的算子 $A_n(f;x)$，等等)，它们对恒等于 1 时的函数值与这函数本身相合，$L(1,x) = 1$. 对于一切这样的算子，不等式

$$|L(f,x) - f(x)| \leq 2M$$

成立，只要连续函数 $f(x)$ 有界，$|f(x)| \leq M$. 其实，由不

第13章 Tschebyscheff 多项式与马尔柯夫定理

等式
$$-M \leqslant f(t) \leqslant M$$
线性正算子的单调性以及由 $L(1,x)=1$，得到
$$-ML(1,x) = -M \leqslant L(f,x) \leqslant M, \quad \|L_n(f,x)\| \leqslant M$$
⑭

从而便推得不等式⑬.

设 $f(x)$ 为 n 阶三角多项式，$f(x) = \dfrac{a_n}{2} + \sum\limits_{k=1}^{n}(a_k \cos kx + b_k \sin kx)$，$|f(x)| \leqslant M$ 并设 $F(n)(f,x)$ 为费耶尔算子，则

$$F_n(f,x) = \frac{a_0}{2} + \sum_{k=1}^{n} \frac{n-k}{n}(a_k \cos kx + b_k \sin kx) =$$

$$\frac{a}{2} + \sum_{k=1}^{n}(a_k \cos kx + b_k \sin kx) -$$

$$\frac{1}{n}\sum_{k=1}^{n} k(a_k \cos kx + b_k \sin kx) =$$

$$f(x) - \frac{1}{n}\sum_{k=1}^{n}(a_k \cos kx + b_k \sin kx)$$

由这一等式以及不等式⑬得到

$$\left|\sum_{k=1}^{n} k(a_k \cos kx + b_k \sin kx)\right| =$$
$$n|F_n(f,x) - f(x)| \leqslant 2nM \qquad ⑮$$

对于每一 n 阶三角多项式 $f(x)$，我们令

$$f(x) = \frac{a}{2} + \sum_{k=1}^{n}(a_k \cos kx + b_k \sin kx)$$

$$D_s f = \sum_{k=1}^{n} k^s(a_k \cos kx + b_k \sin kx)$$

在这些记号下不等式⑮可以改写成

Tschebyscheff 逼近定理

$$|D_1 F| \leqslant 2nM \qquad \text{⑯}$$

显然,$D_{l+m}f = D_l(D_m f)$,所以应用不等式⑯逐步得到

$$|D_2 f| = |D_1(D_1 f)| \leqslant 2n \cdot 2nM = 2^2 n^2 M$$
$$|D_3 f| = |D_1(D_2 f)| \leqslant 2n \cdot 2^2 n^2 M = 2^3 n^3 M$$
$$\vdots$$
$$|D_s f| \leqslant 2^s n^s M$$

运算 $D_s f$ 在 s 为偶数情形与导数 $f^{(s)}(x)$ 相合或与它仅相差一符号。例如 $D_2 f = -f''(x)$,$D_4 f = f^{\text{IV}}(x)$,$D_6 f = -f^{\text{VI}}(x)$ 等。由此有

$$|f^{(s)}(x)| \leqslant 2^s n^s M \qquad \text{⑰}$$

只要 $f(x)$ 为 n 阶三角多项式,$|f(x)| \leqslant M$,s 为偶数。现在我们证明不等式⑰对奇数 s 的正确性。为此预先指出,函数 $|f'(x)|$ 在那样的点 z 取得最大值,于那里三角多项式 $f'(x)$ 有极大或极小值,因而 $f''(z) = 0$。令 $T(x) = f^2(x) - \dfrac{M^2}{2}$。在点 z 有等式

$$T''(z) = 2f'^2(z) + 2f(z)f''(z) = 2f'^2(z) \qquad \text{⑱}$$

然 $T(x)$ 为 $2n$ 阶三角多项式,对于它不等式

$$-\frac{M^2}{2} \leqslant T(x) \leqslant M^2 - \frac{M^2}{2} = \frac{M^2}{2}, \quad |T(x)| \leqslant \frac{M^2}{2}$$

成立,只要 $|f(x)| \leqslant M$。由此以及由关系式⑰与⑱得到

$$|f'(x)| \leqslant |f'(z)| = \sqrt{\frac{1}{2} T''(z)} \leqslant$$

$$\sqrt{\frac{1}{2}(2n)^2 \cdot 2^2 \cdot \frac{M^2}{2}} = 2nM$$

于是,不等式⑰对于 $s = 1$ 获证。由于奇阶导数为偶阶导数的一阶导数,所以不等式⑰对于一切正整数 s 都正确。于是我们证明了

第13章 Tschebyscheff 多项式与马尔柯夫定理

伯恩斯坦第一不等式 若 $f(x)$ 为 n 阶三角多项式，$|f(x)|\leq M$，则对于它的导数，不等式⑮成立.

附注 刚才这里所用的证明伯恩斯坦不等式的方法是头一个证法. 用其他方法可以证明不等式成立且无因子 2^s 出现①.

转向代数多项式的导数增长状态的研究. 设 $P(x)=a_0+a_1x+\cdots+a_nx^n$ 为在区间 $[a,b]$ 上满足不等式 $|P(x)|\leq M$ 的 n 次代数多项式. 令 $T(\theta)=P(x)$，这里 $x=\dfrac{1}{2}\{(b-a)\cos\theta+(b+a)\}$. $T(\theta)$ 为 n 阶偶三角多项式，且由于 θ 由 $-\pi$ 变到 π 时量 x 由 a 变到 b，所以 $|T(\theta)|\leq M$. 据伯恩斯坦第一不等式，有

$$|T'(\theta)|=|P'(x)|\frac{b-a}{2}|\sin\theta|\leq 2nM \qquad ⑮$$

但 $\dfrac{b-a}{2}\cos\theta=x-\dfrac{a+b}{2}$，因而

$$\frac{b-a}{2}|\sin\theta|=\left\{\left(\frac{b-a}{2}-\frac{b-a}{2}\cos\theta\right)\cdot\left(\frac{b-a}{2}+\frac{b-a}{2}\cos\theta\right)\right\}^{\frac{1}{2}}=\sqrt{(b-x)(x-a)}$$

考虑到这一点与不等式⑮，得到

$$|P'(x)|\leq\frac{2nM}{\sqrt{(b-x)(x-a)}} \qquad ⑯$$

函数 $\dfrac{1}{\sqrt{(b-x)(x-a)}}$ 在区间 $[a',b']$，$a<a'<b'<b$ 上

---

① 例如，参看 И·П·纳唐松. 函数构造论，上册[M]. 北京：科学出版社，1958.

Tschebyscheff 逼近定理

有界,因而在这区间上不等式

$$|P'_n(x)| < cnM \qquad ⑯$$

成立,并且常数 c 与 n 无关,而只依赖于量 $a'-a>0$ 与 $b-b'>0$. 现在已经容易在对多项式 $P(x)$ 所作的假定之下证明不等式

$$|P^{(m)}(x)| \leqslant Dn^m M \quad (a' \leqslant x \leqslant b') \qquad ⑯$$

其中常数 D 与 n 无关. 其实,对于 $m=1$ 这个不等式已经证明了. 假定它对于 $m=k$ 为真,亦即假定 $|P^{(k)}(x)| \leqslant c'n^k M, a'' \leqslant x \leqslant b''$,这里数 a'' 与 b'' 这样选择:使得 $a < a'' < a' < b' < b'' < b$. 利用这一假设与不等式⑯,得到

$$|P^{(k+1)}(x)| < c''nc'n^k M = Dn^{k+1}M \quad (a' \leqslant x \leqslant b')$$

这便证明了不等式⑯以及

伯恩斯坦第二不等式 若 $P(x)$ 为 n 次代数多项式,$|P(x)| \leqslant M, a \leqslant x \leqslant b$,则在区间 $[a',b']$ 上,这里 $a < a' < b' < b$,不等式⑯成立,那里常数 D 与 n 无关.

13.2 函数的可微性质的表征

设 $f(x)$ 为连续的周期函数并设 $t_n(x) = t_n(f, x)$ 为与 $f(x)$ 有最小偏差的三角多项式. 我们假定,数 $E_n(f)$ 满足不等式

$$E_n(f) = \|f - t_n\| \leqslant \frac{A}{n^\alpha} \quad (n=1,2,\cdots; 0 < \alpha < 1) \qquad ⑯$$

我们证明,当 $|x-y| \leqslant \delta$ 与 $\frac{1}{2^m} \leqslant \delta \leqslant \frac{1}{2^{m-1}}$ 时,不等式

$$|t_{2^m}(x) - t_{2^m}(y)| \leqslant c'\delta^\alpha \qquad ⑯$$

成立.

为此我们令

第13章 Tschebyscheff 多项式与马尔柯夫定理

$$u_0(x) = t_1(x), u_k(x) = t_2^k(x) - t_2^{k-1}(x) \quad (k=1,2,\cdots)$$

考虑到不等式⑯对于 $k=1,2,3,\cdots$，得到

$$|u_k(x)| = |t_2^k(x) - f(x) + f(x) - t_2^{k-1}(x)| \leqslant$$
$$|t_2^k(x) - f(x)| + |f(x) - t_2^{k-1}(x)| =$$
$$E_2^k(f) + E_2^{k-1}(f) \leqslant \frac{A}{2^{k\alpha}} + \frac{A}{2^{(k-1)\alpha}} < \frac{3A}{2^{k\alpha}} \qquad ⑯$$

对于 $k=0$ 这个估值可能是不对的. 但

$$|u_0(x)| = |t_1(x)| = |t_1(x) - f(x) + f(x)| \leqslant$$
$$|t_1(x) - f(x)| + |f(x)| \leqslant A + M$$

其中 $M = \max|f(x)|$，因而若把常数 A 增大使不等式 $2A \geqslant M$ 成立，则不等式⑯对于 $k=0$ 的情形也适用. 假定如此并利用伯恩斯坦第一不等式（三角多项式 $u_k(x)$ 的阶不超过 2^k），得到

$$|u_k'(x)| = 2 \cdot 2^k \cdot \frac{3A}{2^{k\alpha}} = 6A \cdot 2^{k(1-\alpha)} \qquad ⑯$$

迄今为止我们都是着眼于多项式 $u_k(x)$ 的讨论. 回到多项式 $t_2^m(x)$，首先我们指出

$$t_2^m(x) = \{t_2^m(x) - t_2^{m-1}(x)\} + \{t_2^{m-1}(x) - t_2^{m-2}(x)\} + \cdots +$$
$$\{t_2(x) - t_1(x)\} + t_1(x) = u_m(x) + u_{m-1}(x) + \cdots +$$
$$u_1(x) + u_0(x) = \sum_{k=0}^{m} u_k(x)$$

因而

$$t_2^m(x) - t_2^m(y) = t_2^{m'}(z)(x-y) = (x-y) \sum_{k=0}^{m} u_k'(z)$$
$$(z \in [x,y])$$

从式⑯推出

$$|t_2^m(x) - t_2^m(y)| \leqslant |x-y| \sum_{k=0}^{m} |u_k'(z)| \leqslant$$

Tschebyscheff 逼近定理

$$6A\mid x-y\mid \sum_{k=0}^{m} 2^{k(1-\alpha)} =$$

$$6A\mid x-y\mid \frac{2^{(m+1)(1-\alpha)}-1}{2^{1-\alpha}-1} <$$

$$6A\frac{2^{(m+1)(1-\alpha)}}{2^{1-\alpha}-1} \qquad \circledast$$

最后,注意到后一不等式以及 $\mid x-y\mid \leqslant \delta, 2^{m-1}\leqslant \frac{1}{\delta}$,得到

$$\mid t_2^m(x)-t_2^m(y)\mid <6A\delta\frac{2^{2(1-\alpha)}}{2^{1-\alpha}-1}\delta^{\alpha-1}=c'\delta^\alpha$$

其中假定了 $c'=6A\dfrac{4^{1-\alpha}}{2^{1-\alpha}-1}$.

伯恩斯坦第一定理 若对于连续的周期函数 $f(x)$ 满足不等式⑯,则函数 $f(x)$ 属于类 Lips α.

证明 有

$$\mid f(y)-f(x)\mid = \mid f(y)-t_2^m(y)+t_2^m(y)-t_2^m(x)+t_2^m(x)-f(x)\mid \leqslant \mid f(y)-t_2^m(y)\mid +\mid t_2^m(y)-t_2^m(x)\mid +\mid t_2^m(x)-f(x)\mid \leqslant 2E_2^m(f)+\mid t_2^m(y)-t_2^m(x)\mid$$

这一不等式对任何值 m 都成立. 现在,认为 $\mid x-y\mid \leqslant \delta$, $\delta \leqslant \frac{1}{2}$,我们取这样的 m,使得不等式 $\frac{1}{2^m}\leqslant \delta \leqslant \frac{1}{2^{m-1}}$ 成立. 这时由后一不等式及不等式⑯与⑯便得到

$$\mid f(y)-f(x)\mid \leqslant \frac{2A}{2^{m\alpha}}+c'\delta^\alpha \leqslant (2A+c')\delta^\alpha = D\delta^\alpha$$

这个不等式对于 $\delta \leqslant \frac{1}{2}$ 已证明了. 但如果 $\delta \geqslant \frac{1}{2}$,则

$$\frac{|f(x)-f(y)|}{|x-y|^\alpha} \leqslant 2^n \cdot 2M$$

这里 $M = \|f\|$,因而,取 S 为数 D 与 $2^{\alpha+1}M$ 中的较大者便得到

伯恩斯坦第二定理 若对于周期函数 $f(x)$ 不等式

$$E_n(f) = \|f-t_n\| \leqslant \frac{A}{n^{k+\alpha}} \qquad ⑯⑦$$

成立,其中 k 为自然数且 $0<\alpha<1$,则函数 $f(x)$ 为 k 次可微且 $f^{(k)}(x)$ 属于类 Lips α.

证明 如以前一样,令

$$u_0(x) = t_1(x), u_k(x) = t_2^k(x) - t_2^{k-1}(x) \quad (k=1,2,\cdots)$$

前面曾证明

$$t_2^m(x) = \sum_{v=0}^{+\infty} u_v(x)$$

因而由 $t_2^m(x) \to f(x)$,知

$$f(x) = \sum_{v=0}^{+\infty} u_v(x)$$

成立. 我们证明,从上述级数微分 k 次所得级数

$$\sum_{v=0}^{+\infty} u_v^{(k)}(x) \qquad ⑯⑧$$

一致收敛. 其实,根据伯恩斯坦第一不等式由不等式

$$|u_v(x)| \leqslant |f(x)-t_2^v(x)| + |f(x)-t_2^{v-1}(x)| \leqslant$$

$$E_2^v(f) + E_2^{v-1}(f) < \frac{A+A\cdot 2^{k+\alpha}}{2^{v(k+\alpha)}} = \frac{B}{2^{v(k+\alpha)}}$$

得到

$$|u_v^{(k)}(x)| \leqslant 2^k \cdot 2^{vk} \frac{B}{2^{v(k+\alpha)}} = 2^k \frac{B}{2^{v\alpha}} = \frac{C}{2^{v\alpha}}$$

从而推得级数⑯⑧的一致收敛性以及等式

Tschebyscheff 逼近定理

$$f^{(k)}(x) = \sum_{v=0}^{+\infty} u_v^{(k)}(x)$$

接下来要证明函数 $f^{(k)}(x)$ 属于类 Lips α. 显然，为此只需证明对于函数 $f^{(k)}(x)$ 的伯恩斯坦第一定理的条件满足即可. 设 n 为自然数，取这样的一个自然数 m，使它满足不等式

$$2^{m-1} \leqslant n \leqslant 2^m$$

我们有

$$E_n = (f^{(k)}) = \| f^{(k)}(x) - t_n(f^{(k)}, x) \| \leqslant$$

$$\| f^{(k)}(x) - \sum_{v=0}^{m-1} u_v^{(k)}(x) \| =$$

$$\| \sum_{v=m}^{+\infty} u_v^{(k)}(x) \| \leqslant \sum_{v=m}^{+\infty} \| u_v^{(k)}(x) \|$$

(多项式 $u_{m-1}(x)$ 的阶不大于 $2^{m-1} \leqslant n$) 利用以前对于 $u_v^{(k)}(x)$ 所得的估值，得到

$$E_n(f^{(k)}) \leqslant \sum_{v=m}^{+\infty} \frac{C}{2^{v\alpha}} = \frac{2^\alpha C}{2^\alpha - 1} \cdot \frac{1}{2^{m\alpha}}$$

记住 $\frac{1}{2^m} \leqslant \frac{1}{n}$，并设 $C' = \frac{2^\alpha C}{2^\alpha - 1}$，得到

$$E_n(f^{(k)}) \leqslant \frac{C'}{n^\alpha}$$

由这个不等式与伯恩斯坦第一定理推得伯恩斯坦第二定理.

第 14 章 多元逼近

某些二维区域上的最小零偏差多项式:

用 P_n 表示所有次数不大于 n 的二元实系数多项式构成的线性空间. $D \subset \mathbf{R}^2$ 是有界闭区域. 若 $p^*(x,y)$ 是函数 $x^m y^n$ 在 P_{m+n-1} 中在区域 D 上的最佳逼近多项式, 即

$$\| x^m y^n - p^*(x,y) \| = \inf_{p \in P_{m+n-1}} \| x^m y^n - p(x,y) \|$$

则称 $x^m y^n - p^*(x,y)$ 是区域 D 上以 $x^m y^n$ 为首项的最小零偏差多项式, 并称

$$e_{m,n} = \max_{(x,y) \in D} |x^m y^n - p^*(x,y)|$$

Tschebyscheff 逼近定理

为最小零偏差.

在这里我们仅能针对三角形区域

$$B = \{(x,y) \in \mathbf{R}^2 \mid x \geqslant -1, y \geqslant -1, x+y \leqslant 0\}$$

和圆域

$$C = \{(x,y) \in \mathbf{R}^2 \mid x^2 + y^2 \leqslant 1\}$$

讨论.

下面介绍最小零偏差多项式的表达式.

定理 21 对于 $\forall m, n \in \mathbf{N}$,有

$$B_{m,n}(x,y) = \prod_{i=1}^{m} \left(x + \cos \frac{2i-1}{2(m+n)} \pi \right) \cdot \prod_{j=1}^{n} \left(y + \cos \frac{2j-1}{2(m+n)} \pi \right)$$

是三角形区域 B 上的以 $x^m y^n$ 为首项的最小零偏差多项式. 对应的最小零偏差为 $\min(2^{-m-n+1}, 1)$.

定理 22 对于 $\forall m, n \in \mathbf{N}$,有

$$C_{m,n}(x,y) = \prod_{i=1}^{m} \left(x + \sin \frac{2i-m-1}{2(m+n)} \pi \right) \cdot \prod_{j=1}^{n} \left(y + \sin \frac{2j-n-1}{2(m+n)} \pi \right)$$

是区域 C 上以 $x^m y^n$ 为首项的最小零偏差多项式,其最小零偏差是 $\min(2^{-m-n+1}, 1)$.

证明参见梁学章,李强编著的《多元逼近》(国防工业出版社,p.54~62).

多元逼近问题中的未解决问题

第 15 章

令 D 是一个在 (x,y) 平面中具有光滑边界的有界闭区域,$f(x,y)$ 是一个在 D 中具有连续二阶偏导数的函数,这个函数在 D 的

$$N=(\mu+1)(\gamma+1)+1 \qquad ⑰$$

个不同点上的值通过观测得到,这些点表示为 $P_j=(x_j,y_j)\in D$,设

$$\pi_{\mu\gamma}(x,y)=\sum_{l_1=0}^{\mu}\sum_{l_2=0}^{\gamma}c_{l_1 l_2}x^{l_1}y^{l_2} \qquad ⑰$$

是一个具有性质

$$\pi_{\mu\gamma}(x_j,y_j)=\pi_{\mu\gamma}(p_j)=f(p_j)$$
$$(j=1,2,\cdots,n) \qquad ⑰$$

的多项式.

Tschebyscheff 逼近定理

现在选择这些点 $p_1, p_2, \cdots, p_n$ 使得这个方程组的行列式 $\Delta(p_1, \cdots, p_n)$ 为最大

$$|\Delta(p_1, \cdots, p_n)| = \max_{a_1, \cdots, a_n \in D} |\Delta(Q_1, \cdots, Q_n)| \quad ⑰③$$

容易看到,式⑰③中这个根的最大值是正的. 对于点 p_k 的这种选择,多项式⑰①,⑰② 是唯一确定的. 设 $p = (x, y)$,我们有

$$\pi_{\mu\gamma}(p) = \sum_{j=1}^{n} f(p_j) \cdot \frac{\Delta(p_1, \cdots, p_{j-1}, p_{j+1}, \cdots, p_n)}{\Delta(p_1, \cdots, p_n)} \triangleq$$

$$\sum_{j=1}^{n} f(p_j) l_j(p, p_1, \cdots, p_n) \quad ⑰④$$

设 $\pi_{\mu\gamma}^*(p, f, D)$ 是 f 在 D 中的次数不大于 μ 和 γ 的最佳逼近多项式. 显然

$$\sum_{j=1}^{n} \pi_{\mu\gamma}^*(p_j) l_j(p, p_1, \cdots, p_n) = \pi_{\mu\gamma}^*(p) \quad ⑰⑤$$

我们有

$$\pi_{\mu\gamma}(p) - \pi_{\mu\gamma}^*(p, f, D) =$$
$$\sum_{j=1}^{n} f(p_j) - \pi_{\mu\gamma}^*(p_j) l_j(p, p_1, \cdots, p_n) \quad ⑰⑥$$

令

$$\max_{p \in D} |f(p) - \pi_{\mu\gamma}^*(p)| \overset{\text{def}}{=\!=\!=} d_{\mu\gamma}(f, D) \quad ⑰⑦$$

于是

$$|\pi_{\mu\gamma}(p) - f(p)| \leq |\pi_{\mu\gamma}(p) - \pi_{\mu\gamma}^*(p)| + |\pi_{\mu\gamma}^*(p) - f(p)| \leq d_{\mu\gamma}(f, D) \times$$
$$\left(1 + \sum_{j=1}^{n} |l_j(p, p_1, \cdots, p_n)|\right) \quad ⑰⑧$$

因为由式⑰④可知

$$|l_j(p, p_1, \cdots, p_n)| \leq 1 \quad (j = 1, \cdots, n) \quad ⑰⑨$$

156

第 15 章　多元逼近问题中的未解决问题

从而

$$|f(p)-\pi_{\mu\gamma}(p)| \leq (n+1)d_{\mu\gamma}(f,D) \quad ⑱$$

对于不太大的 μ 和 γ 的值,点 $p_1, p_2, \cdots, p_n$ 可以用数值方法确定.

进一步的研究可见 Journal of Approximation Theory.

非线性 Tschebyscheff 逼近

第 16 章

设 $f(x)$ 与 $F(A,x)$ 为给定的两个实连续函数. 其中 $x \in M$, $A = (a_1, a_2, \cdots, a_n) \in D$. M 表示实轴上的某个有界闭集,D 是 n 维欧氏空间中的子集. Tschebyscheff 逼近问题就是在 D 中确定 $A^* \in D$,使对一切 $A \in D$. 有
$$\rho^* = \max_{x \in M} |F(A^*, x) - f(x)| \leqslant$$
$$\max_{x \in M} |F(A, x) - f(x)|$$
这时,$F(A^*, x)$ 称为 $f(x)$ 在 $F(A, x)$ 中的最优逼近,ρ^* 称为 $f(x)$ 用 $F(A, x)$ 逼近的最小偏差.

关于线性 Tschebyscheff 逼近 ($F(A, x)$ 对于参数 A 的依赖为线性情

第16章 非线性 Tschebyscheff 逼近

形)的理论早已相当成熟和完善. 在计算上也已给出了一些较为有效的算法. 但对非线性 Tschebyscheff 逼近,要想进行一般的研究,仍有许多困难. T. S. Motzkim 和 L. Tornheim 引进的所谓"单解(unisolvence)函数族"的概念就是一个著名的例子. 一个连续函数族类 $F(A,x)(A \in D, x \in M)$ 是单解的,如果:

(1)对于任意给定的 n 对数 $(\xi_i, \eta_i)(i=1,2,\cdots, n)$,方程组
$$F(A, \xi_i) = \eta_i \quad (i=1,2,\cdots,n)$$
总有解 $A \in D$(所谓有解性). 其中 ξ_i 是包含 M 的最小闭区间 $I(M)$ 中任意给定的点.

(2)任意两个相异函数 $F(A', x)$ 与 $F(A'', x)$ 之差在 $I(M)$ 中至多有 $n-1$ 个零点(包括计算重数在内,即两边差值不变号的零点算作两个零点). 由此立即推知(1)的解是唯一的(所谓单解性).

对于单解函数族 $F(A,x)$ 已经证明了如下定理

定理 23 $f(x)$ 在 $F(A,x)$ 中的最优逼近是存在的.

定理 24 $f(x)$ 在 $F(A,x)$ 中的最优逼近是唯一的.

定理 25 $F(A^*, x)$ 是 $f(x)$ 在 $F(A,x)$ 中的最优逼近的充要条件是在 M 中至少有 $n+1$ 个点(称为偏差点)正负交替地取值 $\max_{x \in M} |F(A^*, x) - f(x)|$.

1983 年何新贵得到类似 Vallée-Poussin 的定理.

定理 26 对 $\forall A \in D$,若
$$F(A, x_i) - f(x_i) = (-1)^i \rho_i \quad (i=1,2,\cdots,n+1)$$
其中 ρ_i 或为零或为同号的数,则最小偏差 ρ^* 满足

Tschebyscheff 逼近定理

$$\min_{1\leqslant i\leqslant n+1}|\rho_i|\leqslant \rho^* \leqslant \max_{1\leqslant i\leqslant n+1}|\rho_i|$$

其中两处等号当且仅当

$$|\rho_1|=|\rho_2|=\cdots=|\rho_{n+1}|=\rho^*$$

时成立.

定理 27 函数 $f(x)$ 在整个集合 M 上的最优逼近正好是 $f(x)$ 在使偏差达到最大的 $n+1$ 个点上的最优逼近.

巴拿赫空间中的 Tschebyscheff 多项式

第 17 章

一个相关的有趣问题是寻求马尔柯夫定理在巴拿赫空间中的类似结果,即寻求最小数 $M_{n,k}$ 具有下述性质:若 P 是阶数不大于 n 的任何多项式,把一个实赋范线性空间映入另一个实赋范线性空间,则

$$\sup \| \hat{D}^k P(x) \| \leqslant M_{n,k} \sup \| (x) \|$$

这里, $\hat{D}^k P(x) y = \dfrac{\mathrm{d}^k}{\mathrm{d} t^k} P(x+ty) \Big|_{t=0}$. 容易证明

$$T_n^{(k)}(1) \leqslant M_{n,k} \leqslant 2^{2k-1} T_n^{(k)}(1) \qquad ⑱$$

这里 T_n 是 n 阶 Tschebyscheff 多项式. 事实上,令 P 是实赋范线性空间中阶

Tschebyscheff 逼近定理

数不大于 n 的任何实值多项式,假设 $\|x\| \leq 1$ 蕴涵 $|P(x)| \leq 1$. 令
$$\|x\| \leq 1, \quad \|y\| \leq 1, \quad -1 \leq s \leq 1$$
定义 $q(t) = P(\psi(t))$,这里 $\psi(t) = [x - sy + t(x + sy)]/2$. 注意,当 $-1 \leq t \leq 1$ 时 $\|\psi(t)\| \leq \dfrac{(1+t)}{2} + \dfrac{(1-t)}{2} = 1$. 因此 q 是阶数不大于 n 的多项式,当 $-1 \leq t \leq 1$ 时满足 $|q(t)| \leq 1$,从而 $|q^k(1)| \leq |T_n^{(k)}(1)|$,显然,$q^k(1) = 2^{-k} \hat{D}^k P(x)(x+sy)$. 因此,映象 $s \to \hat{D}^k(x+sy)$ 是阶数不大于 k 的多项式,在 $[-1, 1]$ 上以 $2^k T_n^{(k)}(1)$ 为界,$\hat{D}^k P(x) y$ 是这个多项式中 2^k 的系数. 从而,得
$$|\hat{D}^k P(x) y| \leq 2^{k-1} [2^k T_n^{(k)}(1)]$$
再由罕-巴拿赫定理便推出式⑱.

注意,只考虑 l_2^1 上的实值多项式时,$M_{n,k}$ 的值是不变的. 只考虑实希尔伯特空间时,$M_{n,1} = n^2$,这时,对于 $1 \leq k \leq n$,考虑是否有 $M_{n,k} = T_n^{(k)}(1)$ 是一个有意思的问题,参见 [11].

FIR 数字滤波器设计的 Tschebyscheff 逼近法[①][②]

第 18 章

前面几章所介绍的均为在纯数学方面 Tschebyscheff 多项式的应用,实际上在许多应用领域中它也有意想不到的应用. FIR 数字滤波器的设计有两种常用方法:傅里叶级数法(窗函数法)和频率抽样法,用这两种方法设计出的滤波器的频率特性都是在不同意义上对所给理想频率特性 $H_d(e^{j\omega})$ 的逼近. 从数值逼近的理论来看,对某个函数 $f(x)$ 的逼近一般有三种方法:(1)插值法. (2)最小平方逼

[①] 本章选自胡广书. 数字信号处理[M]. 北京:清华大学出版社,2003.

[②] Tschebyscheff 多项式应用广泛. 康托洛维奇将其应用于统计标志的振动(变异)、广分布曲线的系数、置信度、复回归与曲线方程以及统计参数的误差和平方近似值.

Tschebyscheff 逼近定理

近法.(3)最佳一致逼近法.

所谓插值,即寻找一 n 阶多项式(或三角多项式) $p(x)$,使它在 $n+1$ 个点 $x_0,x_1,\cdots,x_n$ 处满足

$$p(x_k)=f(x_k)\quad(k=0,1,\cdots,n)$$

在非插值点上,$p(x)$ 是 $f(x_k)$ 的某种组合.当然,在非插值点上,$p(x)$ 和 $f(x)$ 存在一定的误差.频率抽样法可以看做插值法,它在抽样点 ω_k 上保证了 $H(e^{j\omega_k})=H_d(e^{j\omega_k})$,而在非抽样点上,$H(e^{j\omega})$ 是插值函数 $S(\omega,k)$ 的线性组合,其权重是 $H_d(e^{j\omega_k})$.这种设计方法的缺点是通带和阻带的边缘不易精确地确定.

最小平方逼近是在所需要的范围内,如在区间 $[a,b]$ 内使积分 $\int_a^b[p(x)-f(x)]^2\mathrm{d}x$ 为最小.这种设计方法是着眼于使整个区间 $[a,b]$ 内的总误差为最小,但它并不一定能保证每个局部位置误差都最小.实际上,在某些位置上,有可能存在着较大的误差.傅里叶级数法就是一种最小平方逼近法.该方法在间断点处出现了较大的过冲(Gibbs 现象).为了减小这种过冲和欠冲,采用了加窗口的方法,当然,加窗以后的设计法,已不再是最小平方逼近.

最佳一致逼近法,是着眼于在所需要的区间 $[a,b]$ 内,使误差函数

$$E(x)=|p(x)-f(x)|$$

较均匀一致,并且通过合理地选择 $p(x)$,使 $E(x)$ 的最大值 E_n 达到最小.Tschebyscheff 逼近理论解决了 $p(x)$ 的存在性、唯一性及如何构造等一系列问题.

麦克莱伦(McClellan),帕克斯(Parks)和拉宾纳(Rabiner)等人应用 Tschebyscheff 逼近理论,提出了一

第18章 FIR 数字滤波器设计的 Tschebyscheff 逼近法

种 FIR 数字滤波器的计算机辅助设计方法. 这种设计方法由于是在一致意义上对 $H_d(\mathrm{e}^{\mathrm{j}\omega})$ 作最佳逼近, 因而获得了较好的通带和阻带性能, 并能准确地指定通带和阻带的边缘, 是一种有效的设计方法.

18.1 Tschebyscheff 最佳一致逼近原理

Tschebyscheff 最佳一致逼近的基本思想是: 对于给定区间 $[a,b]$ 上的连续函数 $f(x)$, 在所有 n 次多项式的集合 $\mathscr{P}_n$ 中, 寻找一多项式 $\hat{p}(x)$, 使它在 $[a,b]$ 上对 $f(x)$ 的偏差和其一切属于 $\mathscr{P}_n$ 的多项式 $p(x)$ 对 $f(x)$ 的偏差相比是最小的, 即

$$\max_{a\leqslant x\leqslant b}|\hat{p}(x)-f(x)|=\min\{\max_{a\leqslant x\leqslant b}|p(x)-f(x)|\}$$

Tschebyscheff 逼近理论指出, 这样的多项式 $\hat{p}(x)$ 是存在的, 且是唯一的, 并指出了构造这种最佳一致逼近多项式的方法, 这就是有名的"交错点组定理", 该定理可描述如下:

设 $f(x)$ 是定义在 $[a,b]$ 上的连续函数, $p(x)$ 为 $\mathscr{P}_n$ 中一个次数不超过 n 的多项式, 并令

$$E_n = \max_{a\leqslant x\leqslant b}|p(x)-f(x)|$$

及

$$E(x)=p(x)-f(x)$$

$p(x)$ 是 $f(x)$ 的最佳一致逼近多项式的充要条件是 $E(x)$ 在 $[a,b]$ 上至少存在 $n+2$ 个交错点 $a\leqslant x_1<x_2<\cdots<x_{n+2}\leqslant b$, 使得

$$E(x_i)=\pm E_n \quad (i=1,2,\cdots,n+2)$$

且

Tschebyscheff 逼近定理

$$E(x_i) = -E(x_{i+1}) \quad (i=1,2,\cdots,n+2)$$

这 $n+2$ 个点即是"交错点组", 显然 $x_1, x_2, \cdots, x_{n+2}$ 是 $E(x)$ 的极值点.

n 阶 Tschebyscheff 多项式

$$C_n(x) = \cos(n\arccos x) \quad (-1 \leqslant x \leqslant 1)$$

在区间 $[-1,1]$ 上存在 $n+1$ 个点 $x_k = \cos\left(\dfrac{\pi}{n}k\right)$ ($k=0,1,\cdots,n$) 轮流使得 $C_n(x)$ 取得最大值 +1 和最小值 -1. $C_n(x)$ 是 x 的多项式, 且最高项 x^n 的系数是 2^{n-1}, 可以证明, 在所有 n 阶多项式中, 多项式 $\dfrac{C_n(x)}{2^{n-1}}$ 和 0 的偏差为最小. 这样, 如果我们在寻找 $p(x)$ 时, 能使误差函数为某一个 $C_n(x)$, 那么, 这样的 $p(x)$ 将是对 $f(x)$ 的最佳一致逼近.

18.2 利用 Tschebyscheff 逼近理论设计 FIR 数字滤波器

设所希望的理想频率响应是

$$H_d(\mathrm{e}^{\mathrm{j}\omega}) = \begin{cases} 1, & 0 \leqslant \omega \leqslant \omega_p \\ 0, & \omega_s \leqslant \omega \leqslant \pi \end{cases} \quad ⑱$$

式中 ω_p 为通带频率, ω_s 为阻带频率. 现在的任务是, 寻找一 $H(\mathrm{e}^{\mathrm{j}\omega})$, 使其在通带和阻带最佳地一致逼近 $H_d(\mathrm{e}^{\mathrm{j}\omega})$. 图 3 给出了对理想频率响应逼近的示意图, 图中, δ_1 为通带纹波峰值, δ_2 为阻带纹波峰值. 这样, 对设计的低通数字滤波器 $H(\mathrm{e}^{\mathrm{j}\omega})$, 共提出了 5 个参数, 即 $\omega_p, \omega_s, \delta_1, \delta_2$ 和相应的单位抽样响应的长度 N.

第18章 FIR 数字滤波器设计的 Tschebyscheff 逼近法

根据上述交错点组定理,可以想象,如果 $H(e^{j\omega})$ 是对 $H_d(e^{j\omega})$ 的最佳一致逼近,那么 $H(e^{j\omega})$ 在通带和阻带内应具有如图所示的等纹波性质,所以最佳一致逼近有时又称等纹波逼近.

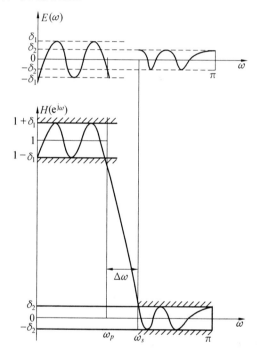

图 3 低通数字滤波器的一致逼近

为了讨论方便,现在先假设 $h(n)$ 为偶对称,N 为奇数,则

$$H(e^{j\omega}) = e^{\frac{-j(N-1)\omega}{2}} H_g(e^{j\omega}) \qquad ⑱$$

式中

$$H_g(e^{j\omega}) = \sum_{n=0}^{M} a(n)\cos(n\omega), \quad M = \frac{(N-1)}{2} \qquad ⑱⑭$$

Tschebyscheff 逼近定理

令 B_p 代表通带$(0 \sim \omega_p)$ 上的频率,B_s 代表阻带$(\omega_s \sim \pi)$ 上的频率. 并令 $F = B_p \cup B_s$ 代表定义在频率范围$(0 \sim \pi)$ 上的频率集合的一个子集,再定义加权函数

$$W(e^{j\omega}) = \begin{cases} \dfrac{1}{k}, & 0 \leq \omega \leq \omega_p, k = \dfrac{\delta_1}{\delta_2} \\ 1, & \omega_s \leq \omega \leq \pi \end{cases} \quad \text{⑱}$$

记误差函数

$$E(e^{j\omega}) = W(e^{j\omega})[H_g(e^{j\omega}) - H_d(e^{j\omega})] =$$
$$W(e^{j\omega}) \left[\sum_{n=0}^{M} a(n)\cos(n\omega) - H_d(e^{j\omega}) \right] \quad \text{⑱}$$

这样,用 $H_g(e^{j\omega})$ 一致逼近 $H_d(e^{j\omega})$ 的问题可表述为: 寻求系数 $a(n), n = 0, \cdots, M$,使加权误差函数 $E(e^{j\omega})$ 的最大值为最小.

上式中使用加权函数 $W(e^{j\omega})$ 是考虑在设计滤波器时对通带和阻带常要求不同的逼近精度,故乘以不同的加权函数,这种逼近又称加权 Tschebyscheff 一致逼近. 为了书写的方便,下面把 $e^{j\omega}$ 的函数都改写为 ω 的函数.

由交错点组定理可知,$H_g(\omega)$ 在子集 F 上是对 $H_d(\omega)$ 的唯一最佳一致逼近的充要条件是误差函数 $E(\omega)$ 在 F 上至少呈现 $M+2$ 个"交错",使得

$$|E(\omega_i)| = |-E(\omega_{i+1})| = E_n$$

其中

$$E_n = \max_{\omega \in F} |E(\omega)|$$

且

$$\omega_0 < \omega_1 < \cdots < \omega_{M+1} \quad (\omega \in F)$$

如果我们已经知道了在 F 上的 $M+2$ 个交错频率,即 $\omega_0, \omega_1, \cdots, \omega_{M+1}$,由式⑱,有

第18章 FIR 数字滤波器设计的 Tschebyscheff 逼近法

$$W(\omega_k)\left[H_d(\omega_k) - \sum_{n=0}^{M} a(n)\cos(n\omega_k)\right] = (-1)^k \rho$$
$$(k = 0, 1, \cdots, M+1) \qquad ⑰$$

式中

$$\rho = E_n = \max_{\omega \in F} |E(\omega)|$$

式⑰可写成矩阵形式,并将 $a(n)$ 写成 a_n,则有

$$\begin{bmatrix} 1 & \cos\omega_0 & \cos(2\omega_0) & \cdots & \cos(M\omega_0) & \dfrac{1}{W(\omega_0)} \\ 1 & \cos\omega_1 & \cos(2\omega_1) & \cdots & \cos(M\omega_1) & \dfrac{-1}{W(\omega_1)} \\ 1 & \cos\omega_2 & \cos(2\omega_2) & \cdots & \cos(M\omega_2) & \dfrac{1}{W(\omega_2)} \\ \vdots & \vdots & \vdots & & \vdots & \vdots \\ 1 & \cos\omega_M & \cos(2\omega_M) & \cdots & \cos(M\omega_M) & \dfrac{(-1)^M}{W(\omega_M)} \\ 1 & \cos\omega_{M+1} & \cos(2\omega_{M+1}) & \cdots & \cos(M\omega_{M+1}) & \dfrac{(-1)^{M+1}}{W(\omega_{M+1})} \end{bmatrix}$$

$$\begin{bmatrix} a_0 \\ a_1 \\ a_2 \\ \vdots \\ a_M \\ \rho \end{bmatrix} = \begin{bmatrix} H_d(\omega_0) \\ H_d(\omega_1) \\ H_d(\omega_2) \\ \vdots \\ H_d(\omega_M) \\ H_d(\omega_{M+1}) \end{bmatrix} \qquad ⑱$$

上式的系数矩阵是 $(M+2)\times(M+2)$ 的方阵,它是非奇异的. 解此方程组,可唯一地求出系数 $a_0, a_1, \cdots, a_M$ 及偏差 ρ,这样最佳滤波器 $H(e^{j\omega})$ 便可构成.

但是,这样做在实际中存在着两个困难:一是交错

Tschebyscheff 逼近定理

点组 $\omega_0, \omega_1, \cdots, \omega_M$ 事先并不知道,当然也就无法求解式⑱,确定一组交错点组并非易事,即使对于较小的 M 也是如此. 二是直接求解方程组⑱比较困难. 为此,麦克莱伦等人利用数值分析中的 Remez 算法,靠一次次的迭代来求得一组交错点组,而且在每一次迭代过程中避免直接求解式⑱. 现把该算法的步骤归纳如下:

第一步,首先在频率子集 F 上等间隔地取 $M+2$ 个频率 $\omega_0, \omega_1, \cdots, \omega_{M+1}$ 作为交错点组的初始猜测位置,然后按公式

$$\rho = \frac{\sum_{k=0}^{M+1} \alpha_k H_d(\omega_k)}{\sum_{k=0}^{M+1} \frac{(-1)^k \alpha_k}{W(\omega_k)}} \qquad ⑱⑨$$

计算 ρ,式中

$$\alpha_k = (-1)^k \prod_{i=0, i \neq k}^{M+1} \frac{1}{\cos \omega_i - \cos \omega_k} \qquad ⑲⓪$$

把 $\omega_0, \omega_1, \cdots, \omega_{M+1}$ 代入上式,可求出 ρ,它是相对第一次指定的交错点组所产生的偏差,实际上就是 δ_2. 求出 ρ 以后,利用重心形式的拉格朗日插值公式,可以在不求出 $a_0, \cdots, a_M$ 的情况下,得到一个 $H_g(\omega)$,即

$$H_g(\omega) = \frac{\sum_{k=0}^{M} \left(\frac{\beta_k}{\cos \omega - \cos \omega_k} \right) C_k}{\sum_{k=0}^{M} \frac{\beta_k}{\cos \omega - \cos \omega_k}} \qquad ⑲①$$

式中

$$C_k = H_d(\omega_k) - (-1)^k \frac{\rho}{W(\omega_k)} \quad (k=0,1,\cdots,M) \qquad ⑲②$$

$$\beta_k = (-1)^k \prod_{i=0, i \neq k}^{M} \frac{1}{(\cos \omega_i - \cos \omega_k)} \qquad ⑲③$$

第 18 章　FIR 数字滤波器设计的 Tschebyscheff 逼近法

把 $H_g(\omega)$ 代入式⑱,可求得误差函数 $E(\omega)$. 如果在子集 F 上,对所有的频率 ω 都有 $|E(\omega)| \leqslant |\rho|$,这说明 ρ 是纹波的极值,初始猜定的 $\omega_0, \omega_1, \cdots, \omega_{M+1}$ 恰是交错点组. 这时,设计工作即可结束. 当然,对第一次猜测的位置,不会恰好如此. 一般在某些频率处,总有 $E(\omega) > |\rho|$,这说明,需要交换上次猜测的交错点组中的某些点,得到一组新的交错点组.

第二步,对上次确定的交错点组 $\omega_0, \omega_1, \cdots, \omega_{M+1}$ 中的每一个点,都在其附近检查是否在某一个频率处有 $|E(\omega)| > \rho$,如若有,再在该点附近找出局部极值点,用这一局部极点代替原来的点. 待这 $M+2$ 个点都检查过后,便得到一组新的交错点组 $\omega_0, \omega_1, \cdots, \omega_{M+1}$,再次利用式⑲~⑲求出 $\rho, H_g(\omega)$ 和 $E(\omega)$,这样就完成一次迭代,也即完成了一次交错点组的交换. 通过交换算法,使得这一次的交错点组中的每一个 ω_i 都是由上一次的交错点组所产生的 $E(\omega)$ 的局部极值频率点,因此,用这次的交错点组求出的 ρ 将增大.

第三步,利用和第二步相同的方法,把在各频率处使 $|E(\omega)| > \rho$ 的点作为新的局部极值点,从而又得到一组新的交错点组.

重复上述步骤. 因为新的交错点组的选择都是作为每一次求出的 $E(\omega)$ 的局部极值点,因此,在迭代中,每次的 $|\rho|$ 都是递增的. ρ 最后收敛到自己的上限,也即 $H_g(\omega)$ 最佳地一致逼近 $H_d(\omega)$ 的解. 因此,若再迭代一次,新的误差曲线 $E(\omega)$ 的峰值将不会大于 $|\rho|$,这时迭代即可结束. 由最后的交错点组可按式⑲得到 $H_g(\omega)$,将 $H_g(\omega)$ 再附上式⑱的线性相位后作逆

Tschebyscheff 逼近定理

变换,便可得到单位抽样响应 $h(n)$.

由于按式⑱定义了 $W(\omega)$,因此最后求出的 ρ 即是阻带的峰值偏差 δ_2,而 $k\delta_2$ 便是通带的峰值偏差. 由上面的讨论可以看出,交错点数 $\omega_0, \omega_1, \cdots, \omega_{M+1}$ 是限制在通带和阻带内的. 因而,上述方法是在通带和阻带内对 $H_d(\omega)$ 的最佳一致逼近,而对过渡带 $(\omega_p \sim \omega_s)$ 内的逼近偏差没提出要求. 过渡带内的 $H_g(\omega)$ 曲线是由通带和阻带内的交错点组插值产生的. 对通带和阻带内的逼近误差 δ_1, δ_2 不需事先指定,而是由 Tschebyscheff 最佳一致逼近理论保证了逼近的最大偏差为最小. 因 ω_p 和 ω_s 在每一次的迭代过程中都始终是对应的极值频率点,所以这种设计方法仅需指定 N, k, ω_p 和 ω_s 这 4 个参数,因而通带和阻带的边缘可准确地确定. 图 4 给出了上述算法的流程图.

18.3 误差函数 $E(\omega)$ 的极值特性

前文已指出,为保证 $H_g(\omega)$ 是对 $H_d(\omega)$ 的最佳一致逼近,误差函数 $E(\omega)$ 必须呈现 $M+2$ 个"交错",即 $E(\omega)$ 至少必须有 $M+2$ 个极值,且交替改变符号. 由于 $E(\omega) = W(\omega) | H_g(\omega) - H_d(\omega) |$,而 $W(\omega)$ 和 $H_d(\omega)$ 都是常数,所以 $E(\omega)$ 的极值也是 $H_g(\omega)$ 的极值. 讨论一下 $H_g(\omega)$ 的极值特性对于我们理解前述算法是很有帮助的. 式⑱的 $H_g(\omega)$ 经展开后可写成幂级数的形式,即

$$H_g(\omega) = \sum_{n=0}^{M} a(n)(\cos \omega)^n \qquad ⑭$$

对上式求导,得

第 18 章 FIR 数字滤波器设计的 Tschebyscheff 逼近法

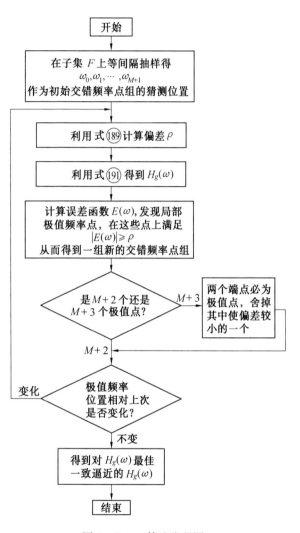

图 4 Remez 算法流程图

$$H_g'(\omega) = -\sin \omega \sum_{n=0}^{M} a(n) n (\cos \omega)^{n-1} \quad ⑲$$

Tschebyscheff 逼近定理

显然,在 $\omega=0$ 和 $\omega=\pi$ 处,$H_g(\omega)$ 必取极大值或极小值. 这样,在闭区间 $[0,\pi]$ 上,$H_g(\omega)$ 至多有 $M+1$ 个极值. 如令 $N=13$,则 $M=6$,$H_g(\omega)$ 的极值点如图 5 所示. 图中,$\omega_0=0$,$\omega_6=\pi$,显然,若不包括 ω_p 和 ω_s,最多可有 $M+1=7$ 个极值频率,包括 ω_p 和 ω_s 时,最多可有 $M+3=9$ 个极值频率.

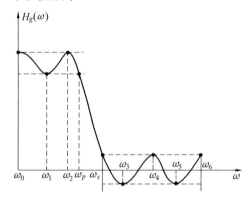

图 5 $M=6$ 时 $H_g(\omega)$ 的极值特性

利用上述的极值特性可得到下述一组方程

$$H_g(\omega_0)=1+\delta_1, H_g(e^{j\pi})=\delta_2$$
$$H_g(\omega_1)=1-\delta_1, H_g'(\omega_1)=0$$
$$H_g(\omega_2)=1+\delta_1, H_g'(\omega_2)=0$$
$$H_g(\omega_3)=-\delta_2, H_g'(\omega_3)=0$$
$$H_g(\omega_4)=\delta_2, H_g'(\omega_4)=0$$
$$H_g(\omega_5)=-\delta_2, H_g'(\omega_5)=0$$

这里总共有 12 个方程. $a_0,a_1,\cdots,a_6$ 是 7 个未知数,加上 5 个未知频率 $\omega_1,\omega_2,\cdots,\omega_5$ 总共 12 个未知数,因此可以求解. 但上述方程都是非线性方程,求解比较困难,

第18章 FIR 数字滤波器设计的 Tschebyscheff 逼近法

因此,这种方法仅用于 M 较小的场合. 此方法是早期用等纹波法设计 FIR 数字滤波器的思路,是由赫尔曼 (Herrmann) 于 1970 年提出的. 尔后霍夫施泰特尔 (Hofstetter) 等人对上述方法作了改进,用迭代的方法确定极值频率,代替上述方法求解非线性方程组. 在用等纹波逼近法设计滤波器时,有 5 个待定的参数,即 M (或 N),$\delta_1, \delta_2, \omega_p$ 和 ω_s. 我们不可能独立地把这 5 个参数全部给定,而只能指定其中的几个,留下其余的几个在迭代中确定. 赫尔曼和霍夫施泰特尔的算法是给定 M, δ_1 和 δ_2,而 ω_p 和 ω_s 是可变的,这样就有一个缺点,即滤波器通带和阻带的边缘频率不能精确地确定. 这两种方法都要求解包括端点 $[0, \pi]$ 在内的 $M+1$ 个极值频率.

前面讨论的麦克莱伦的算法是指定 M, ω_p 和 ω_s,而把 δ_1 和 δ_2 当作可变的,在迭代中确定最佳的 δ_1 和 δ_2. 根据通带和阻带的定义,应有

$$H(\mathrm{e}^{\mathrm{j}\omega_p}) = 1 - \delta_1$$

$$H(\mathrm{e}^{\mathrm{j}\omega_s}) = \delta_2$$

即 ω_p 和 ω_s 也是极值频率点. 这样,对于低通滤波器来说,指定 ω_p 和 ω_s 作为极值频率后,最多将会有 $M+3$ 个极值频率. 但利用交错定理时,只需要 $M+2$ 个极值频率. 因此,在每一次迭代过程中,若出现有 $M+3$ 个极值频率,则在 $\omega=0$ 或 $\omega=\pi$ 处,选取其中呈现最大误差的一个作为极值频率点,形成新的交错点组. 出现 $M+3$ 个极值频率的情况称之为"超纹波",用赫尔曼和霍夫施泰特尔方法设计出的滤波器也称之为"超纹波滤波器". 图 5 中的 ω_p, ω_s 和 $0, \pi$ 处都是极值频率,因此这是一种"超纹波"的情况[5,3].

苏格兰咖啡馆的大本子

第 19 章

在第二次世界大战之前,波兰的一群年轻数学家喜欢聚集在华沙的一家叫苏格兰的咖啡馆中高谈阔论,并将讨论后的结论记录在一个大本子上. 二战爆发后,为了不让德军发现,他们曾将大本子埋藏到一个操场的篮球架之下. 二战结束后又将其挖出,可惜已物是人非,多数年轻的数学家都死于战场,但这本珍贵的文献被留了下来,并且我们从中发现了关于本书所论及问题的进一步背景资料.

例 19(梅热(Mazur),奥里奇(Orlicz)) 设 c_n 是具有下述性质的最小数:若 $F(x_1, \cdots, x_n)$ 是任何对称的 n 元线性算子(定义域是一个 (B) 型空

第19章 苏格兰咖啡馆的大本子

间,值域也是这种空间),则

$$\sup_{\|x_i\|\leqslant 1, i=1,2,\cdots,n} \|F(x_1,\cdots,x_n)\| \leqslant c_n \sup_{\|x\|\leqslant 1} \|F(x,\cdots,x)\|$$

已知(巴拿赫):c_n 存在,可以证明数 c_n 满足不等式

$$\frac{n^n}{n!} \leqslant C_n \leqslant \frac{1}{n!}\sum_{k=1}^{n}\binom{n}{k}\cdot k^n$$

是否有 $c_n = \frac{n^n}{n!}$?

评论(Lawrence A. Harris, Kentucky 大学数学系):对于任何实赋范线性空间,本问题有肯定回答,现在已是无穷维全纯性质领域内的一个基本事实了. 1932年马丁(R. S. Martin)在其学位论文中利用一个 n 维极化公式证明了 $c_n \leqslant \frac{n^n}{n!}$. 几年以后,泰勒(A. E. Taylor)在[15]中发表了他的论证,尽管上述极化公式是梅热与奥里奇得出的[12],不过他们用的似乎是 $x=0$ 的情形,而不是 $x=-\sum_{1}^{n}\frac{h_k}{2}$ 的情形,后一情形给出最好的估值. 在[5],[10]中,给出了 l^1 和 $L[0,1]$ 中 $c_n = \frac{n^n}{n!}$ 这种极端情形的例子. 详细的说明可参看[4]与[13]. 注意,根据罕-巴拿赫定理,如果认为本问题中所有的多线性映象都取复值也无损一般性.

1938 年巴拿赫[1]证明了:只就实希尔伯特空间而言,$c_n = 1$(他还假定了可分性,然而这不是必需的). 巴拿赫的结果很容易从[3]或[9]推出. 至于近代的提法可见[2],[6]或[8]. [6]中指出:巴拿赫的结果与赛戈(Szego)关于三角多项式的伯恩斯坦不等式的改进

Tschebyscheff 逼近定理

结果容易彼此互推. 对于复 L^p 空间, $1 \leqslant p < +\infty$, [6] 中猜测

$$c_n \leqslant \left(\frac{n^n}{n!}\right)^{\frac{|p-2|}{p}}$$

并就 n 为 2 的乘幂时给予了证明. 从 [6] 还可推出: 对 J^* 代数有[7]

$$c_n \leqslant \frac{n^{\frac{n}{2}}(n+1)^{\frac{(n+1)}{2}}}{2^n n!}$$

(特别是, S 为紧致豪斯道夫空间——定义在 S 上的所有连续复值函数构成的空间 $C(S)$, 或者更一般的任何 B^* 代数都是 J^* 代数) 当 S 是两点集时, 对于空间 $C(S)$ 则有 $c_n = 1^{[6]}$, 所以自然要问, 这一结果是否对于任何紧致豪斯道夫空间 S 都成立. 如果成立, 则容易推出: 对于 $C(S)$ 上的多项式, 伯恩斯坦不等式成立 (见 [6]).

例 19 的一种自然推广如下: 设 $k_1, \cdots, k_n$ 是非负整数, 其和为 n. $c(k_1, \cdots, k_n)$ 是具有下述性质的最小数: 若 F 是任何对称的 n 元线性映象, 把一个实赋范线性空间映入另一个实赋范线性空间, 则

$$\sup_{\|x_i\| \leqslant 1, i=1,2,\cdots,n} \|F(x_1^{k_1}, \cdots, x_n^{k_n})\| \leqslant$$
$$c(k_1, \cdots, k_n) \sup_{\|x\| \leqslant 1} \|F(x, \cdots, x)\|$$

这里各指数表示该基本变元以坐标形式出现的次数①. [6] 中指出, 如果只考虑复赋范线性空间与复纯

---

① 这里 $F(x_1^{k_1}, \cdots, x_n^{k_n})$ 是一个简写符号, 表示 $F(x_1, \cdots, x_1, \cdots, x_n)$, 坐标 x_1 有 k_1 个, 坐标 x_2 有 k_2 个, 等等. 例如 $F(x_1^n, x_2^0, \cdots, x_n^0) = F(x_1, x_1, \cdots, x_1)$.

第 19 章　苏格兰咖啡馆的大本子

量,则有

$$c(k_1,\cdots,k_n) = \frac{k_1! \cdot \cdots \cdot k_n!}{k_1^{k_1} \cdot \cdots \cdot k_n^{k_n}} \cdot \frac{n^n}{n!} \quad ⑯$$

(这里 $0^0 = 1$) 但是考虑实赋范空间时,有很多例子说明式⑯不成立.

例 20(梅热,奥里奇)　给了实变量 $t_1,\cdots,t_n$ 的 n 阶齐次多项式

$$W(t_1,\cdots,t_n) = \sum_{k_1+\cdots+k_n=n} a_{k_1,\cdots,k_n} t_1^{k_1} ,\cdots, t_n^{R_n}$$

假设对于所有满足 $|t_1| + \cdots + |t_n| \leq 1$ 的 $t_1,\cdots,t_n$ 有 $|W(t_1,\cdots,t_n)| \leq 1$. 那么,是否有

$$|a_{k_1,\cdots,k_n}| \leq \frac{n^n}{k_1! \cdot \cdots \cdot k_n!}$$

评论(Lawrence A. Harris,Kentucky 大学数学系):这个问题的答案是肯定的,其解易从例 19 的解答推出. 事实上,设 F 是 l_n^1 上的对称 n 元线性映象,使得对所有 $x \in l_n^1$ 有 $W(x) = F(x,\cdots,x)$. 于是对 $x = t_1 e_1 + \cdots + t_n e_n$ 使用多项式定理[12],由于 $W(t_1,\cdots,t_n)$ 的表法的唯一性,我们得到

$$a_{k_1,\cdots,k_n} = \frac{n!}{k_1! \cdot \cdots \cdot k_n!} F(e_1^{k_1},\cdots,e_n^{k_n}) \quad ⑰$$

这里 $e_1,\cdots,e_n$ 是 l_n^1 的标准基底,由此推出所要的估值. 注意,例 20 中确定最佳估值 $a(k_1,\cdots,k_n)$ 的问题等价于确定 $c(k_1,\cdots,k_n)$ 的问题. 因为,若 F 是对称的 n 元线性映象,使得 $\|x\| \leq 1$ 蕴涵 $\|F(x,\cdots,x)\| \leq 1$,又若 $\|x_1\| \leq 1, \cdots, \|x_n\| \leq 1$,则多项式 $W(t_1,\cdots,t_n) = F(x,\cdots,x)$ 满足例 20 的假设,这里 $x = t_1 x_1 + \cdots + t_n x_n$,而

179

Tschebyscheff 逼近定理

$$a_{k_1,\cdots,k_n} = \frac{n!}{k_1! \cdot \cdots \cdot k_n!} F(x_1^{k_1}, \cdots, x_n^{k_n})$$

因此

$$a(k_1,\cdots,k_n) = \frac{n!}{k_1! \cdot \cdots \cdot k_n!} c(k_1,\cdots,k_n) \quad ⑭$$

对 R^m 上满足已知增长条件的 m 个变量的多项式系数,寻求估值的一般问题可利用下面的广义极化公式解决:设 $W(t_1,\cdots,t_m)$ 是任何 n 阶齐次多项式,$a_{k_1,\cdots,k_m}$ 是其展开式中 $t_1^{k_1},\cdots,t_m^{k_m}$ 的系数. 对于每个 $i = 1,\cdots,m$,选取各不相同的实数 $x_{i0},\cdots,x_{ik_i}$,令

$$\Gamma_{ij} = \prod_{l\neq j}(x_{ij} - x_{il}) \quad (0 \leqslant j \leqslant k_i)$$

$k_i = 0$ 时,$\Gamma_{ij} = 1$. 则

$$a_{k_1,\cdots,k_m} = \sum \frac{W(x_{1j_1},\cdots,x_{mj_m})}{\Gamma_{1j_1} \cdot \cdots \cdot \Gamma_{mj_m}} \quad ⑲$$

这里求和是对所有 $0 \leqslant j_1 \leqslant k_1, \cdots, 0 \leqslant j_m \leqslant k_m$ 取的. 我们指出,对于任何 $m-1$ 个变量的阶数不大于 n 的多项式 p,可以定义

$$W(t_1,\cdots,t_m) = t_m^n p\left(\frac{t_1}{t_m},\cdots,\frac{t_{m-1}}{t_m}\right)$$

而得到 m 个变量的 n 阶齐次多项式 W. 求式⑲右端的估值与极小值就可以得到它左端的估值(适当的初始选择是 $x_{ij} = \frac{k_i}{2-j}$). 为了证明式⑲,请注意:若 $p(t)$ 是阶数不大于 k_i 的多项式,则 p 的拉格朗日插值公式中 t^{k_i} 的系数是 $\sum_{j=0}^{k_i} \frac{p(x_{ij})}{\Gamma_{ij}}$,再把这一结果应用到 W 的每一变量. 例如,我们证明,例 20 有更好的估值

第19章 苏格兰咖啡馆的大本子

$$|a_{k_1,\cdots,k_n}| \leqslant \frac{n^n}{k_1! \cdot \cdots \cdot k_n!} r^l \qquad ⑳⓪$$

这里

$$r = \frac{1+e^{-2}}{2}, \ l = \sum_{i=1}^{n}\left[\frac{k_i}{2}\right]$$

事实上,对 $i=1,\cdots,l$,取 $x_{i0}=2, x_{i1}=0, x_{i2}=-2$;对 $i=l+1,\cdots,n-l$ 取 $x_{i0}=1, x_{i1}=-1$;对 $i=n-l+1,\cdots,n$ 取 $x_{i0}=0$. 那么由式⑲得

$$|a_{2,\cdots,2,1,\cdots,1,0,\cdots,0}| \frac{1}{4^l}\sum_{j=0}^{l}\binom{l}{j}(n-2j)^n \leqslant 2^{-l}n^n r^l \qquad ⑳①$$

这里最后一个不等式是从 $(1-\frac{25}{n})^n \leqslant e^{-2j}$ 推出的. 显然

$$c(k_1,\cdots,k_n) \leqslant c(2,\cdots,2,1,\cdots 1,0,\cdots, 0)$$

再加上式⑲⑧与式⑳①便得出式⑳⓪.

一个相关的有趣问题是寻求马尔柯夫定理在巴拿赫空间中的类似结果,即寻求最小数 $M_{n,k}$ 具有下述性质:若 P 是阶数不大于 n 的任何多项式,把一个实赋范线性空间映入另一个实赋范线性空间,则

$$\sup_{\|(x)\|\leqslant 1}\|\hat{D}^k p(x)\| \leqslant M_{n,k}\sup_{\|(x)\|\leqslant 1}\|(x)\|$$

这里,$\hat{D}^k P(x)y = \frac{d^k}{dt^k}P(x+ty)\Big|_{t=0}$. 容易证明

$$T_n^{(k)}(1) \leqslant M_{n,k} \leqslant 2^{2k-1}T_n^{(k)}(1) \qquad ⑳②$$

这里 T_n 是 n 阶 Tschebyscheff 多项式(见[14]). 事实上,令 P 是实赋范线性空间中阶数不大于 n 的任何实值多项式,假设 $\|x\|\leqslant 1$ 蕴涵 $|P(x)|\leqslant 1$. 令 $\|x\|\leqslant 1, \|y\|\leqslant 1, -1\leqslant s\leqslant 1$. 定义 $q(t) = P(\phi(t))$,这里 $\phi(t) = \frac{x-sy+t(x+sy)}{2}$. 注意,当 $-1\leqslant t \leqslant 1$ 时 $\|\phi(t)\| \leqslant$

$\frac{1+t}{2}+\frac{1-t}{2}=1$. 因此 q 是阶数不大于 n 的多项式, 当 $-1\leqslant t\leqslant 1$ 时满足 $|q(t)|\leqslant 1$, 从而由[14], [1], [5], [11] 得 $|q^k(1)|\leqslant|T^k(1)|$, 显然, $q^k(1)=2^{-k}\hat{D}^k P(x)\cdot(x+sy)$. 因此, 映象 $s\to\hat{D}^k(x+sy)$ 是阶数不大于 k 的多项式, 在 $[-1,1]$ 上以 $2^k T_n^{(k)}(1)$ 为界, $\hat{D}^k P(x)y$ 是这个多项式中 2^k 的系数. 从而, 由[14]得

$$|\hat{D}^k P(x)y|\leqslant 2^{k-1}[2^k T_n^{(k)}(1)]$$

再由罕-巴拿赫定理便推出式⑳.

注意, 只考虑 l_2^1 上的实值多项式时, $M_{n,k}$ 的值是不变的. [6]与[9]中指出, 只考虑实希尔伯特空间时, $M_{n,1}=n^2$, 这时, 对于 $1\leqslant k\leqslant n$, 考虑是否有 $M_{n,k}=T_n^{(k)}(1)$ 是一个有意思的问题, 参见[11].

参 考 文 献

[1] BANACH S. Über Homogene Polynome in (L^2) [J]. Studia Math. ,1938(7):36-44.

[2] BOCHNAK J, SICIAK J. Polynomials and Multilinear Mappings in Topological Vector Spaces[J]. Studia Math. ,1971(39):59-76.

[3] VANDER CORPUT J G, SCHAAKE G. Unglei Chungen für Polynome und Trigonometrishe Polynome[J]. Compositio Math. ,1935(2):321-361.

[4] FEDERER H. Geometric Measure Theory[M]. Berlin-Heidelberg-New York: Springer-Verlag, 1969.

[5] GRUNBAUM B. Two Examples in the Theory of Polynomial Functionals[J]. Reveon Lematematika,

第 19 章 苏格兰咖啡馆的大本子

1957(11):56-60.
[6] HARRIS L A. Bounds on the Derivatives of Holomorphic Functions of Vectors[M]. Paris:Hermann, 1975.
[7] HAWIS L A. Bounded Symmetric Homogeneous Domains in Infinite Dimensional Spaces[M]. Berlin: Springer-Verlag,1974.
[8] HÖRMANDER L. On a Theorem of Grace[J]. Math. Scand.,1954(2):55-64.
[9] KELLOGG O D. On Bounded Polynomials in Several Variables[J]. Math. Zeit.,1928(27):55-64.
[10] KOPÉC J, MUSIELAK J. On the Estimation of the Norm of the N-Linear Symmetric Operation[J]. Studia Math.,1955(15):29-30.
[11] WILHELMSEN D R. A Markov Inequality in Several Dimensions[J]. J. Approx. Theory,1974(11):216-220.
[12] MAZUR S, ORLICZ W. Grundlegende Eigenschaften der Polynomischen Operatoren $\text{I}-\text{II}$[J]. Studia Math.,1935(5):50-68,179-189.
[13] NACHBIN L. Topology on Spaces of Holomorphic Mappings[M]. New York:Springer-Verlag,1969.
[14] RIVLIN T J. The Chebyshev Polynomials[M]. New York:Wiley,1974.
[15] TAYLOR A E. Additions to the Theory of Polynomials in Normed Linear Spaces[J]. Tohoku Math. J.,1938(44):302-318.

逼近论中的伯恩斯坦猜测

第20章

20.1 引 言

对于定义在区间 $[-1,+1]$ 上的任一实值函数 $f(x)$ 以及任意的 $\delta>0$, $f(x)$ 的连续模定义为

$$\omega(\delta;f) \triangleq \sup_{\substack{|x_1-x_2|\leqslant\delta \\ x_1,x_2\in[-1,+1]}} |f(x_1)-f(x_2)|$$

⑳

而其在 $[-1,+1]$ 上的一致范数定义为

$$\|f\|_{L_\infty[-1,+1]} \triangleq \sup\{|f(x)| \mid x\in[-1,+1]\}$$

⑳'

用 π_n 代表次数最高为 $n(n=0,1,\cdots)$ 的全体实多项式的集合,下面

第20章 逼近论中的伯恩斯坦猜测

是熟知的杰克逊(Jackson)在[8]中的结果(见[9]，[11]).

定理 28(杰克逊[8]) 若$f(x)$为定义在$[-1,+1]$上的一个连续实值函数，则

$$E_n(f) \leq \sigma\omega\left(\frac{1}{n};f\right) \quad (n=1,2,\cdots) \quad \text{⑳}$$

其中

$$E_n(f) \triangleq \inf\{\|f-g\|_{L_\infty[-1,+1]} \mid g \in \pi_n\} \quad \text{⑳}$$

大家都知道(见[9])，对定义在$[-1,+1]$上的任何连续实值函数$f(x)$，在π_n中存在唯一的多项式$\hat{p}_n(x) = \hat{p}_n(x;f)$使得

$$E_n(f) = \|f-\hat{p}_n\|_{L_\infty[-1,+1]} \quad (n=0,1,\cdots) \quad \text{⑳}'$$

并称$\hat{p}_n(x)$为在$[-1,+1]$上函数$f(x)$在π_n中的最佳一致逼近多项式. 此外，由式⑳知，序列$\{E_n(f)\}_{n=0}^{+\infty}$是一个非负的非增序列. 根据魏尔斯特拉斯逼近定理(见[11])，它趋于零，即

$$\lim_{n\to\infty} E_n(f) = 0 \quad \text{⑳}$$

对于$[-1,+1]$上特殊的连续函数$|x|$，不难看出

$$\omega(\delta;|x|) = \delta \quad (0<\delta\leq 1)$$

从定理 28 中式⑳，可有

$$E_n(|x|) \leq \frac{\sigma}{n} \quad (n=1,2,\cdots) \quad \text{⑳}'$$

对于函数$|x|$说来，这个式子比式⑳更精确.

因为$|x|$是$[-1,+1]$上的连续偶函数，那么对任何$n \geq 0$，它在$[-1,+1]$上在π_n中的最佳一致逼近也是偶函数(见[11]). 把这个结果和最佳一致逼近多项式的 Tschebyscheff 变化特性结合起来，就可进一步得到

Tschebyscheff 逼近定理

(见[11])
$$E_{2n}(|x|) = E_{2n+1}(|x|) \quad (n=1,2,\cdots) \quad ⑳⑦$$

因此,对我们来说,仅考虑序列$\{E_{2n}(|x|)\}_{n=1}^{+\infty}$减到零的情形就足够了. 从式⑳⑥′,我们显然有

$$2nE_{2n}(|x|) \leqslant 6 \quad (n=1,2,\cdots) \quad ⑳⑧$$

为了改进式⑳⑧的上界,在$[-1,+1]$上把$|x|$展成 Tschebyscheff 级数,即

$$|x| = \frac{4}{\pi}\left\{\frac{1}{2} + \sum_{m=1}^{+\infty}\frac{(-1)^{m+1}T_{2m}(x)}{(2m-1)(2m+1)}\right\} \quad ⑳⑨$$

其中$T_n(x)$是n阶 Tschebyscheff 多项式(第一类). 因为对所有在$[-1,+1]$中的x都有$|T_n(x)| \leqslant 1$,去掉式⑳⑨中的前n项和,则式⑳⑨余项的绝对值满足

$$\frac{4}{\pi}\left|\sum_{m=n+1}^{+\infty}\frac{(-1)^{m+1}T_{2m}(x)}{(2m-1)(2m+1)}\right| \leqslant$$

$$\frac{4}{\pi}\left|\sum_{m=n+1}^{+\infty}\frac{1}{(2m-1)(2m+1)}\right|$$

但是

$$\frac{1}{(2m-1)(2m+1)} = \frac{1}{2}\left(\frac{1}{2m-1} - \frac{1}{2m+1}\right)$$

则这个上界可简单地缩小为

$$\frac{4}{\pi} + \sum_{m=n+1}^{+\infty}\frac{1}{(2m-1)(2m+1)} = \frac{2}{\pi}\left\{\left(\frac{1}{2n+1} - \frac{1}{2n+3}\right) + \left(\frac{1}{2n+3} - \frac{1}{2n+5}\right) + \cdots\right\} = \frac{2}{\pi(2n+1)}$$

于是

$$\left| |x| - \frac{4}{\pi}\left\{\frac{1}{2} + \sum_{m=1}^{n} \frac{(-1)^{m+1} T_{2m}(x)}{(2m-1)(2m+1)}\right\} \right| \leq \frac{2}{\pi(2n+1)}$$

㉑

对于在 $[-1,+1]$ 中的所有 x 及任何 $n \geq 1$ 成立. 因为式㉑中给出 $|x|$ 的逼近多项式为一个具体的 $2n$ 次多项式, 从而式㉑隐含着

$$2n E_{2n}(|x|) \leq \frac{4n}{\pi(2n+1)} < \frac{2}{\pi} = 0.63661\cdots$$
$$(n = 1, 2, \cdots)$$

㉑

它又改进了式㉘.

在 1913 年伯恩斯坦[2]进一步把式㉑中的上界 $0.63661\cdots$ 大大精确化了. 从式㉑直接给出

$$\overline{\lim_{n \to \infty}} 2n E_{2n}(|x|) \leq \frac{2}{\pi} = 0.63661\cdots$$

㉒

经过一个冗长复杂的证明, 伯恩斯坦给出下面更进一步的结果.

定理 29(伯恩斯坦[2]) 存在正的常数 β(β 代表伯恩斯坦常数), 使得

$$\lim_{n \to \infty} 2n E_{2n}(|x|) = \beta$$

㉓

其中 β 满足

$$0.278 < \beta < 0.286$$

㉔

除上述结果外, 伯恩斯坦在[2]中还指出常数

$$\frac{1}{2\sqrt{\pi}} = 0.2820947917\cdots$$

㉕

也在式㉔给出的界限内, 并接近于式㉔中 β 的上、下限的平均值 0.282, 这好像是一种奇怪的巧合. 过几年这个发现就成为伯恩斯坦猜测(1913)

Tschebyscheff 逼近定理

$$\beta \stackrel{?}{=} \frac{1}{2\sqrt{\pi}} = 0.282\,094\,917\cdots \qquad ㉑⑥$$

从伯恩斯坦的论文问世以来,尽管有几位作者(见贝尔(Bell)和沙(Shah)[1],博亚尼奇(Bojanic)和埃尔金斯(Elluns)[3],以及萨尔瓦蒂(Salvati)[12])进行了数值上的探索,但这个猜测的真伪尚属未知,这个猜测迄今悬而未决的原因可能是由于如下事实:(i)对于大的 n 值,准确地确定 $E_{2n}(|x|)$ 绝非易事,而且(ii)由式㉑③确立的 $2nE_{2n}(|x|)$ 收敛到 β 的速度太慢.

近年来,瓦尔加(Varga)和卡彭特(Carpenter)[13]在 1985 年指出伯恩斯坦猜测不成立;这个结论是基于[13]中给出的 β 的改进界限(将在下节中讨论)

$$0.280\,168\,546\,0\cdots = l_{20} \leqslant \beta \leqslant 2\mu_{100} = 0.280\,173\,379\,1\cdots$$
㉑⑦

因为式㉑⑦中 β 的上界小于 $\frac{1}{(2\sqrt{\pi})} = 0.282\,094\,791\,7\cdots$,因此伯恩斯坦猜测㉑⑥不成立.下面 20.2 中我们讨论高精度计算 $E_{2n}(|x|)$ 的问题,20.3 和 20.4 介绍在[13]中根据伯恩斯坦方法得到伯恩斯坦常数 β 上、下界的方法.在 20.5 中,我们讨论高精度数 $\{2nE_{2n}(|x|)\}_{n=1}^{52}$ 的理查森(Richaxdson)外插,它给出了有 50 位有效数字的 β 的估计值

$$\beta = 0.280\,169\,499\,023\,869\,133\,036\,436\,491\,230\,672$$
$$ 000\,042\,482\,139\,812\,36 \qquad ㉑⑧$$

在 20.6 中我们讨论关于估计非负量 $\beta(\alpha)$ 这一更一般的问题,其中

$$\beta(\alpha) = \lim_{n\to\infty}(2n)^{2\alpha}E_{2n}(x^\alpha;[0,1]) \quad (\alpha > 0) \qquad ㉑⑨$$

$\beta\left(\dfrac{1}{2}\right)$ 收敛到式 ⑱ 中的常数 β. 最后作为本章的结束,在 20.7 中我们讨论对于 $|x|$ 在 $[-1,+1]$ 上的最佳一致有理逼近问题.

20.2 高精度计算 $\{2nE_{2n}(|x|)\}_{n=1}^{52}$

令 π_{2n} 中的 $\hat{p}_{2n}(x)$ 是 $[-1,+1]$ 上函数 $|x|$ 的最佳逼近多项式,即(见式 ⑳⑤′)

$$\||x|-\hat{p}_{2n}(x)\|_{L_\infty[-1,+1]} = E_{2n}(|x|) \quad (n=1,2,\cdots) \qquad ⑳$$

因为 $|x|$ 在 $[-1,+1]$ 区间中是偶函数,根据 20.1 的讨论,其在 $[-1,+1]$ 中的最佳一致逼近多项式也是偶函数,从而

$$\hat{p}_{2n}(x) = \sum_{j=0}^{n} a_j(n) x^{2j} \quad (n=1,2,\cdots) \qquad ㉑$$

我们作变量变换 $x^2=t, t\in[0,1]$,于是我们的逼近问题变为

$$E_{2n}(|x|) = E_n(\sqrt{t};[0,1]) \triangleq$$
$$\inf\{\|\sqrt{t}-h_n(t)\|_{L_\infty[0,1]} \mid h_n \in \pi_n\} \qquad ㉒$$

如果我们记

$$E_n(\sqrt{t};[0,1]) = \|\sqrt{t}-\hat{h}_n(t)\|_{L_\infty[0,1]} \quad (\hat{h}_n\in\pi_n) \qquad ㉓$$

那么,显然有(见式 ㉑)

$$\hat{p}_{2n}(x) = \hat{h}_n(x^2) \quad (n=1,2,\cdots) \qquad ㉔$$

因此,确定 $E_{2n}(|x|)$ 和 $\hat{p}_{2n}(x)$ 等价于确定 $E_n(\sqrt{t};[0,1])$ 和 $\hat{h}_n(t)$.

Tschebyscheff 逼近定理

利用下面的标准(第二类)Remez 算法就可以解决式㉒的极小化问题(见[9]).

第一步,令 $S \triangleq \{t_j\}_{j=0}^{n+1}$ 是 $[0,1]$ 上 $n+2$ 个离散点构成的集合,它满足

$$0 \leq t_0 < t_1 < \cdots < t_{n+1} \leq 1 \qquad ㉕$$

第二步,找出唯一多项式 $h_n(t)$ 和常数 λ(这是一个线性问题)使得

$$h_n(t_j)+(-1)^j\lambda = \sqrt{t_j} \quad (j=0,1,\cdots,n+1) \qquad ㉖$$

因此,$h_n(t)$ 是在这个离散集 S 内的 π_n 中对于 $\sqrt{t}$ 的最佳一致逼近多项式,而且在 S 的点 t_j 上其交错误差为 $|\lambda|$,使用和式㉒中类似的记号,我们可以写

$$\|\sqrt{t}-h_n(t)\|_{L_\infty(S)} = E_n(\sqrt{t};S) = |\lambda| \qquad ㉗$$

由于 S 是 $[0,1]$ 的子集,故显然有

$$\|\sqrt{t}-h_n(t)\|_{L_\infty[0,1]} - |\lambda| \geq 0 \qquad ㉘$$

第三步,事先给定(小的)$\varepsilon > 0$,如果 $\|\sqrt{t}-h_n(t)\|_{L_\infty[0,1]} - |\lambda| \leq \varepsilon$,则迭代终止. 否则,利用前面第二步,在函数 $\sqrt{t}-h_n(t)$ 位于 $[0,1]$ 区间上交替变号的局部极值点中找出一个新的集合 S',并重复第二和第三步,等等.

从计算的观点说来,知道由重复利用 Remez 算法产生的 $|\lambda|$ 序列是单调非减这一事实是很有用的.

从一个特殊的交错集 $S^{(0)} \triangleq \{t_j^{(0)}\}_{j=0}^{n+1}$,其中

$$t_j^{(0)} \triangleq \frac{1}{2}\left\{1+\cos\left[\frac{(n+1-j)\pi}{n+1}\right]\right\} \quad (j=0,1,2,\cdots,n+1) \qquad ㉙$$

开始,其中 $t_j^{(0)}$ 是在 $[0,1]$ 区间上的 Tschebyscheff 多项式 $T_{n+1}(2t-1)$ 的 $n+2$ 个极值点,并利用布兰特(Brent)

第 20 章 逼近论中的伯恩斯坦猜测

的 MP 软件包[4]在肯特州立大学数学系的 Vax 11/780 上进行高精度计算,当 $\|\sqrt{t}-h_n(t)\|_{L_\infty[0,1]}$ 和 $|\lambda|$ 重合到 100 位有效数字时,Remez 算法迭代终止. 由于已知这种(第二类) Remez 算法有二次收敛性,因此对于每种所考虑的情况,最多需要 9 次迭代. 考虑到保留位和可能存在的小的舍入误差,我们相信我们得到的 $\{2nE_{2n}(|x|)\}_{n=1}^{52}$ 至少有 95 位有效数字.

为了节省篇幅,在表 4 中给出了截断到 20 位有效数字后的 $\{2nE_{2n}(|x|)\}_{n=1}^{52}$ 的值. 为的是表明该序列收敛是很慢的(如果需要的话可以打印出 $\{2nE_{2n}(|x|)\}_{n=1}^{52}$ 至 100 位有效数字).

看来表 4 中的乘积 $\{2nE_{2n}(|x|)\}_{n=1}^{52}$ 已经收敛到 4 位有效数字,按照 20.6 中的渐近估计式,若 $|2nE_{2n}(|x|)-\beta|<10^{-10}$,则必须 $n \geqslant 20\,968$. 这将需要在 $[0,1]$ 上找出 $\sqrt{t}$ 的至少 10 484 阶最佳一致逼近多项式,这是绝对令人生畏的计算工作.

表 4 $\{2nE_{2n}(|x|)\}_{n=1}^{52}$

n	$2nE_{2n}(|x|)$	n	$2nE_{2n}(|x|)$
1	0.250 000 000 000 000 000 00	27	0.280 109 236 522 206 185 25
2	0.270 483 597 111 137 101 07	28	0.280 113 460 889 950 283 84
3	0.275 574 372 401 175 386 04	29	0.280 117 256 249 499 617 92
4	0.277 517 824 675 052 696 46	30	0.280 120 678 772 662 828 33
5	0.278 451 185 535 508 601 52	31	0.280 123 775 731 660 884 50
6	0.278 967 917 464 958 706 36	32	0.280 126 587 138 731 918 44
7	0.279 282 944 958 518 024 60	33	0.280 129 147 043 904 517 20
8	0.279 488 837 594 507 447 71	34	0.280 131 484 570 012 610 69

Tschebyscheff 逼近定理

续表

n	$2nE_{2n}(|x|)$	n	$2nE_{2n}(|x|)$
9	0.279 630 657 410 128 201 25	35	0.280 133 624 744 030 046 76
10	0.279 732 433 771 973 829 68	36	0.280 135 589 169 271 117 13
11	0.279 807 917 288 743 873 83	37	0.280 137 396 572 336 696 62
12	0.278 965 432 123 793 272 79	38	0.280 139 063 250 782 895 91
13	0.279 910 254 315 557 690 36	39	0.280 140 603 441 582 482 18
14	0.279 945 858 485 782 132 47	40	0.280 142 029 625 997 940 87
15	0.279 974 606 686 407 492 31	41	0.280 143 352 783 104 081 69
16	0.279 998 151 956 316 728 27	42	0.280 144 582 601 611 087 07
17	0.280 017 677 133 297 253 79	43	0.280 145 727 657 645 500 97
18	0.280 034 047 414 993 509 64	44	0.280 146 795 564 600 416 24
19	0.280 047 907 285 905 851 56	45	0.280 147 793 099 959 135 46
20	0.280 059 744 760 423 152 65	46	0.280 148 726 313 048 744 46
21	0.280 069 934 831 809 430 67	47	0.280 149 600 616 931 436 84
22	0.280 078 769 475 287 534 23	48	0.280 150 420 867 046 950 23
23	0.280 086 478 757 075 570 49	49	0.280 151 191 428 744 923 26
24	0.280 093 245 938 808 505 47	50	0.280 151 916 235 465 273 55
25	0.280 099 218 452 382 835 58	51	0.280 152 598 839 017 816 32
26	0.280 104 515 986 556 704 89	52	0.280 153 242 453 163 842 49

20.3 计算伯恩斯坦常数 β 的上界

为了得到伯恩斯坦常数 β 的上界和下界,伯恩斯坦引进了下面的特殊函数

$$F(t) \triangleq \sum_{k=0}^{+\infty} \frac{t}{(t+2k+1)^2 - \frac{1}{4}} \quad (t \geq 0) \quad ㉚$$

利用 psi(双 γ 函数)函数

第20章 逼近论中的伯恩斯坦猜测

$$\psi(z) \triangleq \frac{\mathrm{d}}{\mathrm{d}z}(\log \Gamma(z)) = \frac{\Gamma'(z)}{\Gamma(z)} \quad ㉛$$

式㉛中的函数 $F(t)$ 有表达式

$$F(t) = \frac{t}{2}\left\{\psi\left(\frac{t}{2}+\frac{3}{4}\right) - \psi\left(\frac{t}{2}+\frac{1}{4}\right)\right\} \quad (t \geq 0) \quad ㉜$$

$F(t)$ 的其他表达式还包括(见[2])

$$\begin{aligned}F(t) &= \frac{t}{2t+1}\mathrm{H}\left(1,1;t+\frac{3}{2};\frac{1}{2}\right) = t\int_0^1 \frac{z^{\frac{t-1}{2}}\mathrm{d}z}{z+1}\\ &= \frac{1}{2}\int_0^{+\infty} \frac{\mathrm{e}^{-u}\mathrm{d}u}{\cos h\left(\frac{u}{2t}\right)}\end{aligned} \quad ㉝$$

其中 $\mathrm{H}(a,b;c;z)$ 代表古典的超几何函数(见[7]). 在式㉝中最后的积分式表明 $F(t)$ 在 $[0,+\infty)$ 严格增加,且 $F(0)=0$ 和 $F(+\infty)=\frac{1}{2}$.

至此,还看不出为什么式㉚中的函数 $F(t)$ 在确定伯恩斯坦常数 β 中能起作用. 为了说明如何想出这个函数,伯恩斯坦考虑了下面的多项式插值问题. 对于每个固定的正整数 n,考虑在 $[-1,+1]$ 中由下式给定的 $2n+1$ 个离散点

$$x_0 \triangleq x_0(2n)=0; x_k \triangleq x_k(2n) \triangleq \cos\left[\frac{(k-\frac{1}{2})\pi}{2n}\right]$$
$$(k=1,2,\cdots,2n) \quad ㉞$$

因为点 $\{x_k\}_{k=1}^{2n}$ 恰恰是 Tschebyscheff 多项式 $T_{2n}(x)$ 的零点,我们记

$$\omega(x) \triangleq \prod_{j=0}^{2n}(x-x_j) = \frac{xT_{2n}(x)}{2^{2n-1}} \quad ㉟$$

如果 $R_{2n}(x)$ 代表在 π_{2n} 中唯一的多项式,它在式㉞

中给定的点上插值$|x|$,那么$R_{2n}(x)$是x的偶多项式,经过一定演算后,它满足

$$|x|-R_{2n}(x)=\frac{T_{2n}(x)}{n}H_{2n}(x) \quad (x\in[0,1]) \tag{236}$$

其中

$$H_{2n}(x) \triangleq -\sum_{k=0}^{n-1}\frac{(-1)^k\sin\left[\left(k+\frac{1}{2}\right)\frac{\pi}{2n}\right]}{x+\cos\left[\left(k+\frac{1}{2}\right)\frac{\pi}{2n}\right]}$$

$$(x\in[0,1]) \tag{236'}$$

经过很长的证明后,伯恩斯坦[2]得到

$$|x|-R_{2n}(x)=\frac{T_{2n}(x)}{n}\left[F\left(\frac{2nx}{\pi}\right)+\eta_n(x)\right] \quad (x\in[0,1]) \tag{237}$$

其中

$$|\eta_n(x)|\leqslant\frac{4+\pi^2}{2n^{\frac{2}{5}}} \quad (x\in[0,1];n=1,2,\cdots) \tag{238}$$

接下来,按照定义

$$E_{2n}(|x|)\leqslant\||x|-R_{2n}(x)\|_{L_\infty[-1,+1]}=$$
$$\||x|-R_{2n}(x)\|_{L_\infty[0,1]}$$

最后一个不等式是由$|x|$和$R_{2n}(x)$都是偶函数这一事实得出的. 因为在$[0,1]$中$|T_{2n}(x)|\leqslant 1$,而且$F(t)$在$[0,+\infty)$上是增函数,并有$F(+\infty)=\frac{1}{2}$,所以从式(237)和式(238)可知

$$\overline{\lim_{n\to\infty}}\, 2nE_{2n}(|x|)\leqslant 2F(+\infty)=1 \tag{239}$$

注意,当用来改进式(208)时,式(239)不如式(212)好. 此外,式(237)表明:对于大的n,误差$|x|-R_{2n}(x)$在$[-1,+1]$区间

内远不是等幅振荡. 因为 $F(t)$ 的严格递增性质隐含了最大的误差只能出现在 $x=\pm 1$ 的区域内而不可能出现在 $x=0$ 的邻域内.

为了在 $x=0$ 附近有等振荡性, 伯恩斯坦[2] 建议如下, 对于任何偶数 n, 首先记

$$T_{2n}(x) = \cos(2n\arccos x) = \cos(2n\arcsin x) \quad ⑳$$

其次, 对于固定的任意非负整数 m 和所有的 $n>m$, 令 $\left\{\xi_k(2n) \triangleq \sin\left[\dfrac{(2k-1)\pi}{4n}\right]\right\}_{k=1}^m$ 是 $T_{2n}(x)$ 在 $[0,1]$ 中 m 个最小的零点. 显然, 对于每个 $1 \leqslant k \leqslant m$, $\dfrac{T_{2n}(x)}{x^2-\xi_k^2(2n)}$ 是 π_{2n} 中的偶多项式, 于是由

$$Q_{2n}(x) \triangleq R_{2n}(x) + \frac{T_{2n}(x)}{n}\left\{a_0 + \left(\frac{\pi}{2n}\right)^2 \sum_{k=1}^m \frac{a_k}{x^2 - \xi_k^2(2n)}\right\} \quad ㉑$$

定义的多项式 $Q_{2n}(x)$, 对于每个 $n>m$ 也是一个 π_{2n} 中的偶多项式. 除此之外, 令 $x = \dfrac{\pi b}{(2n)}$, 从式 ㉗ 和式 ㉑ 可以得到, 对于每个 $n>m$, 有

$$|x| - Q_{2n}(x) = \frac{T_{2n}(x)}{n} \cdot \left\{F(b) - \left(a_0 + \sum_{k=1}^m \frac{a_k}{b^2 - \left[\dfrac{2n}{\pi}\sin\left(\dfrac{(2k-1)\pi}{4n}\right)\right]^2} + \eta_n(x)\right)\right\} \quad ㉒$$

对每个非负整数 m 现在我们定义

$$\mu_m \triangleq \inf_{\substack{a_0,a_1,\cdots,a_m \\ 实}} \left\{ \left\| \cos(\pi b) \left[F(b) - \left(a_0 + \sum_{k=1} \frac{a^k}{b^2 - \left(\frac{2k-1}{2} \right)^2} \right) \right] \right\|_{L_\infty[0,+\infty)} \right\} \quad (243)$$

对于固定的 $b \geqslant 0$,由式(240)可得

$$T_{2n}(x) = \cos\left[2n \arcsin\left(\frac{\pi b}{2n} \right) \right] \to \cos(\pi b) \quad (n \to \infty)$$

类似于式(239),可以验证

$$\overline{\lim_{n\to\infty}} 2nE_{2n}(|x|) \leqslant 2\mu_m \quad (m=0,1,\cdots) \quad (244)$$

从式(243)可知正常数序列 $\{\mu_m\}_{m=0}^{+\infty}$ 显然是非增的

$$\mu_0 \geqslant \mu_1 \geqslant \mu_2 \geqslant \cdots \quad (245)$$

因此是收敛的. 伯恩斯坦[2]证明了式(213)中的伯恩斯坦常数 β 和序列 $\{\mu_m\}_{m=0}^{+\infty}$ 的极限通过

$$\beta = 2 \lim_{m\to\infty} \mu_m \quad (246)$$

联系起来,由式(245),每个逼近问题(243)的常数 μ_m 给出了如下 β 的上界

$$\beta \leqslant 2\mu_m \quad (m=0,1,\cdots) \quad (247)$$

有趣的是,在 1913 年伯恩斯坦[2]对于 $m=3$ 数值估计了式(243)的解,并发现 $\mu_3 < 0.143$,于是

$$2\mu_3 < 0.286$$

这就是在式(214)中提到的 β 的上界(在表 5 中可以找到 $2\mu_m$ 更精确的估计值).

用来确定常数 μ_m 的式(243)中的极小化问题是一个在 $[0,+\infty)$ 上 $F(b)$ 的特殊加权有理逼近问题,然而式(243)中的权 $\cos(\pi b)$ 在这个区间上肯定不是单一符号

第20章 逼近论中的伯恩斯坦猜测

的. 但是, 正如[13]中指出的, 式㉔中逼近问题的解具有有趣的振荡性质, 这就允许我们使用修正的(第二种)Remez 算法(应该强调伯恩斯坦在 1913 年的工作[2]早于 1934 年出现的 Remez 算法). 利用修改后的(第二种)Remez 算法解决极小化问题㉔:

表 5 $\{2\mu_m\}$

m	$2\mu_m$	m	$2\mu_m$
0	0.500 000 000 000 000	10	0.280 568 148 084 662
1	0.309 816 648 277 486	20	0.280 267 918 128 026
2	0.289 644 642 836 759	30	0.280 213 001 347 551
3	0.284 585 623 264 382	40	0.280 193 895 181 171
4	0.282 681 644 408 752	50	0.280 185 082 723 738
5	0.281 779 992 624 272	60	0.280 180 306 766 681
6	0.281 286 520 869 723	70	0.280 177 431 742 434
7	0.280 988 433 465 837	80	0.280 175 568 033 390
8	0.280 795 058 278 109	90	0.280 174 291 500 582
9	0.280 662 672 087 176	100	0.280 173 379 101 718

第一步. 令 $\widetilde{S} \triangleq \{t_j\}_{j=0}^{m+1} (m \geq 1)$ 是在 $[0, +\infty)$ 中 $m+2$ 个离散点的点集, 满足

$$0 = t_0 < t_1 < \cdots < t_m \leq m - \frac{1}{2} < t_{m+1} \triangleq +\infty \qquad ㉔$$

第二步. 找 $m+2$ 个唯一常数 $\{a_i\}_{i=0}^{m}$ 和 λ (这是个线性问题), 使得

$$\begin{cases} \cos(\pi t_j) \left\{ a_0 + \sum_{k=1}^{m} \dfrac{a_k}{t_j^2 - \left[\dfrac{(2k-1)}{2}\right]^2} + (-1)^{j+1} \lambda \right\} = \\ \cos(\pi t_j) F(t_j) \quad (j = 0, \cdots, m) \\ a_0 + \lambda = \dfrac{1}{2} = F(+\infty) \end{cases} \qquad ㉕$$

Tschebyscheff 逼近定理

线性问题 ㉙ 的解强迫函数

$$R_m(t) \triangleq \cos(\pi t_j)\left[F(t) - \left(a_0 + \sum_{k=1}^{m} \frac{a_k}{t^2 - \left[\frac{(2k-1)}{2}\right]^2}\right)\right]$$

㉚

在 $\widetilde{S}$ 的子集 $\{t_j\}_{i=0}^{m}$ 上等幅振荡,而且在点集 $\{t_j\}_{i=0}^{m}$ 上依次取交替的误差 $|\lambda|$. 此外,强制 $R_m(t)$ 在 $t \to \infty$ 时在 $+|\lambda|$ 和 $-|\lambda|$ 之间振荡. 因此,与式 ㉗ 类似. 我们有

$$\|R_m(t)\|_{L_\infty(\widetilde{S})} = |\lambda|$$

㉛

并因为 $\widetilde{S}$ 是 $[0, +\infty)$ 的子集,那么(见式 ㉘)

$$\|R_m(t)\|_{L_\infty[0,+\infty)} - |\lambda| \geq 0$$

㉜

作为修正 Remez 算法下一步的基础,我们有

$$\frac{d}{dt}\left\{F(t) + \left(a_0 + \sum_{k=1}^{m} \frac{a_k}{t^2 - \left[\frac{(2k-1)}{2}\right]^2}\right)\right\} =$$

$$F'(t) + 2t \sum_{k=1}^{m} \frac{a_k}{\left[t^2 - \left[\frac{(2k-1)}{2}\right]^2\right]^2}$$

㉝

如果式 ㉙ 的解 $a_k(0 \leq k \leq m)$ 和 λ 都是正数,式 ㉝ 及 $F(t)$ 的严格增加性质给出

$$G_m(t) \triangleq F(t) - \left(a_0 + \sum_{k=1}^{m} \frac{a_k}{t^2 - \left[\frac{(2k-1)}{2}\right]^2}\right)$$

在区间 $\left(m-\frac{1}{2}, +\infty\right)$ 上从 $-\infty$ 严格增加到 $+\lambda$. 定义 $\tau_m = \tau_m(\widetilde{S})$ 是 t 在 $\left(m-\frac{1}{2}, +\infty\right)$ 中使 $G_m(\tau_m) = -\lambda$ 成立的唯一解,那么由于 $|R_m(t)| = |\cos(\pi t)| \cdot$

第20章 逼近论中的伯恩斯坦猜测

$G_m(t)|\leq|G_m(t)|\leq\lambda$ 对所有 $t\geq\tau_m$ 成立,并且因为当 $t\to+\infty$ 时 $G_m(t)\to\frac{1}{2}-a_0=\lambda$,可以得到

$$\|R_m(t)\|_{L_\infty[\tau_m,+\infty)}=|\lambda|$$

另一方面,因为 $\{t_j\}_{j=0}^m$ 是 $\left[0,m-\frac{1}{2}\right]$ 的一个子集,从式㉕则有 $\|R_m(t)\|_{L_\infty[0,m-\frac{1}{2}]}|\lambda|$. 又因为 $\left[0,m-\frac{1}{2}\right]$ 是 $[0,\tau_m]$ 的一个子集,则

$$\|R_m(t)\|_{L_\infty[0,\tau_m(\tilde{S})]}=\|R_m(t)\|_{L_\infty[0,+\infty)} \quad ㉕$$

然后更强的命题

$$\|R_m(t)\|_{L_\infty[0,m-\frac{1}{2}]}=\|R_m(t)\|_{L_\infty[0,+\infty)} \quad ㉕'$$

是否也成立可以由数值上确定(迄今为止在确定 β 上界的文献中所处理的特殊情况为:博亚尼奇和埃尔金斯[3]处理了 $m=1$ 的情况. 伯恩斯坦[2]处理了 $m=3$ 的情况. $\tilde{S}$ 集的选法都使得在第二步中的 $a_k(0\leq k\leq m)$ 和 λ 为正的,在这些情况下式㉕'式也成立). 这使我们有

第三步. 预先给定(小的)$\varepsilon>0$,如果第二步的解 $\{a_k\}_{k=0}^m$ 和 λ 全是正数,且式㉕'成立,并有

$$\|R_m(t)\|_{L_\infty[0,m-\frac{1}{2}]}-\lambda<\varepsilon$$

则迭代终止. 否则,在 $\left(0,m-\frac{1}{2}\right]$ 中找一个新的集合 $\tilde{S}'$,它由 $R_m(t)$ 的 m 个局部极值点(交错变号的)再加上 $t_0\triangleq 0$ 和 $t_{m+1}\triangleq\infty$ 构成. 然后再重复第二、第三步,如此继续下去.

从一个具体的集合 $\tilde{S}^{(0)}\triangleq\{t_j^{(0)}\}_{j=0}^{m+1}$ 开始,其中

Tschebyscheff 逼近定理

$$t_0^{(0)} \triangleq 0, \ t_j^{(0)} = \frac{2j-1}{2} \quad (j=1,2,\cdots,m), \ t_{m+1}^{(0)} \triangleq +\infty$$

㉕

用部分主元素的高斯消去法解出相应的线性方程组 ㉙,得到 $\{a_k^{(0)}\}_{k=0}^m$ 和 $\lambda^{(0)}$. 对于表 5 所考虑的每个情况, $\tilde{S}$ 的开始值㉕都足够好,使得前面所述的修正 Remez 算法永远给出正的 $\{a_k\}_{k=0}^m$ 和 λ,以及满足式㉘和式㉔′的交替集. 此外,正如所希望的那样,该算法的收敛性是二次的,在所考虑的情况中最多需要迭代 10 次修正 Remez 算法.

在图 6 中,我们画出式㉚中的函数 $R_5(t)$ 以及式㉓中最佳一致逼近常数 μ_5. 该图中有 6 个交错点(用小黑点标出),并在 $\left(\frac{9}{2}, +\infty\right)$ 上振荡,当 $t \to \infty$ 时它趋于 μ_5.

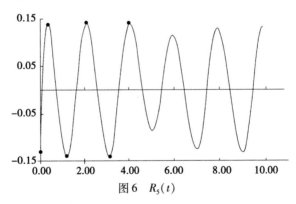

图 6 $R_5(t)$

遗憾的是,上述计算不能和 20.2 中计算 $\{2n \cdot E_{2n}(|x|)_{n=1}^{52}\}$ 具有同样高的精确度(100 位有效数字).

原因之一是把修正 Remez 算法用于极小化问题㉔③时需要反复算出式㉓⓪中函数 $F(t)$ 的值. 这里,我们用了 $F(t)$ 的表达式㉓②,其中用到了 psi 函数 $\psi(x)$. 而 $\psi(x)$ 是用科迪(Cody),Strecok,和撒切尔(Thacher)[6] 给出的近似式

$$\psi(x) \approx (x-x_0)r_{8,8}(x) \quad \left(\frac{1}{2} \leqslant x \leqslant 3\right) \tag{256}$$

其中 $x_0 = 1.461\ 63\cdots$ 是 $\psi(x)$ 的唯一正零点,已知其 40 位有效数字,并且 $r_{8,8}(x)$ 是两个 x 的八次特殊多项式的比(见[6]);还用到了

$$\psi(x) \approx \ln x - \frac{1}{2x} + r_{6,6}\left(\frac{1}{x^2}\right) \quad (x \leqslant 3 < +\infty) \tag{257}$$

其中 $r_{6,6}(u)$ 是两个 u 的六次特殊多项式的比(见[6]). 对于 $0<x\leqslant\frac{1}{2}$,我们利用熟知的递推关系式

$$\psi(x) = \psi(x+1) - \frac{1}{x} \tag{258}$$

和近似式㉕⑥.

上面给出 $\psi(x)$ 的逼近式可以准确到大约 20 位有效数字(见[6]),所以我们估计我们算出的 $\{2\mu_m\}_{m=0}^{+\infty}$ 至少精确到 15 位有效数字. 为了省纸,在表 5 中给出了 $\{2\mu_m\}_{m=0}^{100}$ 的一个子集,它们全都截断到 15 位有效数字.

从表 5 以及式㉑⑤和式㉔⑤,㉔⑥显然可以看出

$$\frac{1}{2\sqrt{\pi}} = 0.282\ 094\ 791\ 7\cdots > 2\mu_5 > \beta \tag{259}$$

这就说明伯恩斯坦猜测不成立.

20.4 计算伯恩斯坦常数 β 的下界

已经看出,在 19.3 中我们计算的伯恩斯坦常数 β 的上界 $\{2\mu_m\}_{\mu=0}^{100}$ 足以否定伯恩斯坦猜测㉑⑥. 因此,就伯恩斯坦猜测本身确立与否来说,显然没有必要再去计算 β 的下界. 但是为了使我们的讨论至臻完美,这里我们从伯恩斯坦另一个明快的方法出发,讨论一下有关 β 下界的计算问题. β 下界的计算证明工作是十分艰难的.

为了叙述确定 β 下界的伯恩斯坦方法[2],我们定义

$$\phi_m \triangleq \prod_{j=1}^{m-1}(x^2 - j^2) \quad (m \geq 1) \qquad ㉖⓪$$

和

$$\psi_m(x) = \psi_m(x;\lambda_1,\lambda_2,\cdots,\lambda_m) \triangleq \prod_{j=1}^{m}(x^2 - \lambda_j^2)$$
$$(m \geq 1) \qquad ㉖①$$

(这里我们采用习惯记法 $\prod_{j=\beta}^{\alpha} \triangleq 1$,如果 $\alpha < \beta$)并假定式 ㉖① 中的参数 λ_j 满足

$$j-1 < \lambda_j < j \quad (j \geq 1) \qquad ㉖②$$

对于每个 $m \geq 1$,设

$$B_m(\lambda_1,\lambda_2,\cdots,\lambda_m) \triangleq$$
$$\dfrac{\sum_{i=1}^{m}\dfrac{\phi_m(\lambda_i)}{\psi'_m(\lambda_i)}\left[1 - \left(\dfrac{2\lambda_i}{\lambda_i + \dfrac{1}{2}}\right)F\left(\lambda_i + \dfrac{1}{2}\right)\right]}{\sum_{i=1}^{m}\dfrac{\phi_m(\lambda_i)}{\psi'_m(\lambda_i)}\left[\dfrac{2}{\pi\lambda_i} + \tan\left(\dfrac{\pi}{2}[\lambda_i - i + 1]\right)\right]} \qquad ㉖③$$

第20章 逼近论中的伯恩斯坦猜测

其中函数 $F(t)$ 的定义在式 ㉚ 中. 从式 ㉖⓪ 和式 ㉖①, 我们可以写

$$\frac{\phi_m(\lambda_i)}{\psi'_m(\lambda_i)} = \frac{\prod_{j=1}^{i-1}(\lambda_i^2 - j^2) \cdot \prod_{j=i}^{m-1}(j^2 - \lambda_i^2)}{2\lambda_i \prod_{j=1}^{i-1}(\lambda_i^2 - \lambda_j^2) \cdot \prod_{j=i+1}^{m}(\lambda_j^2 - \lambda_i^2)}$$

$$(1 \leq i \leq m) \qquad ㉖④$$

再考虑到条件 ㉖②, 我们可以看出上述各比值均为正的. 因为 $F(t)$ 在区间 $[0, +\infty)$ 上是严格增函数, 且 $F(0) = 0, F(+\infty) = \frac{1}{2}$ (见 20.3), 我们导出式 ㉖③ 中每个和式中的诸项必是正的, 于是得到 $B_m(\lambda_1, \lambda_2, \cdots, \lambda_m) > 0$.

如果用式 ㉑③ 中定义的 β, 伯恩斯坦[2] 证明了 $B_m(\lambda_1, \lambda_2, \cdots, \lambda_m)$ 是 β 的一个下界, 即

$$\beta \geq B_m(\lambda_1, \lambda_2, \cdots, \lambda_m) \qquad ㉖⑤$$

对每个正整数 m 和满足式 ㉖② 的 $\{\lambda_j\}_{j=1}^m$, 上式都成立. 对于每个 $m \geq 1, \beta$ 的最好下界显然是

$$l_m \triangleq \max\{B_m(\lambda_1, \lambda_2, \cdots, \lambda_m) \mid \{\lambda_j\}_{j=1}^m \text{满足式} ㉖②\} \qquad ㉖⑥$$

于是

$$\beta \geq l_m > 0 \quad (m \geq 1) \qquad ㉖⑦$$

下面考虑满足式 ㉖② 的任一组参数 $\{\lambda_j\}_{j=1}^{m+1}$, 固定 $\{\lambda_j\}_{j=1}^m$ 并让 λ_{m+1} 减到 m, 从式 ㉖③ 和式 ㉖④ 可以很快地验证

$$\lim_{\lambda_{m+1} \to m} B_{m+1}(\lambda_1, \lambda_2, \cdots, \lambda_m, \lambda_{m+1}) = B_m(\lambda_1, \lambda_2, \cdots, \lambda_m)$$

㉖⑧

最后, 我们可以从式 ㉖⑥ 看到

$$l_{m+1} \geqslant l_m \quad (m \geqslant 1) \tag{269}$$

所以根据式㉖⑦，$\{l_m\}_{m=1}^{+\infty}$是一个有界非减的正数序列.伯恩斯坦[2]进一步证明了这个序列的极限是β

$$\beta = \lim_{m \to +\infty} l_m \tag{270}$$

实际上伯恩斯坦在[2]中数值估计了l_1和l_2，并发现

$$l_1 > 0.27,\ l_2 > 0.278 \tag{271}$$

l_2的估计式作为式㉑④中β的下界(可以在表6中找到l_1和l_2更精确的估计值).

表6　$\{l_m\}_{m=1}^{20}$

m	l_m	m	l_m
1	0. 271 982 359 030 477	11	0. 280 163 464 187 524
2	0. 278 930 922 849 406	12	0. 280 164 893 327 009
3	0. 279 811 000 437 231	13	0. 280 165 905 238 063
4	0. 280 024 333 928 903	14	0. 280 166 641 527 680
5	0. 280 097 791 315 214	15	0. 280 167 189 892 928
6	0. 280 129 183 079 687	16	0. 280 167 606 600 825
7	0. 280 144 691 009 336	17	0. 280 167 928 871 653
8	0. 280 153 187 711 753	18	0. 280 168 181 990 114
9	0. 280 158 217 699 044	19	0. 280 168 383 539 180
10	0. 280 161 379 471 687	20	0. 280 168 546 002 042

现在叙述我们所算的l_m下界.从式㉖③显然看出参数$\lambda_1, \lambda_2, \cdots, \lambda_m$在$B_m(\lambda_1, \lambda_2, \cdots, \lambda_m)$的定义中是非线性地出现.我们用了非常标准的不带导数的优化(极大化)子程序来确定满足约束㉖②的参数$\{\lambda_i\}_{i=1}^{m}$，它们使$B_m(\lambda_1, \lambda_2, \cdots, \lambda_m)$极大化，因此确定了式㉖⑥中的

第 20 章 逼近论中的伯恩斯坦猜测

l_m. 又因为函数 $F(t)$ 显式地出现在式㉖㉛中的 $B_m(\lambda_1, \lambda_2, \cdots, \lambda_m)$ 的定义中, 我们用 psi 函数 $\psi(t)$ 的逼近式 ㉖㉖ ~ ㉖㉘ 和 $\psi(t)$ 表示 $F(t)$ 的表达式 ㉓㉜. 和计算 β 的上界 $2\mu_m$ 时一样(见式 ㉔㉖), 我们计算 β 的下界 l_m 不能和在 20.2 中计算 $2nE_{2n}(|x|)$ 时具有一样高的精确度 (95 位有效数字). 因为用于 20.3 中数值计算结果的类似理由, 对于 $\psi(t)$ 的逼近达到大约 20 位有效数字, 我们估计优化计算 $\{l_m\}_{m=1}^{20}$ 的结果至少有 15 位有效数字. 在表 6 中给出了 $\{l_m\}_{m=1}^{20}$, 它们都被截断到 15 位有效数字.

当比较表 5 中的上界和表 6 中的下界时, 我们看到式 ㉖㉗ 中的下界 l_m 比式 ㉔㉖ 的上界 $2\mu_m$, 对于每个 $1 \leqslant m \leqslant 20$, 都精确得多. 事实上, l_{20} 近似 β 的误差大约是 9.53×10^{-7}, 而 $2\mu_{100}$ 近似 β 的误差仅为 3.88×10^{-6}. 至今, 这种高精度是靠用我们的优化子程序找 l_m 时增加所需的计算机时间取得的. 近似 β 的下界 l_m 有较大的精度也解释了为什么表 6 中 m 的值的精度没有达到像表 5 中算上界 $2\mu_m$ 时那样高.

20.5 数 $\{2n\sum_{2n}(|x|)\}_{n=1}^{52}$ 的理查森外插

在表 4 中 $\{2nE_{2n}(|x|)\}_{n=1}^{52}$ 的值表明这些数收敛到伯恩斯坦常数 β 是很慢的(见式㉑㉓). 改进收敛慢的序列收敛性的典型方法是理查森外插方法(见布热津斯基(Brezinski)[5]). 它可以描述如下, 如果 $\{S_n\}_{n=1}^{N}$ ($N>2$) 为一给定(有限的)实数序列, 设 $T_0^{(n)} \triangleq S_n (1 \leqslant n \leqslant N)$ 并称 $\{T_0^{(n)}\}_{n=1}^{N}$ 为理查森外插表的第零列, 由 N

Tschebyscheff 逼近定理

个数组成. 理查森外插表的第一列由 $N-1$ 个数组成, 其定义如下

$$T_1^{(n)} \triangleq \frac{x_n T_0^{(n+1)} - x_{n+1} T_0^{(n)}}{x_n - x_{n+1}} \quad (1 \leq n \leq N-1) \qquad ⑰$$

按归纳法, 理查森外插表的第 $k+1$ 列由 $N-k-1$ 个数组成, 对每个 $k=0,1,\cdots,N-2$ 定义为

$$T_{k+1}^{(n)} \triangleq \frac{x_n T_k^{(n+1)} - x_{n+k+1} T_k^{(n)}}{x_n - x_{n+k+1}} \quad (1 \leq n \leq N-k-1) \qquad ⑰$$

其中 $\{x_n\}_{n=1}^N$ 是给定的常数. 按这种方法建立了一个由 $\frac{N(N+1)}{2}$ 个元素组成的三角形的表. 在我们计算 $\{2nE_{2n}(|x|)\}_{n=1}^{52}$ 的情况下, 由 1 378 个元素组成的三角形表被算出. 为了选择式 ⑰, ⑰ 中的数 $\{x_n\}_{n=1}^{52}$, 前期计算表明 $2nE_{2n}(|x|) = \beta + \frac{K}{n^2} + $ 低阶项, 所以选 $x_n \triangleq \frac{1}{n^2}$. 我们要说明, 在式 ⑰ 和式 ⑰ 中分数里的分子、分母的减法可能造成精确度的损失, 这件事提醒我们计算 $2nE_{2n}(|x|)$ 时应当以很高的精度完成(95 位有效数字).

$\{2nE_{2n}(|x|)\}_{n=1}^{52}$ 的理查森外插产生了意想不到的好结果. 这里不准备给出整个由 1 378 个数构成的外插表(比如说每个数给出到 95 位有效数字), 看来给出这个表的最后 20 列就足够了. 在这些列中全部 210 个数里除 3 个以外都和式 ⑱ 中的 β 的近似值重合到前 45 位数字

$$\beta \approx 0.280\ 169\ 499\ 023\ 869\ 133\ 036\ 436\ 491$$
$$230\ 672\ 000\ 042\ 482\ 139\ 812\ 36 \qquad ⑭$$

此外,在该理查森外插表中最后 20 列的 210 个数中有 182 个数的前 50 位十进数和式㉔中 β 的近似值完全重合.

20.6　某些未解决的问题

理查森外插$\left(用\ x_n \triangleq \dfrac{1}{n^2}\right)$成功地用于$\{2n E_{2n}(|x|)\}_{n=1}^{52}$. 提示 $2n E_{2n}(|x|)$ 可能有如下形状的渐近展开(见[7])

$$2n E_{2n}(|x|) \stackrel{?}{\approx} \beta - \dfrac{K_1}{n^2} + \dfrac{K_2}{n^4} - \dfrac{K_3}{n^6} + \cdots \quad (n \to \infty) \quad ㉕$$

其中常数 K_j 与 n 无关. 假定式㉕成立,可得出

$$n^2(2n E_{2n}(|x|) - \beta) \approx -K_1 + \dfrac{K_2}{n^2} - \dfrac{K_3}{n^4} + \cdots \quad (n \to \infty)$$

㉖

用已经得到的表 4 中 $2n E_{2n}(|x|)$ 的高精度近似和 $\{2n E_{2n}(|x|)\}_{n=1}^{52}$ 外插表最后一个元素确定的 β 估计值,我们可以再次对 $\{n^2(2n E_{2n}(|x|) - \beta)\}_{n=1}^{52}$ $\left(用\ x_n = \dfrac{1}{n^2}\right)$ 进行理查森外插以便得出式㉖中 K_1 的外插估计值. 这个过程可以通过理查森外插继续得到式㉕中后续的常数 K_j. 正如所担心的那样,在连续确定常数 K_j 的进程中,数值精度逐渐有所损失.

在表 7 中我们列出了估计值 $\{K_j\}_{j=1}^{10}$ 并插入到 10 位有效数字. 正如表 7 所指出的,后面的常数 K_j 增加很快. 因为这些常数都变为正数,我们有下面新的猜测

Tschebyscheff 逼近定理

表 7 $\{K_j\}_{j=1}^{10}$

j	K_j	j	K_j
1	0.043 967 528 88	6	0.595 435 315 1
2	0.026 40 716 87 7	7	2.925 915 470
3	0.031 253 426 46	8	18.494 140 33
4	0.058 890 016 57	9	146.943 012 3
5	0.160 106 997 1	10	1 438.032 717

猜测(瓦尔加和卡彭特[13]) $2nE_{2n}(|x|)$ 有如下形式的渐近展式

$$2nE_{2n}(|x|) \stackrel{?}{\approx} \beta - \frac{K_1}{n^2} + \frac{K_2}{n^4} - \frac{K_3}{n^6} + \cdots \quad (n \to \infty) \qquad ⑦$$

其中常数 K_j(与 n 无关)都是正的.

因为伯恩斯坦常数 β 和函数 $F(t)$ 具体的有理逼近有联系(见式㊺),这里 $F(t)$ 可以用古典的超几何函数表示(见式㉝). 指望着常数 β 以及式⑦中的 K_j 可能有由古典超几何函数和或已知的数学常数给出的封闭表达式并非不合理.

最后,在 1913 年伯恩斯坦的论文[2]的最后,他说:他对 $E_{2n}(|x|;[-1,+1])$ 性态的分析可以推广到对于任何 $\alpha > 0, E_{2n}(|x|^{2\alpha};[-1,+1])$ 的性态上去,即,存在有正常数 $\beta(\alpha)$ 使得

$$\beta(\alpha) = \lim_{n \to \infty} (2n)^{2\alpha} E_{2n}(|x|^{2\alpha};[-1,+1])$$
$$= \lim_{n \to \infty} (2n)^{2\alpha} E_n(x^\alpha;[0,1]) \quad (\alpha > 0) \qquad ⑧$$

和式㉒一样,这里最后一个不等式是从 $|x|^{2\alpha}$ 在 $[-1,+1]$ 上偶性得出的. 因此,用这种符号,本章前面几节

第20章 逼近论中的伯恩斯坦猜测

仅在于估计 $\beta\left(\dfrac{1}{2}\right)$ 并在式㉔中给出.

奇怪的是,在 1913 年发表论文[2]的 25 年之后,于 1938 年伯恩斯坦再次研究了这些问题,某些关于函数 $\beta(\alpha)$ 相当重要的结果发表在[14]中,利用大量有趣的但是困难的技术,在[14]中伯恩斯坦给出了上界

$$\beta(\alpha) < \frac{\Gamma(2\alpha)\,|\sin(\pi\alpha)|}{\pi} \quad (\alpha>0) \qquad ㉙$$

以及下界

$$\frac{\Gamma(2\alpha)\,|\sin(\pi\alpha)|}{\pi}\left(1-\frac{1}{2\alpha-1}\right) < \beta(\alpha) \quad \left(\alpha>\frac{1}{2}\right) \qquad ㉚$$

由这个式子得出,当 $\alpha \to \infty$ 时,精确的渐近式

$$\lim_{\alpha\to\infty} \frac{\beta(\alpha)}{\Gamma(2\alpha)\dfrac{|\sin(\pi\alpha)|}{\pi}} = 1 \qquad ㉛$$

在[14]中他还给出

$$\lim_{\alpha\to 0_+} \beta(2) = \frac{1}{2} \triangleq \beta(0) \qquad ㉜$$

伯恩斯坦给出的结果㉙~㉜可能是他试图找到对于 $\alpha>0$ 的 $\beta(\alpha)$ 的封闭形式. 但除了式㉜中的 $\beta(0)$ 以及式㉔中 $\beta\left(\dfrac{1}{2}\right)$ 这些特殊值之外,在文献中尚没有 $\beta(\alpha)$ 的其他值. 然而近来卡彭特和瓦尔加像上节那样,又利用高精度计算和插值确定了当 α 取某些值时 $\beta(\alpha)$ 的结果,并至少有 40 位十进制数,这些结果列在表 8 中.

尽管对于这个问题进行了很多理论和计算工作,但是当 $\alpha>0$ 时,$\beta(\alpha)$ 的封闭表达式是什么依然是一个悬而未决的问题!

Tschebyscheff 逼近定理

表 8 到 40 位十进数的 $\beta(\alpha)$ 估计值

α	$\beta(\alpha)$
0.000	5.000 000 000 000 000 000 000 000 000 000 000 000 0E-01
0.125	3.921 060 686 524 306 181 028 788 906 500 295 167 407 3E-01
0.250	3.486 482 327 256 100 432 735 006 660 904 270 533 718 1E-01
0.375	3.152 412 741 461 107 187 646 673 856 654 824 993 299 4E-01
0.500	2.801 694 990 238 691 330 364 364 912 306 720 000 424 8E-01
0.625	2.364 447 648 336 463 840 954 677 748 284 406 680 534 7E-01
0.750	1.783 603 316 926 983 670 188 153 355 271 407 477 285 1E-01
0.875	1.008 767 973 591 345 441 686 088 821 616 981 024 648 3E-01
1.000	0.000 000 000 000 000 000 000 000 000 000 000 000 000 0E+00
1.250	2.749 659 750 796 998 831 104 375 161 314 348 009 098 0E-01
1.500	5.911 069 586 273 252 187 191 762 347 676 770 645 623 4E-01
1.750	7.004 367 050 981 847 478 764 148 352 647 247 594 402 3E-01
2.000	0.000 000 000 000 000 000 000 000 000 000 000 000 000 0E+00
2.250	2.482 954 847 765 220 759 098 275 363 988 737 089 887 6E+00
2.500	7.280 319 238 913 027 783 837 144 091 741 948 207 140 6E+00
2.750	1.127 339 025 880 507 558 565 219 844 242 736 479 959 6E+01
3.000	0.000 000 000 000 000 000 000 000 000 000 000 000 000 0E+00

20.7 $|x|$ 在 $[-1,+1]$ 上的有理逼近

因为本章前几节都是研究伯恩斯坦猜测的,即函数 $|x|$ 在 $[-1,+1]$ 上的最佳一致多项式逼近的问题. 在本章我们最后考虑 $|x|$ 在 $[-1,+1]$ 上最佳一致有理逼近的问题是很自然的. 需要新增加一些符号,对任何一个非负整数 n,令 $\pi_{n,n}$ 代表所有有理函数 $r_{n,n}(x)=$

$\frac{p(x)}{q(x)}$ 的集合,而 $p \in \pi_n$ 和 $q \in \pi_n$(这里假定 p 和 q 没有公因子,q 在 $[-1, +1]$ 上不取零值,而且 q 由 $q(0) = 1$ 正规化). 对于定义在 $[-1, +1]$ 上的实函数 $f(x)$, 和式 ⑳类似,我们定义

$$E_{n,n}(f) \triangleq \inf\{\|f - r_{n,n}\|_{L_\infty[-1,+1]} | r_{n,n} \in \pi_{n,n}\} \quad ㉘$$

有趣的是,虽然伯恩斯坦在 1913 年就深入地考虑了 $|x|$ 在 $[-1, +1]$ 上最佳一致逼近多项式的渐近性态,但在 50 年后的 1964 年,纽曼(D. J. Newman)才指出 $|x|$ 在 $[-1, +1]$ 上的最佳一致有理逼近的性质是何等的不同. 纽曼构造性地证明了

$$\frac{1}{2e^{9\sqrt{n}}} \leq E_{n,n}(|x|) \leq \frac{3}{e^{\sqrt{n}}} \quad (n = 4, 5, \cdots) \quad ㉘$$

纽曼按照伯恩斯坦早期关于当 $n \to \infty$ 时 $E_n(|x|)$ 的渐近性态研究的路线提出了不等式㉘的研究课题,这些有价值的研究工作中的一部分集中于当 $n \to \infty$ 时准确地描述 $E_{n,n}(|x|)$ 的渐近性态.

贡恰尔(Gonchar)[8]和其他作者对于 $E_{n,n}(f)$ 的渐近性态的一般理论已经做出了几个重要的贡献. 对于特殊的 $E_{n,n}(|x|)$ 来说,布拉诺夫(Bulanov)[6]已给出了当前最好的结果,他证明了

$$E_{n,n}(|x|) \geq e^{-\pi\sqrt{n+1}} \quad (n = 0, 1, \cdots) \quad ㉘$$

Vjacheslavov 在[18]中证明了存在正常数 M_1 和 M_2 使得

$$M_1 \leq e^{\pi\sqrt{n}} E_{n,n}(|x|) \leq M_2 \quad (n = 1, 2, \cdots) \quad ㉘$$

显然,式㉘和式㉘包含了

$$e^{\pi(1-\sqrt{2})} = 0.272\,18\cdots \leq e^{\pi\sqrt{n}} E_{n,n}(|x|) \leq M_2$$
$$(n = 1, 2, \cdots) \quad ㉘$$

Tschebyscheff 逼近定理

并且,如果

$$\underline{M} \triangleq \varliminf_{n \to \infty} e^{\pi\sqrt{n}} E_{n,n}(|x|)$$

$$\overline{M} \triangleq \varlimsup_{n \to \infty} e^{\pi\sqrt{n}} E_{n,n}(|x|)$$

㉘

则有

$$1 \leqslant \underline{M} \leqslant \overline{M}$$

㉙

式㉖的结果明白地给出了当 $n \to \infty$ 时 $E_{n,n}(|x|)$ 的渐近式中 $\sqrt{n}$ 在渐近意义下的最好的系数,即为 π. 剩下的只是确定式㉙中最佳渐近常数 $\underline{M}$ 和 $\overline{M}$.

为了看到这个问题的内涵,我们在这里介绍由瓦尔加,拉坦(Ruttan)和卡彭特[17]所做的最新的高精度计算 $\{E_{n,n}(|x|)\}_{n=1}^{40}$ 所得到的结果. 和在 20.2 中计算多项式的情况一样,对于每个非负整数 n,在 $\pi_{n,n}$ 中对于 $|x|$ 在区间 $[-1,+1]$ 上的最佳一致逼近是唯一的(见[11]),比如说用 $\hat{r}_{n,n}(x)$ 表示它,于是

$$E_{n,n}(|x|) = \||x| - \hat{r}_{n,n}(x)\|_{L_\infty[-1,+1]} \quad (n=1,2,\cdots)$$

㉚

又因为 $|x|$ 在 $[-1,+1]$ 上是偶函数,所以 $\hat{r}_{n,n}(x)$ 也是偶函数,并可以证明(见式㉗)

$$E_{2n,2n}(|x|) = E_{2n+1,2n+1}(|x|) \quad (n=1,2,\cdots)$$

㉛

于是,对我们来说,只考虑 $\{E_{2n,2n}(|x|)\}_{n=0}^{+\infty}$ 减小到零的方式就足够了.

如果对于每个 $n=1,2,\cdots,\hat{h}_{n,n}(t) \in \pi_{n,n}$ 是 $\pi_{n,n}$ 中在 $[0,1]$ 区间上 $\sqrt{t}$ 的最佳一致逼近,也就是说,如果

$$E_{n,n}(\sqrt{t};[0,1]) = \inf_{\tau_{n,n} \in \pi_{n,n}} \|\sqrt{t} - \tau_{n,n}(t)\|_{L_\infty[0,1]} =$$

$$\|\sqrt{t} - \hat{h}_{n,n}(t)\|_{L_\infty[0,1]}$$

㉜

第20章 逼近论中的伯恩斯坦猜测

那么容易证明(见式㉓)

$$E_{2n,2n}(|x|) = E_{n,n}(\sqrt{t};[0,1]) \quad (n=1,2,\cdots) \quad ㉝$$

其中

$$\hat{r}_{2n,2n}(x) = \hat{h}_{n,n}(x^2) \quad (n=1,2,\cdots) \quad ㉞$$

根据式㉞可以直接从高精度计算 $\{E_{n,n}(\sqrt{t};[0,1])\}_{n=1}^{40}$ 得出 $\{E_{2n,2n}(|x|)\}_{n=1}^{40}$ 的估计值,和20.2中一样,这样的计算又涉及(第二种)Remez 算法并用到 Brent MP 软件包中直到250位有效数字的运算,又考虑到保留位和可能出现的舍入误差,我们相信 $\{E_{n,n}(\sqrt{t};[0,1])\}_{n=1}^{40}$ 准确到200位有效数字. 截断到25位后的 $\{E_{2n,2n}(|x|)\}_{n=21}^{40}$ 和 $\{e^{\pi\sqrt{2n}}E_{2n,2n}(|x|)\}_{n=21}^{40}$ 的数值如表9所示.

表9

n	$E_{2n,2n}(|x|;[-1,+1])$	$e^{\pi\sqrt{2n}}E_{2n,2n}(|x|;[-1,+1])$
21	9.601 122 612 842 236 480 898 718 4E-9	6.675 616 512 649 122 885 656 417 9
22	5.970 823 398 705 558 055 298 613 7E-9	6.703 214 288 224 997 725 642 425 7
23	3.752 381 381 641 316 369 086 450 2E-9	6.729 109 963 476 020 911 052 099 8
24	2.381 499 690 721 783 089 227 969 4E-9	6.753 473 365 851 186 986 196 498 3
25	1.525 473 289 510 979 374 814 720 7E-9	6.776 451 379 185 256 903 334 534 8
26	9.856 763 349 496 352 995 813 741 3E-10	6.798 171 795 031 113 669 577 074 1
27	6.421 358 050 726 624 692 365 324 8E-10	6.818 746 400 291 279 675 079 678 8
28	4.215 884 842 992 714 575 828 506 1E-10	6.838 273 474 222 969 818 037 143 6
29	2.788 324 165 133 927 541 106 021 4E-10	6.856 839 824 093 862 326 770 263 4
30	1.857 072 001 162 821 795 312 570 7E-10	6.874 522 457 133 671 117 247 554 0
31	1.245 078 325 074 423 591 090 236 0E-10	6.891 389 963 299 101 763 905 461 5
32	8.400 599 755 776 278 634 321 604 9E-11	6.907 503 666 267 325 308 041 961 3

Tschebyscheff 逼近定理

续表

n	$E_{2n,2n}(\|x\|;[-1,+1])$	$e^{\pi\sqrt{2n}}E_{2n,2n}(\|x\|;[-1,+1])$
33	5.702 211 575 728 862 026 377 444 7E-11	6.922 918 587 292 003 040 007 665 6
34	3.892 950 581 599 345 944 390 982 3E-11	6.937 684 256 909 916 668 184 585 7
35	2.672 443 556 645 653 736 397 589 4E-11	6.951 845 402 139 240 175 290 985 3
36	1.844 299 509 252 544 160 250 377 7E-11	6.965 442 531 166 209 461 463 720 4
37	1.279 244 840 924 708 988 199 301 0E-11	6.978 512 433 145 669 705 344 080 0
38	3.916 358 294 918 686 087 120 193 9E-12	6.991 088 607 329 832 331 986 247 5
39	6.243 828 154 996 281 262 473 042 4E-12	7.003 201 633 058 588 770 167 246 1
40	4.392 048 409 181 786 189 839 103 7E-12	7.014 879 490 023 366 905 666 533 7

和 20.5 一样,我们对数 $\{e^{\pi\sqrt{2n}}E_{2n,2n}(|x|)\}_{n=1}^{21}$ 采用了几种不同的外插技术,如理查森外插、艾特肯(Aitken)Δ^2 外插等等(见[5]). 我们的最好结果是取 $x_n = \dfrac{1}{\sqrt{n}}$ 并采用理查森外插得到的. 在表 10 中给出了 $\{e^{\pi\sqrt{2n}}E_{2n,2n}(|x|)\}_{n=21}^{40}$ 的第 9 和第 10 次理查森外插的结果. 表 10 中第 9 次和第 10 次理查森外插结果分别为严格减和严格增. 根据表 10 的外插结果,瓦尔加、拉坦和卡彭特在[17]中做了一个数值上很合逻辑的猜测

$$\lim_{n\to\infty} e^{\pi\sqrt{2n}}E_{2n,2n}(|x|) \stackrel{?}{=} 8 \qquad ㉕$$

非常有趣的是,希尔伯特·施塔尔(Herbert Stahl)教授曾向文献[17]的作者索取他们的数值结果,这些结果显示出最佳一致有理逼近的误差,即在[0,1]上 $E_{n,n}(\sqrt{x};[0,1])$ 的极值点的分布. 这些数值结果看来对他很有用处;施塔尔从理论上确认式㉕中的猜测是正确的.

第 20 章　逼近论中的伯恩斯坦猜测

最后,完全类似 20.6 中的讨论,在式㉕之后自然要问对于任何 $\alpha > 0$ 是否有当 $n \to \infty$ 时关于 $E_{n,n}(x^\alpha;[0,1])$ 最佳性质的结果.

表 10

$\{\tau_n\}_{n=21}^{40}$的第 9 次 理查森外插值	$\{\tau_n\}_{n=21}^{40}$的第 10 次 理查森外插值
8.000 000 000 481 851 385 215 090 4	7.999 999 999 337 057 595 765 302 2
8.000 000 000 279 285 724 220 520 5	7.999 999 999 617 491 916 900 985 5
8.000 000 000 166 222 353 765 899 2	7.999 999 999 776 667 141 544 846 1
8.000 000 000 101 886 184 628 378 6	7.999 999 999 867 392 459 685 919 8
8.000 000 000 064 406 595 405 800 2	7.999 999 999 919 459 784 465 717 9
8.000 000 000 041 962 198 441 058 3	7.999 999 999 949 641 968 875 029 9
8.000 000 000 028 099 077 551 120 7	7.999 999 999 967 380 838 908 659 9
8.000 000 000 019 248 920 434 609 9	7.999 999 999 977 999 240 078 618 9
8.000 000 000 013 407 762 553 032 5	7.999 999 999 984 506 879 210 164 9
8.000 000 000 009 428 580 853 842 8	7.999 999 999 988 612 955 024 803 5
8.000 000 000 006 639 815 788 423 1	

参 考 文 献

[1] BELL R A, SHAH S M. Oscillating Polynomials and Approximations to $|x|$ [J]. Publ. of the Ramanujan Inst, 1969(1):167-177.

[2] BERNSTEIN S. Sur la Meilleure Approximation de $|x|$ par des Polynômes de Degré Donnés [J]. Acta Math, 1913(37):1-57.

[3] BOJANIC R, ELKINS J M. Bernstein's Constant and Best Approximation on $[0, \infty)$ [J]. Publ. de l'Inst. Mat., Nouvelle série, 1975, 18(32):19-30.

[4] RICHARD P B. A Fortran Multiple-Precision Arithmetic Package [J]. Assoc. Comput. Mach. Trans. Math. Software, 1978(4):57-70.

[5] BREZINSKI C. Algorithms d'Accélération de la Convergence [M]. Paris: Éditions Technip, 1978.

[6] BULANOV A P. Asymptotics for Least Deviation of $|x|$ from Rational Functions [J]. Mat. Sbornik, 1968, 76(118):288-303.

[7] CODY W J, STRECOK A J, THACHER H C. Chebyshev Approximations for the psi Function [J]. Math. Comp., 1973(27):123-127.

[8] GONCHAR A A. Estimates of the Growth of Rational Functions and Some of Their Applications [J]. Mat. Sbornik, 1967, 72(114):489-503.

[9] HENRICI P. Applied and Computational Complex A-

nalysis,vol. 2. [M]. New York:John Wiley & Sons, 1974.

[10] BERNSTEIN S N. Collected Works [C]. Akad. Nauk SSSR,1954(2):262-272.

[11] MEINARDUS G, Approximation of Functions: Theory and Numerical Methods [M]. New York: Springer-Verlag,1967.

[12] NEWMAN D J. Rational Approximation to $|x|$ [J]. Michigan Math. J. ,1964(11):11-14.

[13] REMEZ E Y. Sur le Calcul Effectiv des Polynômes d'Approximation de Tchebycheff [J]. C. R. Acad. Sci. Paris,1934(199): 337-340.

[14] BERNSTEIN S N. Sur la Meilleure Approximation de $|x|^p$ par des Polynômes de Degrés trés élevés [J]. Bull. Acad. Sci. USSR, Cl. Sci. Math. Nat. ,1938(2):181-190.

[15] WHITTAKER E T, WATSON G N, A Course of Modern Analysis, 4th ed. [M]. Cambridge: Cambridge University Press,1962.

[16] VARGA R S, CARPENTER A J. On the Bernstein Conjecture in Approximation Theory [J]. Constr. Approx,1985(1):333-348.

[17] VARGA R S, RUTTAN A, CARPENTER A J. Numerical Results on Best Uniform National Approximations to $|x|$ on $[-1,+1]$ [J]. Mat. Sbornik, 1991,182(11):1523-1541.

[18] VJACHESLAVOV N S. On the Least Deviations of

Tschebyscheff 逼近定理

the Function Sign x and its Primitives From the Rational Functions in the L_p-Metrics, $0<p\leq\infty$ [J]. Mat. Sbornik, 1977, 103(145):24-36.

关于非线性 Tschebyscheff 逼近的几点注记

附录 I

本附录讨论了一类非线性 Tschebyscheff 逼近问题[1]的两个具体算法,证明了它们的收敛性. 实际计算表明,效果很好.

设 $f(x)$ 与 $F(A,x)$ 为给定的两个实连续函数,其中 $x \in M, A = (a_1, a_2, \cdots, a_n) \in \mathscr{D}, M$ 表示实轴上的某个有界闭集,$\mathscr{D}$ 是 n 维欧氏空间中的子集. Tschebyscheff 逼近问题就是在 $\mathscr{D}$ 中确定 $A^* \in \mathscr{D}$,使对一切 $A \in \mathscr{D}$,有

$$p^* = \max_{x \in M} |F(A^*,x) - f(x)| \leqslant \max_{x \in M} |F(A,x) - f(x)|$$

Tschebyscheff 逼近定理

这时，$F(A^*, x)$ 称为 $f(x)$ 在 $F(A, x)$ 中的最优逼近，ρ^* 称为 $f(x)$ 用 $F(A, x)$ 逼近的最小偏差. 关于线性 Tschebyscheff 逼近 ($F(A, x)$ 对于参数 A 的依赖为线性情形) 理论早已相当成熟和完善, 在计算上也已给出了一些较为有效的算法. 但对非线性 Tschebyscheff 逼近, 要想泛泛地研究, 仍有许多困难. 在做理论上的研究时, 常常需要对 $F(A, x)$ 加上各种限制. 莫兹金 (T. S. Motzkin)[1] 和 L. Tornheim[2] 引进的所谓 "单解 (unisolvence) 函数族"① 的概念, 就是一个著名的例子. 一个连续函数族 $F(A, x)$② $(A \in \mathscr{D}, x \in M)$ 是单解的, 如果

(i) 对于任给的 n 对数 (ξ_i, η_i) $(i=1,2,\cdots,n)$, 方程组

$$F(A, \xi_i) = \eta_i \quad (i=1,2,\cdots,n) \qquad ㉖$$

总有解 $A \in \mathscr{D}$ (所谓有解性), 其中 ξ_i 是包含 M 的最小闭区间 $I(M)$ 中的任意给定的点.

(ii) 任意两个相异函数 $F(A', x)$ 与 $F(A'', x)$ 之差在 $I(M)$ 中至多有 $n-1$ 个零点 (包括计算重数在内, 即两边差值不变号的零点算作两个零点). 由此立即推知, 式 ㉖ 的解是唯一的 (所谓单解性).

对单解函数族 $F(A, x)$, 已经证明了如下定理.

定理 1.1 $f(x)$ 在 $F(A, x)$ 中的最优逼近是存在的.

---

① L. Tornheim 称之为 "含 n 个参数的函数族", 实际上与单解函数族是等价的.

② 当 $F(A, x)$ 的参数 A 在 $\mathscr{D}$ 中取值时, 就得到一族函数, 这里仍用记号 $F(A, x)$ 表示这个函数族. 以后我们用记号 $F(A, x)$ 表示函数族, 有时也可表示该族中的某个函数.

附录Ⅰ 关于非线性Tschebyscheff逼近的几点注记

定理 1.2 $f(x)$在$F(A,x)$中的最优逼近是唯一的.

定理 1.3 $F(A^*,x)$是$f(x)$在$F(A,x)$中的最优逼近的充要条件是在M中至少有$n+1$个点(称为偏差点)正负交替地取值$\max\limits_{x\in M}|F(A^*,x)-f(x)|$.

在实际应用中,$f(x)$往往是列表函数,即M是由实轴上有限个点组成的情形. 为明确起见,设
$$M=\{x_1,x_2,\cdots,x_m\mid x_1<x_2<\cdots<x_m\}\quad(n+1\leqslant m)$$
本文将讨论两种求这类最优逼近的算法,并证明其收敛性.

首先讨论$m=n+1$的情形(称之为简单情形). 我们将看到,与求线性Tschebyscheff逼近一样,这种简单情形是解决一般情形($m\geqslant n+1$)的关键. 根据定理1.3,M中的全部$n+1$个点,对其最优逼近而言,都应该是偏差点. 所以,最优逼近$F(A^*,x)$与最小偏差ρ^*应满足非线性方程组
$$F(A^*,x_i)=f(x_i)+(-1)^i\rho,\ \rho^*=|\rho|$$
$$(i=1,2,\cdots,n+1)\qquad\text{㊗}$$

定理1.1和定理1.2保证了方程组㊗的解存在且唯一. 在此,我们不讨论非线性方程组的具体解法,而认定总能得到它的解. 因而,简单情形的计算归结为求非线性方程组的解的问题.

类似于证明Vallée-Poussin定理,不难证明[3]

定理 1.4 对任意的$A\in\mathscr{D}$,若
$$F(A,x_i)-f(x_i)=(-1)^i\rho_i\quad(i=1,2,\cdots,n+1)$$
其中ρ_i或为零或为同号的数,则最小偏差ρ^*满足
$$\min_{1\leqslant i\leqslant n+1}|\rho_i|\leqslant\rho^*\leqslant\max_{1\leqslant i\leqslant n+1}|\rho_i|\qquad\text{㊘}$$

Tschebyscheff 逼近定理

其中两处等号当且仅当
$$|\rho_1| = |\rho_2| = \cdots = |\rho_{n+1}| = \rho^*$$
时成立.

式㉘可用来估计最小偏差.

现在来讨论一般情形($m \geq n+1$). 我们给出两个算法,均收敛且效果好.

算法1

(i) 在 M 中任取 $n+1$ 个点
$$x_1^{(0)} < x_2^{(0)} < \cdots < x_{n+1}^{(0)}$$
作为偏差点的初始估计,令 $k=0$.

(ii) 解方程组
$$F(A, x_i^{(k)}) = f(x_i^{(k)}) + (-1)^i \rho^{(k)} \quad (i=1,2,\cdots,n+1)$$
得 $A^{(k)} \in \mathscr{D}$ 与 $\rho^{(k)}$.

(iii) 计算误差
$$\Delta_k(x_j) = F(A^{(k)}, x_i) - f(x_i) \quad (i=1,2,\cdots,m)$$
从中选取一个满足
$$|\Delta_k(x_i)| = \max_{1 \leq i \leq m} |\Delta_k(x_i)|$$
的 x_j,用 x_j 换掉 $x_1^{(k)} < x_2^{(k)} < \cdots < x_{n+1}^{(k)}$ 中的一个点,使得在重新得到的点 $x_1^{(k+1)} < x_2^{(k+1)} < \cdots < x_{n+1}^{(k+1)}$ 上, $\Delta(x_i^{(k+1)})$ 仍然保持正负交替.

(iv) 用 $x_i^{(k+1)}$ 代替 $x_i^{(k)}(i=1,2,\cdots,n+1)$,然后计算过程重复(ii).

定理 1.5 不管偏差点的初始估计如何选取,经算法1迭代有限步后,都能使 $F(A^{(k)}, x)$ 收敛到 $f(x)$ 的唯一最优逼近 $F(A^*, x)$,而且 $|\rho^{(k)}|$ 严格递增且到达最小偏差 ρ^*.

证明 显然,每次迭代总有

附录 I 关于非线性 Tschebyscheff 逼近的几点注记

$$\max_{1\leq i\leq m}|\Delta_k(x_i)|\geq|\rho^{(k)}|$$

若其中等号成立,根据定理 1.3,$F(A^{(k)},x)$ 已经是 $f(x)$ 的最优逼近. 若不等号成立,根据定理 1.4,经算法 1 迭代一步,应该有

$$|\rho^{(k+1)}|>|\rho^{(k)}| \qquad ㉙$$

式㉙告诉我们,在迭代过程中,偏差点的估计不会重复出现,而从 m 中取出 $n+1$ 个组合只有有限种可能,故迭代必然在进行有限步后达到

$$\max_{1\leq i\leq m}|\Delta_k(x_i)|=|\rho^{(k)}|$$

这时 $|\rho^{(k)}|=\rho^*$,$F(A^{(k)}x)\equiv F(A^*,x)$. 由定理 1.2 知,这种极限是唯一的,不依赖偏差点初始估计的选取. 定理证毕.

从上述证明,我们看到了下列有趣的结论.

定理 1.6 函数 $f(x)$ 在整个集合 M 上的最优逼近正好是 $f(x)$ 在使偏差达到最大的 $n+1$ 个点上的最优逼近.

算法 2

只需将算法 1 的第(iii)步改写如下:

(iii′) 在 M 中选取 $\Delta_k(x)=F(A^{(k)},x)-f(x)$ 的 $n+1$ 个正负相间的极值点 $x_1^{(k+1)}<x_2^{(k+1)}<\cdots<x_{n+1}^{(k+1)}$,使满足

$$\max_{1\leq j\leq n+1}|\Delta_k(x_j^{(k+1)})|=\max_{1\leq i\leq m}|\Delta_k(x_i)|$$

和

$$\min_{1\leq j\leq n+1}|\Delta_k(x_j^{(k+1)})|\geq|\rho^{(k)}|$$

由定理 1.4 显见,这种算法在迭代时也能保证

$$|\rho^{(k+1)}|>|\rho^{(k)}|$$

推理过程同定理 34,算法 2 也在迭代有限步后收敛.

上述两种算法与求线性 Tschebyscheff 逼近的交换

法和 Peuez 第二算法类似. 实际计算表明, 对非线性 Tschebyscheff 逼近, 也很有实用价值.

参 考 文 献

[1] MOTZKIN T S. Approximation by Curves of a Unisolvent Family[J]. Bull. Amer. Math. Soc. ,1949, 55(8):789-793.

[2] TORNKEIM L T. On N-Parameter Family of Functions and Associated Convex Functions[J]. Trans. Amer. Math. Soc. ,1950,69(3):457−467.

[3] 阿赫叶惹尔. 逼近论讲义[M]. 北京:科学出版社, 1957.

几个多项式问题

附录 II

1. 全 k 次方值蕴涵 k 次方式

给定正整数 $k \geq 2$ 和实系数多项式 $f(x)$，如果对每个充分大的正整数 n，$f(n)$ 都是整数的 k 次方，则必有整值多项式 $g(x)$，使得 $f(x) = g(x)^k$.

证明 不妨设 $\deg f = m \geq 1$，$f(x) = \sum_{i=0}^{m} a_i x^{m-i}$，记 $r = \left[\dfrac{m}{k}\right]$，有

$$u(x) = \sqrt[k]{f(x)} = \sqrt[k]{a_0 x^m} \sqrt[k]{1 + \dfrac{a_1}{a_0 x} + \cdots}$$

（当 k 为偶数时取算术根）
利用函数 $(1 + c_1 t + \cdots)^{1/k}$ 在 $t = 0$ 处的麦克劳林（Maclaurin）展开，我们有

Tschebyscheff 逼近定理

$$u(x) = \sum_{i=0}^{r} a_i^{(0)} x^{\frac{m}{k}-i} + \varepsilon$$

其中 ε 表示 $x \to \infty$ 时的无穷小量.

注意到对 $t > 0$,差分函数

$$\Delta x^t = (x+1)^t - x^t = x^t((1+\frac{1}{x})^t - 1) = \sum_{i=1}^{[t]} c_i x^{t-i} + \varepsilon$$

故 $\quad \Delta u(x) = \sum_{i=1}^{r} a_i^{(1)} x^{\frac{m}{k}-i} + \varepsilon$

归纳可证 $\quad \Delta^j u(x) = \sum_{i=j}^{r} a_i^{(j)} x^{\frac{m}{k}-i} + \varepsilon \quad (j \geq 1)$

最后有 $\quad \Delta^{r+1} u(x) = \varepsilon$

也就是 $\quad \lim\limits_{x \to +\infty} \Delta^{r+1} u(x) = 0$

现在根据已知条件,对充分大的正整数 n,$u(n)$ 都是整数,由差分的定义 $\Delta f(n) = f(n+1) - f(n)$ 可知,$\Delta^j u(n)$ 都是整数,特别的,$\Delta^{r+1} u(n)$ 都是整数. 再由上面的极限式得到:存在 n_0,使得当 $n \geq n_0$ 时,$\Delta^{r+1} u(n) = 0$. 于是存在 (至多) r 次的整值多项式 $g(x)$,使得当 $n \geq n_0$ 时,$u(n) = g(n)$,从而

$$f(n) = u(n)^k = g(n)^k$$

最后,由于 f 与 g^k 都是多项式,根据多项式恒等定理得到 $f(x) = g(x)^k$.

单壿老师在《解题研究》(南京师范大学出版社,2002)一书中用初等方法证明了 $k = m = 2$ 的情形. 对一般的 k, m 恐怕很难用初等方法证明.

2. Tschebyscheff 多项式引申出的几个问题

n 次 Tschebyscheff 多项式 $T_n(x)$ 归纳定义为：$T_0(x)=1, T_1(x)=x$，而对于 $n \geqslant 2$，有
$$T_n(x) = 2xT_{n-1}(x) - T_{n-2}(x)$$
容易归纳证明：当 $n \geqslant 1$ 时，$T_n(x)$ 是首项系数为 2^{n-1} 的 n 次整系数多项式.

归纳可证 $T_n(\cos\theta) = \cos(n\theta)$. 由此可知，当 $|x| \leqslant 1$ 时，$|T_n(x)| \leqslant 1$. $T_n(x)$ 有 n 个不同实根 $u_k = \cos\dfrac{(2k+1)\pi}{2n} (0 \leqslant k \leqslant n-1)$ 和 $n-1$ 个极值点 $v_k = \cos\dfrac{k\pi}{n}(1 \leqslant k \leqslant n-1)$，$T_n(v_k) = (-1)^k$，它们在 $[-1,1]$ 中交错分布：$v_0 > u_0 > v_1 > u_1 > \cdots$

同样归纳可证 $T_n(\dfrac{u+u^{-1}}{2}) = \dfrac{1}{2}(u^n + u^{-n})$，由此有当 $|x| > 1$ 时，$|T_n(x)| > 1$.

利用特征方程可求出显式
$$T_n(x) = \dfrac{1}{2}((x+\sqrt{x^2-1})^n + (x-\sqrt{x^2-1})^n) = \sum_{k=0}^{[\frac{n}{2}]} C_n^{2k}(x^2-1)^k x^{n-2k}$$

Tschebyscheff 定理 给定正整数 n、区间 $[\alpha,\beta]$ 和非零实数 a，首项系数为 a 的 n 次实系数多项式 $f(x)$ 在区间 $[\alpha,\beta]$ 上的振幅
$$M(f,[\alpha,\beta]) = \max_{x \in [\alpha,\beta]} |f(x)| \geqslant \dfrac{|a|(\beta-\alpha)^n}{2^{2n-1}}$$
达到最小值的多项式只有一个

Tschebyscheff 逼近定理

$$f(x) = \frac{a(\beta-\alpha)^n}{2^{2n-1}} T_n\left(\frac{2x-\alpha-\beta}{\beta-\alpha}\right) =$$

$$\frac{a}{2^{n-1}} \sum_{k=0}^{\left[\frac{n}{2}\right]} C_n^{2k}(x-\alpha)^k(x-\beta)^k\left(x-\frac{\alpha+\beta}{2}\right)^{n-2k}$$

令 A 为实系数多项式的一个集合,当 f 取遍 A 时, $M(f,[\alpha,\beta])$ 的最小值记做 $A([\alpha,\beta])$. 达到这个最小值的 f 称为极值函数.

Tschebyscheff 定理的结论及证明方法对于研究多项式族在区间上振幅的极值问题有重要的启发. 它的基本证法是先猜测构造极值函数,然后用同一法给予证明. 证明中用到多项式的唯一性定理(又称恒等定理):如两个至多 n 次的多项式在 $n+1$ 个不同点处的值分别相等,则它们必恒等. 循着这一思路,改变问题的条件和提法,可以引申出一些有趣的极值问题.

引申问题 1 给定正整数 n、区间 $[\alpha,\beta]$ 以及不属于 $[\alpha,\beta]$ 的实数 u,令 $A = \{f(x) | \deg f \leq n, M(f,[\alpha,\beta]) \leq 1\}$. f 取遍 A 时求 $|f(u)|$ 的最大值.

解 记

$$v_k = \frac{\alpha+\beta}{2} + \frac{\beta-\alpha}{2}\cos\frac{k\pi}{n} \quad (0 \leq k \leq n)$$

由拉格朗日插值公式

$$f(x) = \sum_{k=0}^{n} f(v_k) \prod_{i \neq k} \frac{u - v_i}{v_k - v_i}$$

注意到所有的 $u - v_i$ 彼此同号(因为 v_i 都在区间 $[\alpha,\beta]$ 中,而 u 在区间外), v_k 关于 k 递减,且 $|f(v_k)| \leq 1$,故

$$|f(u)| \leq \sum_{k=0}^{n} |f(v_k)| \prod_{i \neq k} \left|\frac{u - v_i}{v_k - v_i}\right| \leq$$

附录 Ⅱ　几个多项式问题

$$\left|\sum_{k=0}^{n}(-1)^{k}\prod_{i\neq k}\frac{u-v_{i}}{v_{k}-v_{i}}\right|=$$

$$\left|\sum_{k=0}^{n}T_{n}(\cos\frac{k\pi}{n})\prod_{i\neq k}\frac{\frac{2u-\alpha-\beta}{\beta-\alpha}-\cos\frac{i\pi}{n}}{\cos\frac{k\pi}{n}-\cos\frac{i\pi}{n}}\right|=$$

$$\left|T_{n}(\frac{2u-\alpha-\beta}{\beta-\alpha})\right|$$

（最后一步是对 $T_n(x)$ 用拉格朗日插值公式）这个上界又可写成

$$\frac{2^{n}}{(\beta-\alpha)^{n}}\sum_{k=0}^{n}C_{n}^{2k}(u-\alpha)^{k}(u-\beta)^{k}\left|u-\frac{\alpha+\beta}{2}\right|^{n-2k}$$

当且仅当 $f(v_k)=(-1)^k(0\leqslant k\leqslant n)$ 或 $f(v_k)=(-1)^{k+1}(0\leqslant k\leqslant n)$ 时，$f(x)$ 为极值函数．由于 $n+1$ 个不同点处的值唯一确定 n 次多项式，故极值函数只有两个，即

$$g(x)=\pm T_{n}(\frac{2x-\alpha-\beta}{\beta-\alpha})$$

本题也可用同一法讨论．设 $f\in A$ 使 $|f(u)|$ 不小于上面给出的值，选取 g 的符号使 $g(u)$ 与 $f(u)$ 同号．不妨设 $u>\beta$ 且 $f(u)>1$．作 $\varphi(x)=g(x)-f(x)$．注意到

$$g(u)\leqslant f(u),\ g(v_{k})=(-1)^{k},\ |f(v_{k})|\leqslant 1$$

就得到：$\varphi(x)$ 在 $v_n,v_{n-1},\cdots,v_0,u$ 处交替变号，而且极值点 $v_k(1\leqslant k\leqslant n-1)$ 若是 $\varphi(x)$ 的根就是二重根，又如果 $\varphi(x)$ 在 (v_1,u) 中只有一个根 v_0，它也是二重的．因此 $\varphi(x)$ 至少有 $n+1$ 个实根（计算重数），故 $\varphi(x)=0$，即 $f=g$．

引申问题 2　给定正整数 n、区间 $[\alpha,\beta]$ 和非零实数 c，$A=\{a_nx^n+\cdots+a_1x+c\mid a_n,\cdots,a_1\in\mathbf{R}\}$，求 $A([\alpha,$

Tschebyscheff 逼近定理

$\beta])$.

解 (1) 如 $0 \in [\alpha, \beta]$,则 $M(f, [\alpha, \beta]) \geqslant |f(0)| = |c|$,$f(x) = c$ 达到,故 $A([\alpha, \beta]) = |c|$.

(2) 设 $0 \notin [\alpha, \beta]$,即 $\alpha\beta > 0$,此时 $\left|\dfrac{\alpha+\beta}{\beta-\alpha}\right| > 1$,可令

$$\frac{\alpha+\beta}{\beta-\alpha} = \left(\frac{u+u^{-1}}{2}\right), \quad u = \frac{\alpha+\beta+2\sqrt{\alpha\beta}}{\beta-\alpha}$$

猜测极值函数为 $g(x) = mT_n\left(\dfrac{\alpha+\beta-2x}{\beta-\alpha}\right)$. 代入 $x=0$,求 m,则

$$g(0) = c = mT_n\left(\frac{u+u^{-1}}{2}\right) = m\left(\frac{u^n+u^{-n}}{2}\right)$$

故

$$m = \frac{2c}{u^n + u^{-n}} = \frac{2c(\beta-\alpha)^n}{(\alpha+\beta+2\sqrt{\alpha\beta})^n + (\alpha+\beta-2\sqrt{\alpha\beta})^n} = \frac{c(\beta-\alpha)^n}{\displaystyle\sum_{i=0}^{[\frac{n}{2}]} C_n^{2i}(4\alpha\beta)^i(\alpha+\beta)^{n-2i}}$$

而

$$g(x) = mT_n\left(\frac{\alpha+\beta-2x}{\beta-\alpha}\right) = \frac{c(-2)^n}{\displaystyle\sum_{i=0}^{[\frac{n}{2}]} C_n^{2i}(4\alpha\beta)^i(\alpha+\beta)^{n-2i}} \cdot$$

$$\sum_{i=0}^{[\frac{n}{2}]} C_n^{2k}(x-\alpha)^k(x-\beta)^k\left(x-\frac{\alpha+\beta}{2}\right)^{n-2k}$$

由以上构造自然有

$$M(g,[\alpha,\beta]) = |m| = \frac{|c|(\beta-\alpha)^n}{\sum_{i=0}^{[\frac{n}{2}]} C_n^{2i}(4\alpha\beta)^i |\alpha+\beta|^{n-2i}} \quad (g \in A)$$

设有 $f \in A$ 使得 $M(f,[\alpha,\beta]) \le |m|$,作 $\varphi(x) = g(x) - f(x)$,则

$$\varphi(0) = 0, \quad g(v_k) = (-1)^k m, \quad |f(v_k)| \le |m|$$

故对 $0 \le k \le n, \varphi(v_k)$ 的值交替地 ≥ 0 和 ≤ 0,而且在极值点 ($1 \le k \le n-1$) 处若有 $\varphi(v_k) = 0, v_k$ 必是重根. 因此 $\varphi(x)$ 至少有 $n+1$ 个实根(计算重数), 但 $\deg \varphi(x) \le n$, 故 $\varphi(x) = 0$, 即 $f = g$.

这样就用同一法证明了 $A([\alpha,\beta]) = |m|$,而且极值函数只有一个,即 $g(x)$.

顺便指出: $0 \in [\alpha,\beta]$ 时的极值函数不唯一,事实上,任意取定正整数 $n \ge 2$ 和 k ($1 \le k \le n-1$),令 $f(x) = (-1)^k c T_n(tx + \cos\frac{k\pi}{n})$. 显然 $f \in A$, 只要选取 t 使得

$$t\alpha, t\beta \in [-1-\cos\frac{k\pi}{n}, 1-\cos\frac{k\pi}{n}]$$

就有 $M(f,[\alpha,\beta]) = |c|$. 符合要求的 t 有无穷多个.

引申问题 3 给定正整数 $n \ge 2$、正数 m 以及非零实数 b, c ($m \ge |c|$). $A = \{a_n x^n + \cdots + a_2 x^2 + bx + c \mid a_n, \cdots, a_2 \in \mathbf{R}\}$. 求最大的 β, 使得 $A([0,\beta]) \le m$.

解 猜测

$$g(x) = \varepsilon m T_n(px + q) \quad (\varepsilon = \pm 1)$$
$$c = g(0) = \varepsilon m T_n(q)$$

Tschebyscheff 逼近定理

记 $\theta = \arccos \dfrac{c}{\varepsilon m}$,则 $q = \cos \dfrac{\theta}{n}$. 再令 $p\beta + q = -1$,得到

$$p = -\dfrac{1+\cos \dfrac{\theta}{n}}{\beta}$$

$$b = g'(0) = \varepsilon m(-\sin \theta) n(-\dfrac{1}{\sqrt{1-q^2}}) p =$$

$$-\varepsilon \dfrac{n\sqrt{m^2-c^2}}{\beta} \cos \dfrac{\theta}{2n}$$

故 $\varepsilon = -\operatorname{sgn} b$.

(1)当 $m > |c|$ 时,$\beta = \dfrac{n\sqrt{m^2-c^2}}{|b|} \cos \dfrac{\theta}{2n}$,显然 $g \in A$ 且 $M(g, [0, \beta]) = m$.

如果有 $f \in A$ 使 $M(f, [0, \beta]) \leqslant m$,作 $\varphi(x) = g(x) - f(x)$. 则 $x^2 | \varphi(x)$,$\varphi(x)$ 有二重根 $x = 0$,对于 $1 \leqslant k \leqslant n$,$\varphi(x)$ 在 $x_k = \dfrac{\cos \dfrac{k\pi}{n} - q}{p} \in (0, \beta]$ 处交替变号,而且极值点如果是根必为重根,故 $\varphi(x)$ 至少有 $n+1$ 个实根(计算重数),但 $\deg \varphi(x) \leqslant n$,因此 $\varphi(x) = 0$,$f = g$.

注意到 $x > \beta$ 时,$px + q < -1$,$|g(x)| > m$,于是当 $m > |c|$ 时,所求的最大 β 是

$$\dfrac{n\sqrt{m^2-c^2}}{|b|} \cot \dfrac{\theta}{2n}$$

其中 $\theta = \arccos(-\operatorname{sgn} b \dfrac{c}{m})$,并且极值函数只有一个

$$g(x) = -(\operatorname{sgn} b) m T_n(\cos \dfrac{\theta}{n} - \dfrac{1+\cos \dfrac{\theta}{n}}{\beta} x)$$

(2) 当 $m=|c|$ 且 $bc>0$ 时，$f(0+)>m$，故 $\beta=0$.

(3) 当 $m=|c|$ 且 $bc<0$ 时
$$\theta=0$$
$$g(x)=cT_n(1-\frac{2x}{\beta})=$$
$$\frac{c}{\beta^n}\sum_{k=0}^{[\frac{n}{2}]}C_n^{2k}(4x)^k(x-\beta)^k(\beta-2x)^{n-2k}$$
$$b=\frac{c}{\beta^n}(-2n\beta^{n-1}+4C_n^2(-\beta)\beta^{n-2})=-\frac{2n^2c}{\beta}$$

故 $\beta=-\dfrac{2n^2c}{b}$，和（1）取 $\theta\to 0$ 的极限相同. 而且（1）的唯一性讨论也成立. 因此这时极值函数也唯一.

笔者曾为1998年全国高中数学联赛提供试题：

设函数 $f(x)=ax^2+8x+3$ $(a<0)$ 对于给定的负数 a，有一个最大的正数 $l(a)$ 使得在整个区间 $[0,l(a)]$ 上，不等式 $|f(x)|\leq 5$ 都成立.

问：a 为何值时 $l(a)$ 最大？求出这个最大的 $l(a)$. 证明你的结论.

它就是引申问题3的特例：$n=2,b=8,c=3,m=5$. 此时

$$\cos\theta=-\frac{3}{5},\ \cos\frac{\theta}{4}=\frac{\sqrt{5}+1}{2},\ \beta=\frac{\sqrt{5}+1}{2}$$
$$g(x)=-5T_2(\frac{1-2x}{\sqrt{5}})=-8x^2+8x+3$$

公布的官方解答和学生实际作的解答都是讨论二次函数区间极值的常规方法，要求较强的运算能力. 下面给出用同一法的讨论.

猜测极值函数 $\qquad g(x)=a(x-x_0)^2+5$

代入 $g(0) = 3 = ax_0^2 + 5$, $b = 8 = -2ax_0$, 解得 $x_0 = \frac{1}{2}$, $a = -8$.

令 $g(l(a)) = -8(l(a) - \frac{1}{2})^2 + 5 = -5$, 得 $l(a) = \frac{\sqrt{5}+1}{2}$. 由二次函数的图像可知

$$g(x) = -8x^2 + 8x + 3 = -8(x - \frac{1}{2})^2 + 5$$

在区间 $[0, \frac{\sqrt{5}+1}{2}]$ 中满足 $|g(x)| \leq 5$. 且当 $x > \frac{\sqrt{5}+1}{2}$ 时, $g(x) < -5$.

设有 $f(x) = ax^2 + 8x + 3$ 在区间 $[0, u]$ 中满足 $|f(x)| \leq 5$, 其中 $u \geq \frac{\sqrt{5}+1}{2}$. 令

$$\varphi(x) = f(x) - g(x) = (a+8)x^2$$

由于 $f(\frac{1}{2}) \leq 5 = g(\frac{1}{2})$, $f(\frac{\sqrt{5}+1}{2}) \geq -5 = g(\frac{\sqrt{5}+1}{2})$, 故 $\varphi(\frac{1}{2}) \leq 0$, $\varphi(\frac{\sqrt{5}+1}{2}) \geq 0$, $\varphi(x)$ 在 $[0, \frac{\sqrt{5}+1}{2}]$ 中有一个根, 又有二重根 $x = 0$, 所以 $\varphi(x)$ 恒等于 0, 即 $f = g$, 而且 $u = \frac{\sqrt{5}+1}{2}$.

因此只有 $a = -8$ 时给出最大的 $l(a) = \frac{\sqrt{5}+1}{2}$.

3. 二次函数的几个问题

1. 给定非零实数 b 和区间 $[\alpha, \beta]$, $A = \{ax^2 + bx +$

$c \mid a, c \in \mathbf{R}\}$. 求 $A([\alpha, \beta])$.

解 令

$$g(x) = mT_2\left(\frac{2x-\alpha-\beta}{\beta-\alpha}\right) = m\left(2\left(\frac{2x-\alpha-\beta}{\beta-\alpha}\right)^2 - 1\right) =$$

$$\frac{m}{(\beta-\alpha)^2}(8x^2 - 8(\alpha+\beta)x + \alpha^2 + 6\alpha\beta + \beta^2)$$

当 $\alpha+\beta \neq 0$ 时,应取

$$m = -\frac{b(\beta-\alpha)^2}{8(\alpha+\beta)} \cdot g(x) =$$

$$-\frac{b}{\alpha+\beta}\left(x^2 - (\alpha+\beta)x + \frac{\alpha^2+6\alpha\beta+\beta^2}{8}\right)$$

$\varphi(x) = g(x) - f(x)$ 在 $[\alpha, \frac{\alpha+\beta}{2}]$,$[\frac{\alpha+\beta}{2}, \beta]$ 中各有一个根. 又因 $\varphi(x)$ 为偶函数(不含一次项),它的根成对互为相反数.

(1) 当 $\frac{\alpha+\beta}{2} \geq -\alpha$ (即 $3\alpha+\beta \geq 0$, 此时 $\alpha+\beta > 0$) 或 $\frac{\alpha+\beta}{2} \leq -\beta$ (即 $\alpha+3\beta \leq 0$, 此时 $\alpha+\beta < 0$) 时, $\varphi(x)$ 有 3 个实根,从而

$$A([\alpha+\beta]) = \frac{|b|(\beta-\alpha)^2}{8|\alpha+\beta|}$$

且极值函数只有一个,即 $g(x)$.

(2) 当 $-\alpha/3 < \beta < -3\alpha$ (此时 $\alpha < 0 < \beta$) 时,令 α,β 中绝对值较小者为 u, 由于 $u, -u \in [\alpha, \beta]$, 故对任意 $f \in A$, 有

$$M(f,[\alpha,\beta]) \geq \max\{|au^2+c+bu|, |au^2+c-bu|\} =$$

$$|au^2+c| + |bu| \geq |bu|$$

极值函数 g 应当满足 $au^2+c = 0$, 即 $g(x) = ax^2 + bx - au^2$.

Tschebyscheff 逼近定理

此时 $g(u) = bu, g(-u) = -bu$,故 g 的顶点 $-\dfrac{b}{2a}$ 不能在区间内.

如果 $\beta \neq -\alpha$, $-u$ 是区间的内点,应当为顶点,唯一的极值函数 $g(x) = \dfrac{b}{2u}(x^2 + 2ux - u^2)$ 达到 $A([\alpha, \beta]) = |bu| = |b| \cdot \{-\alpha, \beta\}$.

如果 $\beta = -\alpha$, $g(x) = ax^2 + bx - a\beta^2$ 达到 $A([\alpha, \beta]) = |b|\beta$ 当且仅当 $|a| \leqslant \dfrac{|b|}{2\beta}$,此时有无穷多个极值函数.

2. 给定正数 k 和区间 $[m, n]$, $A = \{f(x) = ax^2 + bx + c \mid a, b, c \in \mathbf{R}, M(f, [m, n]) \leqslant k\}$. f 取遍 A 时求 $S = |a| + |b| + |c|$ 的最大值.

解 最早见到这种题目是 1990 年之前的苏联 8~9 年级(相当于我国初中)竞赛题:$f(x) = ax^2 + bx + c$ 在 $[0, 1]$ 上满足 $|f(x)| \leqslant 1$,求 $|a| + |b| + |c|$ 的最大值. 解出后自然想推广到任意的区间 $[m, n]$, m 与 n 同号时较简单.

记
$$u = f(m) = am^2 + bm + c$$
$$v = f(n) = an^2 + bn + c$$
$$w = f\left(\dfrac{m+n}{2}\right) = \dfrac{a(m+n)^2}{4} + \dfrac{b(m+n)}{2} + c$$

联立解出
$$a = \dfrac{2v + 2u - 4w}{(n-m)^2}$$
$$b = \dfrac{-(n+3m)v - (3n+m)u + (4n+4m)w}{(n-m)^2}$$
$$c = \dfrac{(m^2+mn)v + (n^2+mn)u - 4mnw}{(n-m)^2}$$

(1) 若 m 与 n 同号 ($m, n \geq 0$ 或 $m, n \leq 0$). 注意到 $|u|, |v|, |w| \leq k$, 故有

$$|a| \leq \frac{|2v| + |2u| + |4w|}{(n-m)^2} \leq \frac{8k}{(n-m)^2}$$

$$|b| \leq \frac{|(n+3m)v| + |(3n+m)u| + |(4n+4m)w|}{(n-m)^2} \leq \frac{|8n+8m|k}{(n-m)^2}$$

$$|c| \leq \frac{|(m^2+mn)v| + |(n^2+mn)u| + |4mnw|}{(n-m)^2} \leq \frac{(m^2+n^2+6mn)k}{(n-m)^2}$$

于是

$$|a| + |b| + |c| \leq \frac{(m^2+n^2+6mn+8|n+m|+8)k}{(n-m)^2}$$

等号当且仅当 $u = v = -w = \pm k$, 即

$$f(x) = \pm \frac{k}{(n-m)^2} \cdot (8(x - \frac{m+n}{2})^2 - (n-m)^2) =$$

$$\pm \frac{k}{(n-m)^2} \cdot$$

$$(8x^2 - 8(m+n)x + (m^2+n^2+6mn))$$

时成立.

(2) 若 $-m = n \leq 1$, 则

$$a = \frac{v+u-2w}{2n^2}, \quad b = \frac{v-u}{2n}, \quad c = w$$

$$|a| + |b| + |c| \leq \frac{|v+u| + |v-u|n}{2n^2} + \frac{|w|(n^2+1)}{n^2} \leq \frac{(n^2+2)k}{n^2}$$

Tschebyscheff 逼近定理

此处用到

$$|v+u|+|v-u|n = \max\{|v+u+(v-u)n|$$
$$|v+u-(v-u)n|\} = \max\{|(1+n)v+(1-n)u|$$
$$|(1-n)v+(1+n)u|\} \leqslant (1+n+1-n)k = 2k$$

最大值也是在 $u=v=-w=\pm k$，即 $f(x)=\pm\dfrac{k}{n^2}\cdot(2x^2-n^2)$ 时达到.

（3）若 $-m=n>1$，由于 $\pm f(\pm x)$ 与 $f(x)$ 情况一样，不妨设 $a,b\geqslant 0$. 此时若 $c\geqslant 0$，则

$$|a|+|b|+|c|=a+b+c=f(1)\leqslant k$$

若 $c<0$，记 $t=\dfrac{n-1}{n+1}$，有

$$v-\dfrac{(n+1)^2}{2}\cdot f(-t)=a(n^2-\dfrac{(n-1)^2}{2})+$$
$$b(n+\dfrac{n^2-1}{2})-$$
$$c(\dfrac{(n+1)^2}{2}-1)=$$
$$\dfrac{n^2+2n-1}{2}\cdot(a+b-c)$$

故

$$|a|+|b|+|c|=a+b-c=\dfrac{2v-(n+1)^2\cdot f(-t)}{n^2+2n-1}\leqslant$$
$$\dfrac{(2+(n+1)^2)k}{n^2+2n-1}=\dfrac{(n^2+2n+3)k}{n^2+2n-1}$$

等号成立当且仅当

$$f(x)=\pm\left(\dfrac{k(n+1)^2}{(n^2+2n-1)^2}\right)\cdot\left(2\left(\dfrac{x\pm(n-1)}{n+1}\right)^2-\right.$$

附录Ⅱ 几个多项式问题

$$\left.\frac{(n^2+2n-1)^2}{(n+1)^2}\right)$$

由于 $\frac{n^2+2n+3}{n^2+2n-1}>1$,最大值就是 $\frac{(n^2+2n+3)k}{n^2+2n-1}$.

剩下 $mn<0$ 且 $-m$ 不等于 n 的情形十分棘手.

离散逼近论

附录 Ⅲ

在计算算子 T 的本征值的数值方法中,逼近不总是在 X 中或 X 的子空间 X_n 中定义,它可以在一个与 X 不同的空间中定义. 微分方程的有限差分法就是一个很好的例子,在那里逼近是在某个 $\mathbb{C}^n$ 中定义的. 离散逼近论的这一观点首先得到了发展,因为它对差分方法的研究是最自然的. 我们在下面给出 Stummel(1970,1972)与 Vainikko(1974,1978a)的公理化描述的一个浏览.

1. Banach 空间的离散逼近

X 与 $\mathscr{X}_n$ ($n \in \mathbf{N}$) 是 Banach 空间,分别赋以范数 $\|\cdot\|$ 与 $\|\cdot\|_n$.

记号 在下文中,拉丁字母 $x, y, \cdots$ 表示 X 的元素,希腊字母 $\xi, \eta, \cdots$ 表示 $\mathscr{X}_n$ 的元素.

序列 $\{\xi_n\}_\mathbf{N}$ 与 $\{\eta_n\}_\mathbf{N}$ 是等价的,当且仅当
$$\|\xi_n - \eta_n\|_n \to 0 \quad (n \to \infty)$$
等价关系记作 $\{\xi_n\} \sim \{\eta_n\}$.

令 $\mathscr{X} := \prod_{n \in \mathbf{N}} \mathscr{X}_n$. $\sim$ 是空间 $\mathscr{X}$ 上的一个等价关系,并把等价的有界序列的商空间记作 $\mathscr{Y}$.

定义 空间序列 $\{\mathscr{X}_n\}_\mathbf{N}$ 离散地逼近空间 X,当且仅当存在线性映射 $\mathscr{P}: X \to \mathscr{Y}$ 使得若
$$x \in X \text{ 及 } \{\xi_n\}_\mathbf{N} \in \mathscr{P}x$$
则当 $n \to \infty$ 时,$\|\xi_n\|_n \to \|x\|$.

设 $\mathscr{Y}$ 赋以商范数 $\|\cdot\|_\mathscr{Y}$,定义为
$$\|\zeta\|_\mathscr{Y} = \inf_{\{\xi_n\} \in \zeta} \sup_{x \in \mathbf{N}} \|\xi_n\|_n, \quad \zeta \in \mathscr{Y}$$
因为对所有 $x \in X$,$\|x\| = \|\mathscr{P}x\|_\mathscr{Y}$,所以,如上定义的映射 $\mathscr{P}$ 是 X 到 $\mathscr{Y}$ 中的等距映射.

在每一个等价类 $\mathscr{P}x$ 中选出一个序列 $\{r_n(x)\}_\mathbf{N}$,我们得到以下等价的定义:$\{\mathscr{X}_n\}_\mathbf{N}$ 离散地逼近 X,当且仅当存在(可能地)非线性映射 $r_n: X \to \mathscr{X}_n$,$n \in \mathbf{N}$,使得

(i) 对于任意一个 $x \in X$,当 $n \to \infty$ 时,$\|r_n(x)\|_n \to \|x\|$.

(ii) 对于任意的 $x, y \in X$ 及 $\alpha, \beta \in \mathbb{C}$
$$\|r_n(\alpha x + \beta y) - \alpha r_n(x) - \beta r_n(y)\|_n \to 0$$

r_n 称为限制算子. 由映射 $\mathscr{P}$ 定义的算子 r_n 不是唯一的. 若 $\{r_n\}$ 与 $\{r'_n\}$ 是对应于 $\mathscr{P}$ 的两组限制算子, 则对任一 $x \in X$, 当 $n \to \infty$ 时 $\|r_n(x) - r'_n(x)\|_n \to 0$. r_n 不必线性, 仅仅渐近于线性, 即满足(ii).

若 $r_n \in \mathscr{L}(X, \mathscr{X}_n)$, 则 r_n 是一致有界的. 但是, 即使存在一个线性的 r_n, 正如我们将在例 3.2 中所看到的, 未必是最易于利用的一个.

例 3.1 设 Ω 是 $\mathbf{R}^k$ 中的有界域, $X = C(\Omega)$. 并设 Ω_n 是在 Ω 内的网格点 $\{t_1^{(n)}, t_2^{(n)}, \cdots, t_{N(n)}^{(n)}\}$ 的集合, 使得任意 $t \in \Omega$ 到 Ω_n 的距离趋向于 0 (当 $n \to \infty$). 我们定义 $\mathscr{X}_n = \mathbb{C}^{N(n)}$, 赋以最大范数 $\|\cdot\|_n$, 而且

$$r_n : x \in X \mapsto r_n(x) = (x(t_1^{(n)}), \cdots, x(t_{N(n)}^{(n)}))^T$$

r_n 是在 X 上线性以及对于所有连续的 x, $\|r_n(x)\|_n \to \|x\|$.

例 3.2 在实践中, 常常存在着与离散化方法内在联系的十分"自然的"限制, 它在稠密于 X 的子空间 X' 上线性, 并且使得对于任一 $x \in X'$, 成立 $\|r_n(x)\|_n \to \|x\|$. 相应的映射 $\mathscr{P}$, 它是 X' 到 $\mathscr{Y}$ 中的等距映射, 而且能够唯一地延拓到 X 到 $\mathscr{Y}$ 中的等距离映射. 然而, 延拓的限制 r_n 通常仅是在 X 上渐近线性的.

我们以 $X = L^2(\Omega)$ 给出一个例子. 正如我们在例 3.1 所见到的, 自然限制 r_n 作为像近似求积或有限差分那样的离散化方法来考虑, 它是一个算子, 把连续函数映射到这个连续函数在网格点取值的向量. 但是, r_n 作为 $L^2(\Omega)$ 中的算子不是有界的.

我们考虑 $\mathscr{X}_n = \mathbb{C}^{N(n)}$, 具有由下式定义的范数

$\|\cdot\|_n$.

$$\|\xi_n\|_n : \left(\sum_{i=1}^{N(n)} W_{in} |\xi_{in}|^2\right)^{1/2}$$

它对于所有的连续函数 x，与收敛于 $\int_\Omega x(t)\,dt$ 的求积公式 $\sum_{i=1}^{N(n)} W_{in} x(t_i^{(n)})$ 相联系。所以，对于 $x \in C(\Omega)$，$\|r_n(x)\|_n \to \|x\|$，因而子空间 X' 是 $C(\Omega)$。

正如已经指出的，对于 $x \notin X'$，$r_n(x)$ 的值可以通过 $\mathscr{P}$ 来定义，而对于 $\mathscr{P}$ 也对应一个限制 r_n'，r_n' 在 X 上是线性的。设 $\{\sigma_i^{(n)}\}_1^{N(n)}$ 是 Ω 的基本子集，$t_i^{(n)} \in \sigma_i^{(n)}$。于是

$$r_n' : x \mapsto (r_n'x)_i = \frac{1}{\operatorname{meas} \sigma_i^{(n)}} \int_{\sigma_i^{(n)}} x(t)\,dt$$

$$i = 1, \cdots, N(n)$$

可以证明：对 $x \in L^2(\Omega)$，$\|r_n'x\|_n \to \|x\|$，以及对 $x \in X'$，$\|r_n(x) - r_n'x\|_n \to 0$。所以，$r_n$ 及 r_n' 都相应于 X' 中相同的 $\mathscr{P}$。而且

$$r_n' \in \mathscr{L}(X, \mathscr{X}_n)$$

对于 $x \notin X'$，我们何以令 $r_n(x) := r_n'x$。

例 3.3 在 Aubin(1972,P.5) 中，一个略为不同的观点是不仅给出线性限制算子 $r_n \in \mathscr{L}(X, \mathscr{X}_n)$ 的应用，而且也给出线性拓展算子 $p_n \in \mathscr{L}(\mathscr{X}_n, X)$。$p_n$ 是从 $\mathscr{X}_n$ 到 X 的闭子空间 X_n 的同构映射。p_n 与 r_n 由性质

$$p_n r_n x \to x \quad (x \in X)$$

相联结。

作为一个例子，我们选取 $X = C(a,b)$（参阅第三章例 3.1）。把 $[a,b]$ 分为 $n-1$ 个区间，分点为 $t_i^{(n)}$，$i = 1, \cdots, n$，$t_1^{(n)} = a$，$t_n^{(n)} = b$。$\mathscr{X}_n = \mathbb{C}^n$ 赋以最大范数，定义

$$r_n : x \mapsto (x(t_1^{(n)}), x(t_2^{(n)}), \cdots, x(t_n^{(n)}))^T$$

Tschebyscheff 逼近定理

$p_n : (x(t_i^{(n)})) \mapsto$ 在 $(t_i, x(t_i^{(n)}))_1^n$ 处的逐段线性插值. 若设 $\max_{i=1,\cdots,n-1} |t_{i+1}^{(n)} - t_i^{(n)}| \to 0$, 则对 X 中的所有 x, $p_n r_n x \to x$. p_n 是从 $\mathbb{C}^n$ 到逐段线性连续函数的子空间 X_n 的同构映射.

例 3.4 在例 3.3 中有性质 $r_n p_n = 1_n$, 其中 1_n 是 $\mathscr{X}_n$ 上的恒同算子. 因而 $\pi_n = p_n r_n$ 是从 X 到 $X_n = p_n \mathscr{X}_n$ 上的投影, 使得在 X 中 $\pi_n \to 1$ 逐点地成立. 当空间 $\mathscr{X}_n$ 是有限维时, 子空间 X_n 也是有限维的.

2. 闭算子的离散逼近

序列 $\{\xi_n\}_N \{\xi_n \in \mathscr{X}_n\}$ 离散收敛 (简短写为 d-收敛) 于 $x \in X$, 当且仅当 $n \to \infty$ 时, $\|r_n x - \xi_n\|_n \to 0$, 记作 $\xi_n \xrightarrow{d} x$. 序列 $\{\xi_n\}_N \{\xi_n \in \mathscr{X}_n\}$ 是离散相对紧的 (简短写为 d-相对紧), 当且仅当每个子序列 $\{\xi_n\}_{N_1 \subset N}$ 有一个 d-收敛子序列 $\{\xi_n\}_{N_2 \subset N_1}$.

设 T 是 X 中的闭线性算子, 其定义域 $D \subseteq X$. 为了简单起见, 我们假定离散逼近 X 的子空间 $\mathscr{X}_n (n \in \mathbf{N})$ 是有限维的. 这几乎是实用中经常出现的情形. $\mathscr{T}_n$ 是从 $\mathscr{X}_n$ 到其自身的线性算子, 其定义域 $\mathscr{D}_n \subseteq \mathscr{X}_n$. 我们假定, 对所有 n, $r_n D \subseteq \mathscr{D}_n$. 若 T (或 $\mathscr{T}$) 在 X (或 $\mathscr{X}_n$) 上有界, 则 $D = X$ (或 $\mathscr{D}_n = X_n$).

现在, 我们定义 $\{\mathscr{T}_n\}_N$ 趋向于 T 的各种类型的收敛性, 它们是:

离散逐点收敛 $\mathscr{T}_n \xrightarrow{d-p} T$, 当且仅当对 D 中所有的 x

$$\|r_n T x - \mathscr{T}_n r_n x\|_n \to 0, \text{当} n \to \infty$$

离散稳定收敛 $\mathscr{T}_n \xrightarrow{d\text{-}s} T$,当且仅当

(i) $\mathscr{T}_n \xrightarrow{d\text{-}p} T$;

(ii) $\exists N$:对 $n>N, \mathscr{T}_n^{-1} \in \mathscr{L}(\mathscr{X}_n)$ 且 $\|\mathscr{T}_n^{-1}\| \leqslant M$.

对于 T 是紧的,离散紧收敛 $\mathscr{T}_n \xrightarrow{d\text{-}c} T$,当且仅当

(i) $\mathscr{T}_n \xrightarrow{d\text{-}p} T$;

(ii) 对任意一个满足 $\xi_n \in \mathscr{X}_n$,且 $|\xi_n|_n \leqslant C$ 的序列 $\{\xi_n\}_{\mathbf{N}}$,序列 $\{\mathscr{T}_n \xi_n\}_{\mathbf{N}}$ 是 d-相对紧的.

离散正则收敛 $\mathscr{T}_n \xrightarrow{d\text{-}r} T$,当且仅当

(i) $\mathscr{T}_n \xrightarrow{d\text{-}p} T$;

(ii) 任意一个序列 $\{\xi_n\}_{\mathbf{N}}$,其中 $\xi_n \in \mathscr{D}_n$, $\|\xi_n\|_n \leqslant c$,且该序列使得对于 $n \in N_1 \subset \mathbf{N}, \mathscr{T}_n \xi_n \xrightarrow{d} y$,则这个序列自身使得对于 $n \in N_2 \subset N_1, \xi_n \xrightarrow{d} x \in D$,且 $Tx = y$.

Tschebyscheff 正交多项式

附录 IV

1. Tschebyscheff 正交多项式问题

上述用来逼近被观察数据的最小二乘法有以下易见的缺点. 我们假设, 对所得的实验数据已建立了二次多项式 $y = a_0 + a_1 x + a_2 x^2$, 它的系数是用最小二乘法计算出来的, 但是发现了, 用它计算出的值 y 与被观察值有很大差别. 在这种情形下, 必须提高多项式的次数, 我们说提高到三次, 即寻找形如 $y = a_0 + a_1 x + a_2 x^2 + a_3 x^3$ 的多项式.

但是这里不仅要求求出系数 a_3, 而且还要重新计算系数 a_0, a_1, a_2, 因为求这些系数的方程组改变了.

附录 Ⅳ Tschebyscheff 正交多项式

本节与下节叙述的 Tschebyscheff 方法容许大大简化这个程序. 这里以提高次数的各项式之和的形式寻求出近似多项式,并且增加的新被加数不改变以前的系数.

这样逐项增加后,可以看出,用所求公式算出的值与被观察数据的偏差平方和减小了. 因此近似多项式次数的选择也大大简化了.

设作 n 次观察,它们给出结果:

x_1	x_2	x_3	$\cdots$	x_{n-1}	x_n
y_1	y_2	y_3	$\cdots$	y_{n-1}	y_n

我们将寻求含 $m+1$ 个未知系数 $a_0, a_1, \cdots, a_m$ 的 m 次多项式,并且照例设多项式次数比观察次数小得多.

Tschebyscheff 方法的本质是,不是直接以 x 的各幂和的形式,而是以用特殊形式选出的各多项式组合的形式,求出近似多项式. 我们把未知多项式写成形式

$$y = a_0 \varphi_0(x) + a_1 \varphi_1(x) + \cdots + a_m \varphi_m(x) \qquad (300)$$

其中 $\varphi_0(x) = 1, \varphi_1(x) = x + \alpha_1, \cdots$,一般地,$\varphi_e(x)$ 是以下形式的 e 次多项式

$$\varphi_e = x^e + \alpha_e^{(1)} x^{e-1} + \cdots \qquad (301)$$

其最高项系数为 1.

稍迟指出建立多项式 $\varphi_e(x)(e = 0, 1, 2, \cdots, m)$ 的方法. 首先设这些多项式以任一方式选出,开始研究寻求系数 $a_0, a_1, \cdots, a_m$ 的最可能值的问题. 为此必须用最小二乘法寻求以下函数的最小值

$$\Phi = \sum_{i=1}^{n} [y_i - a_0 \varphi_0(x_i) - a_1 \varphi_1(x_i) - \cdots - a_m \varphi_m(x_i)]^2$$

Tschebyscheff 逼近定理

这化为方程组

$$\begin{cases} [\varphi_0,\varphi_0]a_0 + [\varphi_0,\varphi_1]a_1 + \cdots + [\varphi_0,\varphi_m]a_m = [y,\varphi_0] \\ [\varphi_1,\varphi_0]a_0 + [\varphi_1,\varphi_1]a_1 + \cdots + [\varphi_1,\varphi_m]a_m = [y,\varphi_1] \\ \quad\quad\quad\quad\quad\quad\quad\quad\quad \vdots \\ [\varphi_m,\varphi_0]a_0 + [\varphi_m,\varphi_1]a_1 + \cdots + [\varphi_m,\varphi_m]a_m = [y,\varphi_m] \end{cases}$$

我们现在写出展开形式. 因为在本节与以下各节中,符号 $\sum$ 表示从 $i=1$ 到 $i=n$ 的求和,所以为简单起见,不写出求和指标

$$\begin{cases} a_0 \sum [\varphi_0(x_i)]^2 + a_1 \sum \varphi_0(x_i)\varphi_1(x_i) + \\ \quad a_2 \sum \varphi_0(x_i)\varphi_2(x_i) + \cdots + \\ \quad a_m \sum \varphi_0(x_i)\varphi_m(x_i) = \sum y_i\varphi_0(x_i) \\ a_0 \sum \varphi_0(x_i)\varphi_1(x_i) + a_1 \sum [\varphi_1(x_i)]^2 + \\ \quad a_2 \sum \varphi_1(x_i)\varphi_2(x_i) + \cdots + \\ \quad a_m \sum \varphi_1(x_i)\varphi_m(x_i) = \sum y_i\varphi_1(x_i) \\ \quad\quad\quad\quad\quad \vdots \\ a_0 \sum \varphi_0(x_i)\varphi_l(x_i) + a_1 \sum \varphi_1(x_i)\varphi_l(x_i) + \cdots + \\ \quad a_l \sum [\varphi_l(x_i)]^2 + \cdots + a_m \sum \varphi_l(x_i)\varphi_m(x_i) = \\ \quad \sum y_i\varphi_l(x_i) \\ \quad\quad\quad\quad\quad \vdots \\ a_0 \sum \varphi_0(x_i)\varphi_m(x_i) + a_1 \sum \varphi_1(x_i)\varphi_m(x_i) + \cdots + \\ \quad a_m \sum [\varphi_m(x_i)]^2 = \sum y_i\varphi_m(x_i) \end{cases}$$

㉛

方程组 ㉛ 的解也将是多项式 ㉚ 系数的最可能值.

附录 IV Tschebyscheff 正交多项式

我们回忆，我们目前完全可以选择多项式 $\varphi_0(x),\cdots,\varphi_m(x)$. 自然这样选择，使求系数 $a_0,\cdots,a_m$ 的方程组 (3) 可能有更简单的形式. 为此我们将这样选择这些多项式，使得满足条件

$$\begin{cases} \sum \varphi_l(x_i)\varphi_k(x_i) = 0 & (l \neq k) \\ \sum [\varphi_l(x_i)]^2 \neq 0 & (l = 0,1,\cdots,m) \end{cases} \quad ⑶$$

后一条件表示，多项式 $\varphi_e(x)$ 至少在点 $x_1, x_2,\cdots,x_n$ 中一点上不等于零. 这样的多项式称为 Tschebyscheff 正交多项式.

在满足条件 ⑳ 时，方程组 ⑳ 中每个方程的左边只留下一项，所以我们立即可以写出系数表达式

$$a_l = \frac{\sum y_i \varphi_l(x_i)}{\sum [\varphi_l(x_i)]^2} \quad (l = 0,1,\cdots,m) \quad ⑭$$

剩下只要证实，条件 ⑳ 是满足的，并在给定点 $x_1, x_2,\cdots,x_n$ 上求出 Tschebyscheff 正交多项式的表达式.

我们已经取 $\varphi_0(x) = 1$. 在条件 ⑳ 的多项式 $\varphi_1(x)$ 中设 $l = 0, k = 1$, 由此由 ⑳ 得出

$$\sum \varphi_1(x_i) = 0 \quad ⑮$$

因为根据 ⑳, 多项式 $\varphi_1 = x + a_1$, 所以 ⑮ 可以改写为

$$\sum (x_i + \alpha_1) = 0$$

从而

$$\sum x_i + n\alpha_1 = 0 \quad 或 \quad \alpha_1 = -\frac{1}{n}\sum x_i$$

即 a_i 的符号与值 $x_1, x_2,\cdots,x_n$ 的算术平均值的符号相反. 最后

$$\varphi_1(x) = x - \frac{1}{n}\sum x_i \quad ⑯$$

Tschebyscheff 逼近定理

为建立多项式 $\varphi_2(x)$，在 ③ 中依次令 $l = 0, k = 2$，然后令 $l = 1, k = 2$. 我们得出两个方程

$$\begin{cases} \sum \varphi_2(x_i) = 0 \\ \sum \varphi_1(x_i)\varphi_2(x_i) = 0 \end{cases} \quad ③⑦$$

多项式 $\varphi_2(x)$ 是含量高项系数为 1 的二次多项式. 因此可以把它写为

$$\varphi_2(x) = (x + \beta_2)\varphi_1(x) + \gamma_2 \varphi_0(x) \quad ③⑧$$

我们想起 $\varphi_0(x) = 1$. 把表达式 ③⑧ 代入方程组 ③⑦，则有

$$\sum x_i \varphi_1(x_i) + \beta_2 \sum \varphi_1(x_i) + n\gamma_2 = 0$$

$$\sum x_i [\varphi_1(x_i)]^2 + \beta_2 \sum [\varphi_1(x_i)]^2 + \gamma_2 \sum \varphi_1(x_i) = 0$$

或者，考虑到 ③⑤

$$\sum x_i \varphi_1(x_i) + n\gamma_2 = 0$$

$$\sum x_i [\varphi_1(x_i)]^2 + \beta_2 \sum [\varphi_1(x_i)]^2 = 0 \quad ③⑨$$

方程组 ③⑨ 给出

$$\beta_2 = \frac{-\sum x_i [\varphi_1(x_i)]^2}{\sum [\varphi_1(x_i)]^2}, \quad \gamma_2 = -\frac{1}{n}\sum x_i \varphi_1(x_i) \quad ③⑩$$

因此求出了多项式 $\varphi_2(x)$ 的系数 β_2 与 γ_2. 在用公式 ③⑩ 计算时，只要知道 x_i 的各幂之和. 实际上（在计算时考虑到 $\sum \varphi(x_i) = 0$）

$$\begin{cases} \sum x_i \varphi_1(x_i) = \sum x_i(x_i + \alpha_1) = \sum x_i^2 + \alpha_1 \sum x_i \\ \sum [\varphi_1(x_i)]^2 = \sum (x_i + \alpha_1)\varphi_1(x_i) = \sum x_i \varphi_1(x_i) \\ \sum x_i [\varphi_1(x_i)]^2 = \sum (x_i^2 + \alpha_1 x_i)\varphi_1(x_i) = \\ \sum x_i^3 + \alpha_1 \sum x_i^2 + \alpha_1 \sum x_i \varphi_1(x_i) \end{cases} \quad ③⑪$$

附录 Ⅳ Tschebyscheff 正交多项式

于是多项式 $\varphi_0(x), \varphi_1(x), \varphi_2(x)$ 已经建立起来了.

我们现在导出一个递推公式,它容许由上述两个多项式计算以下多项式. 这将证明了所有多项式 $\varphi(x)$ 的存在.

我们首先建立以下一个简单命题.

对于所有的 $k < r$ 和 $\sum x_i^k \varphi_r(x_i) = 0$.

假设所有多项式 $\varphi_0(x), \varphi_1(x), \cdots, \varphi_r(x)$ 已被建立起来了,把它们写成下式

$$\begin{cases} \varphi_0(x) = 1 \\ \varphi_1(x) = x + \alpha_1 \\ \varphi_2(x) = x^2 + \alpha_2^{(1)} + \alpha_2^{(2)} \\ \qquad \vdots \\ \varphi_k(x) = x^k + \alpha_k^{(1)} x^{k-1} + \cdots + \alpha_k^{(k)} \\ \qquad \vdots \\ \varphi_r(x) = x^r + \alpha_r^{(1)} x^{r-1} + \cdots + \alpha_r^{(r)} \end{cases} \quad ⑫$$

由此依次求出

$$x = \varphi_1(x) - \alpha_1 \varphi_0(x)$$
$$x^2 = \varphi_2(x) - \alpha_2^{(1)} \varphi_1(x) + (\alpha_2^{(1)} \alpha_1 - \alpha_2^{(2)}) \varphi_0(x)$$
$$\vdots$$

结果 x^k 可表示为含常系数的多项式 $\varphi_0(x), \varphi_1(x), \cdots, \varphi_k(x)$ 的线性组合形式

$$x^k = \varphi_k(x) + b_{k-1} \varphi_{k-1} + \cdots + b_0 \varphi_0(x)$$

($\varphi_k(x)$ 的系数总是等于1). 但是此时,由于 $\varphi(x)$ 的定义(见 ⑬),故有

$$\sum x_i^k \varphi_r(x_i) = \sum \varphi_k(x_i) \varphi_r(x_i) + b_{k-1} \sum \varphi_{k-1}(x_i) \varphi_r(x_i) + \cdots +$$

Tschebyscheff 逼近定理

$$b_0 \sum \varphi_0(x_i)\varphi_r(x_i) = 0$$

由此立即推出,若 $\psi(x)$ 为次数小于 r 的任一多项式,则

$$\sum \psi(x_i)\varphi_r(x_i) = 0 \qquad ⑬$$

我们现在证明,多项式 $\varphi_{r+1}(x)$ 可以用以下递推公式表示

$$\varphi_{r+1}(x) = (x + \beta_{r+1})\varphi_r(x) + \gamma_{r+1}\varphi_{r-1}(x) \qquad ⑭$$

并导出计算系数 β_{r+1} 与 γ_{r+1} 的公式. 首先显然, $\varphi_{r+1}(x)$ 是含量最高项系数 1 的 $r+1$ 次多项式. 然后,若 $k < r-1$, 则由 ⑬, 有

$$\sum \varphi_k(x_i)\varphi_{r+1}(x_i) = \sum (x_i + \beta_{r+1})\varphi_k(x_i)\varphi_r(x_i) + \gamma_{r+1}\sum \varphi_k(x_i)\varphi_{r-1}(x_i) = 0$$

因为 $(x + \beta_{r+1})\varphi_k(x)$ 是次数不大于 r 的多项式. 因此应当只满足 2 个条件

$$\sum \varphi_{r-1}(x_i)\varphi_{r+1}(x_i) = \sum x_i\varphi_{r-1}(x_i) + \gamma_{r+1}\sum [\varphi_{r-1}(x_i)]^2 = 0$$

$$\sum \varphi_r(x_i)\varphi_{r+1}(x_i) = \sum x_i[\varphi_r(x_i)]^2 + \beta_{r+1}\sum [\varphi_r(x_i)]^2 = 0$$

从而

$$\beta_{r+1} = -\frac{\sum x_i[\varphi_r(x_i)]^2}{\sum [\varphi_r(x_i)]^2}$$

$$\gamma_{r+1} = -\frac{\sum x_i\varphi_{r-1}(x_i)\varphi_r(x_i)}{\sum [\varphi_{r-1}(x_i)]^2} \qquad ⑮$$

附录 Ⅳ Tschebyscheff 正交多项式

从而求出了多项式 $\varphi_{r+1}(x)$ 的系数 β_{r+1} 与 γ_{r+1}. 显然, 公式 ⑩ 是公式 ⑮ 在 $r = 1$ 时的特殊情形.

将公式 ⑮ 的计算再次归结为 x_i 的各幂和的计算. 考虑到公式 ⑫ 与等式 ⑬, 我们求出

$$\begin{cases} \sum [\varphi_r(x_i)]^2 = \sum x_i^r \varphi_r(x_i) = \sum x_i^{2r} + \\ \qquad \alpha_r^{(1)} \sum x_i^{2r-1} + \cdots + \alpha_r^{(r)} \sum x_i^r \\ \sum x_i \varphi_{r-1}(x_i) \varphi_r(x_i) = \sum x_i^r \varphi_r(x_i) \\ \sum x_i [\varphi_r(x_i)]^2 = \sum (x_i^{r+1} + \alpha_r^{(1)} x_i^r) \varphi_r(x_i) = \\ \qquad \sum x_i^{2+1} + \alpha_2^{(1)} \sum x_i^{2r} + \cdots + \\ \qquad \alpha_r^{(r)} \sum x_i^{r+1} + \alpha_r^{(1)} \sum x_i^r \varphi_r(x_i) \end{cases} \quad ⑯$$

显然, 公式 ⑪ 由公式 ⑯ 在 $r = 1$ 时得出. 若值 x_i 形成等差数列 (含固定步长的表), 则计算特别简单, 设 $u = x - x_1$, 得 $u_1 = 0, u_2 = h, \cdots, u_n = (n-1)h$, 则

$$\sum_{i=1}^n u_i^r = h^r \sum_{i=1}^{n-1} i^r$$

可以预先对不同的值 n 计算自然数的幂之和.

若表格没有固定步长, 则选择新变量 $u = x - \bar{x}$ 是方便的. 于是 $\sum u_i = \sum x_i - n\bar{x} = 0$. 此外, 此时 u_i 的幂和的计算被简化了.

2. 用 Tschebyscheff 方法逼近函数

这样, 我们确定了, 在给定点 $x_1, x_2, \cdots, x_n$ 上, Tschebyscheff 正交多项式可以依次建立起来.

若现在要寻找以下形式的近似多项式

Tschebyscheff 逼近定理

$$y = a_0\varphi_0(x) + a_1\varphi_1(x) + \cdots + a_m\varphi_m(x) \quad ⑰$$

则正如上节所确定的那样,系数 $a_0, a_1, \cdots, a_m$ 的最可能值用以下公式求出

$$a_r = \frac{\sum y_i\varphi_i(x_i)}{\sum [\varphi_i(x_i)]^2} \quad (r = 0,1,2,\cdots) \quad ⑱$$

此公式的分母用上节公式 ⑯ 求出,分子利用形如 $\sum y_i x_i^r$ 的和求出

$$\sum y_i \varphi_r(x_i) = \sum y_i x_i^r + \alpha_r^{(1)} \sum y_i x_i^{r-1} + \cdots + \alpha_r^{(r)} \sum y_i \quad ⑲$$

若近似多项式 ⑰ 已经建立起来了,即多项式 $\varphi_0(x), \cdots, \varphi_m(x)$ 与系数 $a_0, a_1, \cdots, a_m$ 已求出了,但它的准确度不能满足我们,则必须求以下的项: $a_{m+1}\varphi_{m+1}(x)$. 为此用上节公式 ⑭ 与 ⑮ 建立多项式 $\varphi_{m+1}(x)$,并用本节公式 ⑱ 计算系数 a_{m+1}. 还要指出,偏差的平方和可以这样计算

$$\sum_{i=1}^n \left[y_i - \sum_{r=0}^m a_r\varphi_r(x_i) \right]^2 =$$
$$\sum_{i=1}^n y_i^2 - 2\sum_{r=0}^m a_r \sum_{i=1}^n y_i\varphi_r(x_i) +$$
$$\sum_{r=0}^m a_r^2 \sum_{i=1}^n [\varphi_r(x_i)]^2 \quad ⑳$$

例 3.1 现在我们考虑用 Tschebyscheff 方法逼近的例子. 根据下表所列的数据,寻求一次与二次多项式,用来逼近 x 与 y 之间的依赖关系式. 这里为了计算,我们列出 x 的幂与乘积.

附录 Ⅳ Tschebyscheff 正交多项式

x_i	y_i	x_i^2	x_i^3	x_i^4	$x_i y_i$	$x_i^2 y_i$	y_i^2
0.154 11	19.47	0.023 75	0.003 66	0.000 56	3.000 52	0.462 41	379.08
0.195 16	21.83	0.038 09	0.007 43	0.001 45	4.260 34	0.831 50	476.55
0.221 43	23.11	0.049 03	0.010 86	0.002 40	5.117 25	1.133 08	534.07
0.288 02	26.11	0.082 96	0.023 89	0.006 88	7.520 20	2.166 09	681.73
0.328 08	27.60	0.107 64	0.035 31	0.011 59	9.055 01	2.970 86	761.76
0.381 83	28.89	0.145 79	0.055 67	0.021 26	11.031 07	4.211 87	834.63
0.455 17	33.17	0.207 18	0.094 30	0.042 92	15.097 99	6.872 16	1 100.25
0.570 12	33.38	0.325 04	0.185 31	0.105 65	19.030 61	10.849 84	114.22
0.759 30	32.31	0.576 54	0.437 76	0.332 39	24.532 98	18.628 01	1 043.94
0.910 75	31.88	0.829 47	0.755 44	0.688 01	29.034 71	26.443 50	1 016.33
1.138 95	25.46	1.297 21	1.477 45	1.682 75	28.997 67	33.026 97	648.21
∑ 5.402 92	303.21	3.682 70	3.087 08	2.895 86	156.678 35	107.596 29	8 590.77

Ⅰ. 一次多项式的建立

因为 $n = 11$，所以由上节公式 ⑯

$$\varphi_1(x) = x - \frac{5.402\,92}{11} = x - 0.491\,17$$

可见 $\alpha_1 = -0.491\,17$.

由本节公式 ⑱

$$\alpha_0 = \frac{\sum y_i}{n} = \frac{303.21}{11} = 27.564\,5$$

为了确定 a_1，首先由上节公式 ⑪ 求出

$$\sum [\varphi_1(x_i)]^2 = \sum x_i^2 + \alpha_1 \sum x_i =$$

$3.682\,70 + (-0.491\,17) \cdot 5.402\,92 = 1.028\,95$

由本节公式 ⑲

$$\sum y_i \varphi_1(x_i) = \sum y_i x_i + \alpha_1 \sum y_i =$$

$156.678\,35 + (-0.491\,17) \times 303.21 = 7.750\,69$

Tschebyscheff 逼近定理

于是

$$a_1 = \frac{7.75069}{1.02895} = 7.5315$$

因此,上表实验数据的一次多项式有形式

$$y = 27.5645 + 7.5315\varphi_1(x) =$$
$$27.5645 + 7.5315(x - 0.49117)$$

最后有

$$y = 23.8653 + 7.5315x$$

Ⅱ. 二次多项式的建立

现在我们转向寻求二次近似多项式. 为此必须求多项式 $\varphi_2(x)$ 与系数 a_2 的表达式. 首先计算 β_2 与 γ_2. 为此用上节公式 ⑪ 求出

$$\sum x_i\varphi_1(x_i) = \sum [\varphi_1(x_i)]^2 = 1.02895$$

(这个和已经被计算出了)

$$\sum x_i[\varphi_1(x_i)]^2 = \sum x_i^3 + \alpha_1 \sum x_i^2 + \alpha_1 \sum x_i\varphi_1(x_i) =$$
$$3.08708 + (-0.49117) \times 3.68270 +$$
$$(-0.49117) \cdot 1.02895 = 0.77286$$

于是按照上节公式 ⑩

$$\beta_2 = -\frac{0.77286}{1.02895} = 0.75112$$

$$\gamma_2 = -\frac{1.02895}{11} = -0.09354$$

因此多项式 $\varphi_2(x)$ 有形式

$$\varphi_2(x) = (x - 0.75112)(x - 0.49117) - 0.09354 =$$
$$x^2 - 1.24229x + 0.27539$$

剩下只要用本节公式 ⑱ 求系数 a_2. 由本节公式 ⑲,得出

附录 Ⅳ　Tschebyscheff 正交多项式

$$\sum y_i \varphi_2(x_i) = 107.596\,29 - 1.242\,29 \times 5.402\,92 +$$
$$0.275\,39 \times 303.21 = -3.542\,63$$

其次用上节公式 ⑯ 求出

$$\sum [\varphi_2(x_i)]^2 = 2.895\,86 - 1.242\,29 \cdot 3.087\,08 +$$
$$0.275\,39 \cdot 3.682\,70 = 0.074\,98$$

于是，对 a_2 求出值

$$a_2 = -\frac{3.542\,63}{0.074\,98} = 47.247\,67$$

因此，二次近似多项式等于

$$y = 27.564\,5 + 66.222\,6x - 47.247\,67x^2$$

这里代入已计算出的 $\varphi_1(x)$ 与 $\varphi_2(x)$ 表达式，我们有

$$y = 27.564\,5 + 7.531\,5(x - 0.491\,17) -$$
$$47.247\,67(x^2 - 1.242\,2x + 0.275\,39)$$

或者作代数变换，有

$$y = 10.853\,7 + 66.222\,6x - 47.247\,67x^2$$

为了估计所求出的多项式逼近实验数据好到什么程度，首先利用均方差。计算了一次与二次多项式的值，且求出偏差平方和之后，我们求出了，对于一次多项式，均方差为 175，可是对于二次多项式，均方差只有 7。若考虑到二次多项式所达到的准确度是不够的，则可以继续选择三次多项式，等等。

联合最佳 L_p 逼近

附录 V

1. 引　言

在理论问题和实际问题中,常常提出需要用一个函数同时逼近几个函数、甚至是一族函数的问题,即所谓联合逼近的问题.

例如,1934 年 Remes(参看[1],也可参看[2])在研究对有界不连续函数的逼近问题时,证明了该问题等价于对两个函数(分别是上半连续函数和下半连续函数)的联合逼近问题;又如,近年来提出了对函数本身及其各阶导函数的联合逼近问题[3];再如,在实际问题中,被逼近的函数有时是依赖于参数的,而它的参数值不能准确地知道,只是知道其变化的范围,如此也归结为不是逼近一个函数而是逼近一族函数的问题.

附录 V 联合最佳 L_p 逼近

联合逼近问题的提法是各种各样的. 本文将讨论如下比较常用的联合逼近问题.

设 (X, Σ, μ) 是 σ - 有穷测度空间, $L_p \equiv L_p(X, \Sigma, \mu) (1 \leq p < \infty)$ 表示在空间 (X, Σ, μ) 上所有可测且 p 次幂可积的复值函数的全体组成的线性赋范空间 ([4], 41). 其范数定义为

$$\|f\| = \left\{ \int_X |f(x)|^p \mathrm{d}\mu \right\}^{\frac{1}{p}}$$

又设 $f_i \in L_p (i = 1, 2, \cdots), \lambda_i \geq 0 (i = 1, 2, \cdots)$, $\sum_{i=1}^{\infty} \lambda_i = 1$, 且满足条件

$$\sum_{i=1}^{\infty} \lambda_i \|f_i\|^p < \infty \qquad ㉛$$

令 $K \subset L_p$, 称 $h_0 \in K$ 是 $F \equiv \{f_i\}$ 在 K 中的联合最佳 L_p 逼近, 如果

$$\sum_{i=1}^{\infty} \lambda_i \|f_i - h_0\|^p = \inf_{h \in K} \sum_{i=1}^{\infty} \lambda_i \|f_i - h\|^p \qquad ㉜$$

第二节将证明当 K 是局部紧闭子集 ([9], 12) 时联合最佳 L_p 逼近的存在性.

第三节将导出任意准凸集上联合最佳 L_1 逼近的特征定理.

第四节将导出任意准凸集上联合最佳 L_p 逼近的特征定理, 这里 $1 < p < \infty$.

在讨论之前, 我们引入下列记号

$$\operatorname{sign} x = \begin{cases} \dfrac{x}{|x|}, & x \neq 0 \\ 0, & x = 0 \end{cases}$$

$Z(f) \equiv \{x \in X : f(x) = 0\}$

$\operatorname{Re} \alpha$ 表示复数 α 的实数部分

Tschebyscheff 逼近定理

$\bar{\alpha}$ 表示复数 α 的共轭复数

同时,为简便起计,约定在下面所有出现的符号 $\sum$ 一律表示 $\sum_{i=1}^{\infty}$;且 $F \equiv \{f_i\}$ 总满足条件 ㉑.

定义 集合 $K \subset L_p$ 称为准凸子集,如果对任意 $h_1, h_2 \in K$,都存在一数列 $t_n > 0 (n = 1,2,\cdots), t_n \to 0 (n \to \infty)$,使

$$t_n h_1 + (1 - t_n) h_2 \in \bar{K} \quad (n = 1,2,\cdots)$$

这里 $\bar{K}$ 表示 K 之闭包.

我们看到,准凸集的概念是与集合的拓扑有关的一个概念. 从这个定义看出,凸集、开集以及其闭包是凸集的集,都是准凸集.

2. 存在定理

定理 5.1 设 K 是 $L_p(1 \leq p < \infty)$ 的局部紧闭子集,$F \subset L_p$,则必存在一个函数 $h_0 \in K$,使 ㉒ 成立.

证 设 $\{h^j\}_{j=1}^{\infty}$ 是 K 中的序列,使

$$\lim_{j \to \infty} \sum \lambda_i \|f_i - h_j\|^p = \inf_{h \in K} \sum \lambda_i \|f_i - h\|^p \quad ㉓$$

首先指出,上式右端为有穷. 事实上,对任一函数 $h \in K$,总有①

$$\sum \lambda_i \|f_i - h\|^p \leq \sum \lambda_i (\|f_i\| - \|h\|)^p \leq$$
$$2^p (\sum \lambda_i \|f_i\|^p + \|h\|^p) < \infty$$

因而 ㉓ 右端为有穷. 而对于每一个 j,有

---

① 对任意数 a, b,成立不等式
$$|a + b|^p \leq [2\max(|a|, |b|)]^p \leq 2^p(|a|^p + |b|^p)$$

附录 V　联合最佳 L_p 逼近

$$0 \leqslant \|h_j\|^p = \sum \lambda_i \|h_j\|^p \leqslant$$
$$\sum \lambda_i (\|f_i - h_j\| + \|f_i\|)^p \leqslant$$
$$2^p (\sum \lambda_i \|f_i - h_j\|^p + \sum \lambda_i \|f_i\|^p)$$

上式右端括号内第一项是收敛序列的第 j 项，从而有界. 第二项依假设 ㉒ 也有界. 故 $\{\|h_j\|^p\}$ 有界，从而 $\{\|h_j\|\}$ 有界. 由 K 之局部紧性，存在一个在 L_p 的度量下的收敛子列. 不失一般性，可以假设存在 $h_0 \in K$，使

$$\lim_{j \to \infty} \|h_j - h_0\| = 0 \qquad ㉔$$

而

$$0 \leqslant \sum \lambda_i \|f_i - h_0\|^p - \inf_{h \in K} \sum \lambda_i \|f_i - h\|^p =$$
$$(\sum \lambda_i \|f_i - h_0\|^p - \sum \lambda_i \|f_i - h_j\|^p) +$$
$$(\sum \lambda_i \|f_i - h_j\|^p - \inf_{h \in K} \sum \lambda_i \|f_i - h\|^p) \qquad ㉕$$

由 ㉔，对任何 i 都有

$$\lim_{j \to \infty} \|f_i - h_j\| = \|f_i - h_0\|$$

这样，对每个 i 成立

$$\lim_{j \to \infty} \|f_i - h_j\|^p = \|f_i - h_0\|^p$$

同时容易看出，级数 $\sum \lambda_i \|f_i - h_j\|^p$ 关于 j 是一致收敛的，因而

$$\lim_{j \to \infty} \sum \lambda_i \|f_i - h_j\|^p = \sum \lambda_i \|f_i - h_0\|^p \qquad ㉖$$

而式 ㉕ 右端第二项由 ㉓，在 $j \to \infty$ 时也应趋于 0. 从而

$$\sum \lambda_i \|f_i - h_0\|^p = \inf_{h \in K} \sum \lambda_i \|f_i - h\|^p$$

证毕.

注　注意到式 ㉔，式 ㉖ 实际上表明泛函 $\Phi_p(h) =$

Tschebyscheff 逼近定理

$\sum \lambda_i \|f_i - h\|^p$ 是连续的.

3. L_1 逼近的特征定理

引理 5.2 设 $F \subset L_1, h \in L_1$, 及实数 t, 则成立

$$\lim_{t \to 0} \frac{1}{t} \sum \lambda_i [\, \|f_i + th\| - \|f_i\| - |t| \int_{Z(f_i)} |h| \, \mathrm{d}\mu \,] =$$

$$\sum \lambda_i (\operatorname{Re} \int_X h \, \overline{\operatorname{sign} f_i} \, \mathrm{d}\mu) \tag{327}$$

证 [5] 中证明了, 对每个 i 成立

$$\frac{1}{t} [\, \|f_i + th\| - \|f_i\| - |t| \int_{Z(f_i)} |h| \, \mathrm{d}\mu \,] =$$

$$\int_{X-Z(f_i)} \frac{|f_i + th| - |f_i|}{t} \mathrm{d}\mu$$

及

$$\lim_{t \to 0} \frac{1}{t} [\, \|f_i + th\| - \|f_i\| - |t| \int_{Z(f_i)} |h| \, \mathrm{d}\mu \,] =$$

$$\operatorname{Re} \int_X h \, \overline{\operatorname{sign} f_i} \, \mathrm{d}\mu \tag{328}$$

但

$$\left| \int_{X-Z(f_i)} \frac{|f_i + th| - |f_i|}{t} \mathrm{d}\mu \right| \leq$$

$$\int_{X-Z(f_i)} \left| \frac{|f_i + th| - |f_i|}{t} \right| \mathrm{d}\mu \leq$$

$$\int_{X-Z(f_i)} |h| \, \mathrm{d}\mu \leq \int_X |h| \, \mathrm{d}\mu = \|h\|$$

因而

$$\sum \lambda_i \left| \int_{X-Z(f_i)} \frac{|f_i + th| - |f_i|}{t} \mathrm{d}\mu \right| \leq$$

$$\sum \lambda_i \|h\| = \|h\|$$

即级数

$$\frac{1}{t} \sum \lambda_i [\|f_i + th\| - \|f_i\| - |t| \int_{Z(f_i)} |h| \, d\mu] = \sum \lambda_i \int_{X-Z(f_i)} \frac{|f_i + th| - |f_i|}{t} d\mu$$

关于 t 一致收敛. 所以考虑到 ㉘,由函数项级数的逐项求极限定理

$$\lim_{t \to 0} \frac{1}{t} \sum \lambda_i [\|f_i + th\| - \|f_i\| - |t| \int_{Z(f_i)} |h| \, d\mu] = \sum \lambda_i (\operatorname{Re} \int_{X-Z(f_i)} h \, \overline{\operatorname{sign} f_i} \, d\mu)$$

定理 5.3 使关系式

$$\sum \lambda_i \|f_i - h_0\| \leqslant \sum \lambda_i \|f_i - (h_0 + ch)\| \quad ㉙$$

对所有 $c \geqslant 0$ 都成立的充要条件是

$$\sum \lambda_i (\operatorname{Re} \int_X h \, \overline{\operatorname{sign}(f_i - h_0)} \, d\mu) \leqslant \sum \lambda_i \int_{Z(f_i - h_0)} |h| \, d\mu) \quad ㉚$$

证 充分性. 设 $0 \leqslant t \leqslant 1$,由三角不等式

$$\|f + th\| = \|t(f + h) + (1-t)f\| \leqslant t\|f + h\| + (1-t)\|f\|$$

或写成

$$\|f + th\| - \|f\| \leqslant t[\|f + h\| - \|f\|]$$

将 f 换成 $f_i - h_0$,h 换成 $-ch$,两边同乘以 λ_i,并对 i 求和,得到

$$\sum \lambda_i [\|f_i - h_0 - tch\| - \|f_i - h_0\|] \leqslant$$

Tschebyscheff 逼近定理

$$t \sum \lambda_i [\ \|f_i - h_0 - ch\| - \|f_i - h_0\|\ ] \quad (0 \leqslant t \leqslant 1)$$

㉛

再在 ㉗ 中将 f_i 换成 $f_i - h_0$, h 换成 $-ch(c \geqslant 0)$, 同时取单侧极限, 得到

$$c \sum \lambda_i \int_{Z(f_i - h_0)} |h|\,\mathrm{d}\mu -$$

$$c \sum \lambda_i \mathrm{Re} \int_X h\, \overline{\mathrm{sign}\,(f_i - h_0)}\,\mathrm{d}\mu \leqslant$$

$$\lim_{t \to 0^+} \frac{1}{t} \sum \lambda_i [\ \|f_i - h_0 - tch\| - \|f_i - h_0\|\ ] \quad ㉜$$

而由 ㉛ 得

$$\lim_{t \to 0^+} \frac{1}{t} \sum \lambda_i [\ \|f_i - h_0 - tch\| - \|f_i - h_0\|\ ] \leqslant$$

$$\sum \lambda_i [\ \|f_i - h_0 - ch\| - \|f_i - h_0\|\ ] \quad ㉝$$

因此, 若 ㉚ 成立. 由 $c \geqslant 0$. 则式 ㉜ 的左端是非负的, 从而 ㉝ 的右端也就非负, 此即式 ㉙ 成立.

必要性. 若式 ㉚ 不成立, 这时在 ㉜ 中令 $c = 1$ 得到

$$\lim_{t \to 0^+} \frac{1}{t} \sum \lambda_i [\ \|f_i - h_0 - tch\| - \|f_i - h_0\|\ ] =$$

$$\sum \lambda_i \int_{Z(f_i - h_0)} |h|\,\mathrm{d}\mu -$$

$$\sum \lambda_i \mathrm{Re} \int_X h\, \overline{\mathrm{sign}\,(f_i - h_0)}\,\mathrm{d}\mu < 0$$

从而对充分小的正数 t 有

$$\frac{1}{t} \sum \lambda_i [\ \|f_i - h_0 - th\| - \|f_i - h_0\|\ ] < 0$$

或

$$\sum \lambda_i \|f_i - h_0 - th\| \leqslant \sum \lambda_i \|f_i - h_0\|$$

这与假设 ㉙ 矛盾.

附录 V 联合最佳 L_p 逼近

证毕.

定理 5.3 是证明联合最佳 L_1 逼近的特征定理的有力工具. 下面就是本节的主要结果.

定理 5.4 设 K 是 L_1 中的准凸子集, $F \subset L_1$, $h_0 \in K$. 则 h_0 是 F 在 K 中的联合最佳 L_1 逼近的充要条件是

$$\sum \lambda_i \mathrm{Re} \int_X (h-h_0) \overline{\mathrm{sign}\,(f_i-h_0)} \mathrm{d}\mu \leqslant$$
$$\sum \lambda_i \int_{Z(f_i-h_0)} |h-h_0|\, \mathrm{d}\mu \qquad \text{⟨334⟩}$$

对所有 $h \in K$ 都成立.

证 充分性. 对任何 $h \in K$, 应用定理 5.3, 由式 ⟨334⟩ 得出, 不等式

$$\sum \lambda_i \|f_i - h_0\| \leqslant \sum \lambda_i \|(f_i-h_0) - c(h-h_0)\|$$

对所有 $c \geqslant 0$ 都成立. 特别当 $c=1$ 时, 将得出不等式

$$\sum \lambda_i \|f_i - h_0\| \leqslant \sum \lambda_i \|f_i - h\|$$

对任何 $h \in K$ 都成立.

必要性. 当 $\lambda > 0$ 时, 对每个 i, 注意到 $Z(f_i-h_0)$ 的定义, 我们有

$$\mathrm{Re}\int_X (h-h_0)\overline{\mathrm{sign}[f_i-h_0-\lambda(h-h_0)]}\mathrm{d}\mu =$$
$$\mathrm{Re}\int_{X-Z(f_i-h_0)} (h-h_0)\overline{\mathrm{sign}[f_i-h_0-\lambda(h-h_0)]}\mathrm{d}\mu +$$
$$\mathrm{Re}\int_{Z(f_i-h_0)} (h-h_0)\overline{\mathrm{sign}[f_i-h_0-\lambda(h-h_0)]}\mathrm{d}\mu =$$
$$\mathrm{Re}\int_{X-Z(f_i-h_0)} (h-h_0)\overline{\mathrm{sign}[f_i-h_0-\lambda(h-h_0)]}\mathrm{d}\mu -$$
$$\mathrm{Re}\int_{Z(f_i-h_0)} (h-h_0)\overline{\mathrm{sign}[(h-h_0)]}\mathrm{d}\mu =$$
$$\mathrm{Re}\int_{X-Z(f_i-h_0)} (h-h_0)\overline{\mathrm{sign}[f_i-h_0-\lambda(h-h_0)]}\mathrm{d}\mu -$$

Tschebyscheff 逼近定理

$$\int_{Z(f_i-h_0)} |h-h_0| \, d\mu$$

故

$$\lim_{\lambda \to 0^+} \mathrm{Re} \int_X (h-h_0) \overline{\mathrm{sign}[f_i - h_0 - \lambda(h-h_0)]} \, d\mu =$$

$$\mathrm{Re} \int_{X-Z(f_i-h_0)} (h-h_0) \overline{\mathrm{sign}(f_i-h_0)} \, d\mu -$$

$$\int_{Z(f_i-h_0)} |h-h_0| \, d\mu$$

但显然级数

$$\sum \lambda_i \mathrm{Re} \int_X (h-h_0) \overline{\mathrm{sign}[f_i - h_0 - \lambda(h-h_0)]} \, d\mu$$

关于 $\lambda \geqslant 0$ 是一致收敛的,所以

$$\lim_{\lambda \to 0^+} \sum \lambda_i \mathrm{Re} \int_X (h-h_0) \overline{\mathrm{sign}[f_i - h_0 - \lambda(h-h_0)]} \, d\mu =$$

$$\sum \lambda_i \mathrm{Re} \int_{X-Z(f_i-h_0)} (h-h_0) \overline{\mathrm{sign}(f_i-h_0)} \, d\mu -$$

$$\sum \lambda_i \int_{Z(f_i-h_0)} |h-h_0| \, d\mu \qquad \text{㉟}$$

这样,若 ㉞ 对某个 $h \in K$ 不成立,即

$$\sum \lambda_i \mathrm{Re} \int_X (h-h_0) \overline{\mathrm{sign}(f_i-h_0)} \, d\mu >$$

$$\sum \lambda_i \int_{Z(f_i-h_0)} |h-h_0| \, d\mu \qquad \text{㊱}$$

而由 $Z(f_i-h_0)$ 的定义,上式左端项应等于式 ㉟ 右端的第一项. 于是式 ㉟ 的右端应大于零. 从而当 $\lambda > 0$ 充分小时,特别当 $\lambda < 1$ 时,由式 ㉟ 有

$$\sum \lambda_i \mathrm{Re} \int_X (h-h_0) \overline{\mathrm{sign}[f_i - h_0 - \lambda(h-h_0)]} \, d\mu > 0$$

对这样的 λ,进一步可得

$$\sum \lambda_i \|f_i - h_0\| \geqslant$$

附录 V 联合最佳 L_p 逼近

$$\sum \lambda_i \operatorname{Re} \int_X (f_i - h_0) \overline{\operatorname{sign}[f_i - h_0 - \lambda(h - h_0)]} \mathrm{d}\mu =$$

$$\sum \lambda_i \int_X | f_i - h_0 - \lambda(h - h_0) | \mathrm{d}\mu +$$

$$\lambda \sum \lambda_i \operatorname{Re} \int_X (h - h_0) \overline{\operatorname{sign}[f_i - h_0 - \lambda(h - h_0)]} \mathrm{d}\mu >$$

$$\sum \lambda_i \int_X | f_i - h_0 - \lambda(h - h_0) | \mathrm{d}\mu =$$

$$\sum \lambda_i \| f_i - h_0 - \lambda(h - h_0) \|$$

此即只要 $\lambda > 0$ 充分小,就有

$$\sum \lambda_i \| f_i + h_0 - \lambda(h - h_0) \| < \sum \lambda_i \| f_i - h_0 \|$$

注意到 K 的准凸性,存在 $\lambda^* > 0$ 满足上式且同时有

$$h_1 \equiv h_0 + \lambda^* (h - h_0) = \lambda^* h + (1 - \lambda^*) h_0 \in \bar{K}$$

再考虑到泛函 $\Phi_1(h) = \sum \lambda_i \| f_i - h \|$ 的连续性(第二节末尾的注),存在 $h_2 \in K$ 使 $\| h_2 - h_1 \|$ 充分小,从而 $\Phi_1(h_2) < \Phi_1(h_0)$,此即

$$\sum \lambda_i \| f_i - h_2 \| < \sum \lambda_i \| f_i - h_0 \|$$

这就违反了 h_0 是 F 在 K 中的联合最佳 L_1 逼近的假定.

证毕.

由定理 5.4 容易推出当常用的 K 为 L_1 的子空间时的特征定理.

定理 5.5 设 K 是 L_1 中的子空间,$F \subset L_1, h_0 \in K$. 则 h_0 是 F 在 K 中的联合最佳 L_1 逼近的充分充要条件是

$$\left| \sum \lambda_i \operatorname{Re} \int_X h \overline{\operatorname{sign}(f_i - h_0)} \mathrm{d}\mu \right| \leqslant$$

$$\sum \lambda_i \int_{Z(f_i - h_0)} | h | \mathrm{d}\mu$$

Tschebyscheff 逼近定理

对所有 $h \in K$ 都成立.

证 在 K 是子空间的情形,㉞对所有 $h \in K$ 都成立这一条件等价于

$$\sum \lambda_i \mathrm{Re} \int_X h \,\overline{\mathrm{sign}(f_i - h_0)}\,\mathrm{d}\mu \le$$

$$\sum \lambda_i \int_{Z(f_i - h_0)} |h|\,\mathrm{d}\mu$$

对所有 $h \in K$ 都成立. 但当 $h \in K$ 时,必有 $-h \in K$. 从而上面的条件又等价于㉝对所有 $h \in K$ 都成立.

证毕.

当 $L_1 \equiv L_1(E)$ 表示实轴上 Lebesgue 可测紧子集 E 上 Lebesgue 可积实值函数空间,且当 K 是次数不超过 n 的实多项式集合(是局部紧的)时,令

$$\lambda_1 = \cdots = \lambda_m = \frac{1}{m}$$

$$\lambda_{m+1} = \lambda_{m+2} = \cdots = 0$$

存在定理 2.1(当 $p=1$ 时) 和特征定理 3.4 即是 Carroll 和 McLaughlin 在 [6] 中证明的定理 1 和定理 2.

4. L_p 逼近的特征定理

引理 5.6 设 $F \subset L_p, h \in L_p(1 < p < \infty)$ 满足条件

$$\sum \lambda_i \mathrm{Re} \int_X h\,|f_i|^{p-1}\,\overline{\mathrm{sign}\,f_i}\,\mathrm{d}\mu > 0$$

则存在 $\lambda_0 > 0$,使当 $0 < \lambda \le \lambda_0$ 时成立

$$\sum \lambda_i \|f_i - \lambda h\|^p < \sum \lambda_i \|f_i\|^p \qquad ㉝$$

证 我们首先指出,对每个 i,当 $\lambda \to 0$ 时

$$\lim_{\lambda \to 0} \mathrm{Re}(h\,|f_i - \lambda h|^{p-1}\,\overline{\mathrm{sign}(f_i - \lambda h)}) =$$

$$\mathrm{Re}(h\mid f_i\mid^{p-1}\overline{\mathrm{sign}\,f_i})$$

在 X 上几乎处处成立. 由控制收敛定理([4],640)

$$\lim_{\lambda\to 0}\mathrm{Re}\int_X h\mid f_i-\lambda h\mid^{p-1}\overline{\mathrm{sign}(f_i-\lambda h)}\mathrm{d}\mu =$$
$$\mathrm{Re}\int_X h\mid f_i\mid^{p-1}\overline{\mathrm{sign}\,f_i}\mathrm{d}\mu$$

但当 $\mid\lambda\mid\leqslant 1$ 时

$$\sum\lambda_i\left|\mathrm{Re}\int_X h\mid f_i-\lambda h\mid^{p-1}\overline{\mathrm{sign}(f_i-\lambda h)}\right|\mathrm{d}\mu\leqslant$$
$$\sum\lambda_i\int_X\mid h\mid\cdot\mid f_i-\lambda h\mid^{p-1}\mathrm{d}\mu\leqslant$$
$$2^{p-1}\left(\sum\lambda_i\int_X\mid h\mid\cdot\mid f_i\mid^{p-1}\mathrm{d}\mu+\parallel h\parallel^p\right)$$

令 $q=1/\left(1-\dfrac{1}{p}\right)$,由 Hölder 不等式([4],42)

$$\sum\lambda_i\int_X\mid h\mid\cdot\mid f_i\mid^{p-1}\mathrm{d}\mu\leqslant$$
$$\sum\lambda_i\left\{\int_X\mid h\mid^p\mathrm{d}\mu\right\}^{\frac{1}{p}}\left\{\int_X\mid f_i\mid^{(p-1)q}\mathrm{d}\mu\right\}^{\frac{1}{q}}=$$
$$\sum\lambda_i\parallel h\parallel\cdot\parallel f_i\parallel^{\frac{p}{q}}$$

再利用级数的 Hölder 不等式,并考虑到 $\sum\lambda_i=1$,有

$$\sum\lambda_i\parallel h\parallel\cdot\parallel f_i\parallel^{\frac{p}{q}}=\sum(\lambda_i^{\frac{1}{p}}\parallel h\parallel)(\lambda_i^{\frac{1}{q}}\parallel f_i\parallel^{\frac{p}{q}})\leqslant$$
$$\{\sum(\lambda_i^{\frac{1}{p}}\parallel h\parallel)^p\}^{\frac{1}{p}}\{\sum(\lambda_i^{\frac{1}{q}}\parallel f_i\parallel^{\frac{p}{q}})^q\}^{\frac{1}{q}}=$$
$$\{\sum\lambda_i\parallel h\parallel)^p\}^{\frac{1}{p}}\{\sum\lambda_i\parallel f_i\parallel)^p\}^{\frac{1}{q}}=$$
$$\parallel h\parallel\{\sum\lambda_i\parallel f_i\parallel)^p\}^{\frac{1}{q}}<\infty$$

从而当 $\mid\lambda\mid\leqslant 1$ 时,级数

$$\sum\lambda_i\mathrm{Re}\int_X h\mid f_i-\lambda h\mid^{p-1}\overline{\mathrm{sign}(f_i-\lambda h)}\mathrm{d}\mu$$

Tschebyscheff 逼近定理

关于 λ 一致收敛. 故

$$\lim_{\lambda \to 0} \sum \lambda_i \mathrm{Re} \int_X h \mid f_i - \lambda h \mid^{p-1} \overline{\mathrm{sign}(f_i - \lambda h)} \mathrm{d}\mu =$$

$$\sum \lambda_i \mathrm{Re} \int_X h \mid f_i \mid^{p-1} \overline{\mathrm{sign} f_i} \mathrm{d}\mu > 0$$

所以存在 $\lambda_0 > 0$, 使当 $0 < \lambda \leqslant \lambda_0$ 时有

$$\sum \lambda_i \mathrm{Re} \int_X h \mid f_i - \lambda h \mid^{p-1} \overline{\mathrm{sign}(f_i - \lambda h)} \mathrm{d}\mu > 0$$

这时

$$\sum \lambda_i \| f_i - \lambda h \|^p < \sum \lambda_i \| f_i - \lambda h \|^p +$$
$$\lambda \sum \lambda_i \mathrm{Re} \int_X h \mid f_i - \lambda h \mid^{p-1} \overline{\mathrm{sign}(f_i - \lambda h)} \mathrm{d}\mu =$$
$$\mathrm{Re} \sum \lambda_i \int_X \mid f_i - \lambda h \mid^{p-1} [\mid f_i - \lambda h \mid +$$
$$\lambda h \overline{\mathrm{sign}(f_i - \lambda h)}] \mathrm{d}\mu =$$
$$\mathrm{Re} \sum \lambda_i \int_X \mid f_i - \lambda h \mid^{p-1} f_i \overline{\mathrm{sign}(f_i - \lambda h)} \mathrm{d}\mu \leqslant$$
$$\sum \lambda_i \int_X \mid f_i - \lambda h \mid^{p-1} \mid f_i \mid \mathrm{d}\mu$$

利用 Hölder 不等式

$$\sum \lambda_i \int_X \mid f_i - \lambda h \mid^{p-1} \mid f_i \mid \mathrm{d}\mu \leqslant$$
$$\sum \lambda_i \left\{ \int_X \mid f_i - \lambda h \mid^{(p-1)q} \mathrm{d}\mu \right\}^{\frac{1}{q}} \left\{ \int_X \mid f_i \mid^p \mathrm{d}\mu \right\}^{\frac{1}{p}} =$$
$$\sum \lambda_i \| f_i - \lambda h \|^{\frac{p}{q}} \| f_i \|$$

与前面一样, 再利用级数的 Hölder 不等式, 可以得到

$$\sum \lambda_i \| f_i - \lambda h \|^p < \sum \lambda_i \| f_i - \lambda h \|^{\frac{p}{q}} \| f_i \| \leqslant$$
$$\left\{ \sum \lambda_i \| f_i - \lambda h \|^p \right\}^{\frac{1}{q}} \left\{ \sum \lambda_i \| f_i \|^p \right\}^{\frac{1}{p}}$$

或

$$\{\sum \lambda_i \|f_i - \lambda h\|^p\}^{\frac{1}{p}} < \{\sum \lambda_i \|f_i\|^p\}^{\frac{1}{p}}$$

它等价于 ㉞. 证毕.

定理 5.7 设 $F \subset L_p, h \in L_p (1 < p < \infty)$,则

$$\sum \lambda_i \|f_i\|^p \leqslant \sum \lambda_i \|f_i - \lambda h\|^p \qquad ㉞$$

对所有 $\lambda \geqslant 0$ 都成立的充要条件是

$$\sum \lambda_i \mathrm{Re} \int_X h \|f_i\|^{p-1} \overline{\mathrm{sign}\, f_i} \mathrm{d}\mu \leqslant 0 \qquad ㉞$$

证 充分性. 若 ㉞ 成立,则对 $\lambda \geqslant 0$,首先有

$$\sum \lambda_i \|f_i\|^p \leqslant$$

$$\sum \lambda_i \|f_i\|^p - \lambda \sum \lambda_i \mathrm{Re} \int_X h |f_i|^{p-1} \overline{\mathrm{sign}\, f_i} \mathrm{d}\mu =$$

$$\mathrm{Re} \sum \lambda_i \int_X |f_i|^{p-1} [|f_i| - \lambda h \overline{\mathrm{sign}\, f_i}] \mathrm{d}\mu =$$

$$\mathrm{Re} \sum \lambda_i \int_X |f_i|^{p-1} (f_i - \lambda h) \overline{\mathrm{sign}\, f_i}] \mathrm{d}\mu \leqslant$$

$$\sum \lambda_i \int_X |f_i|^{p-1} |f_i - \lambda h| \mathrm{d}\mu$$

与上面的引理 4.1 的证明一样,两次利用 Hölder 不等式,最后得到

$$\sum \lambda_i \|f_i\|^p \leqslant \sum \lambda_i \|f_i - \lambda h\|^p$$

必要性. 若 ㉞ 不成立,即

$$\sum \lambda_i \mathrm{Re} \int_X h |f_i|^{p-1} \overline{\mathrm{sign}\, f_i}] \mathrm{d}\mu > 0$$

由引理 4.1,存在正数 λ,使

$$\sum \lambda_i \|f_i - \lambda h\|^p < \sum \lambda_i \|f_i\|^p$$

这与 ㉞ 相矛盾. 证毕.

定理 4.3 设 $F \subset L_p, h \in L_p (1 < h < \infty)$,则 ㉞ 对所有 λ(当 L_p 是实函数空间时 λ 为实数,当 L_p 是复

Tschebyscheff 逼近定理

函数空间时 λ 为复数)都成立的充要条件是

$$\sum \lambda_i \int_X h \mid f_i \mid^{p-1} \overline{\operatorname{sign} f_i} \, \mathrm{d}\mu = 0 \qquad ㉑$$

证 我们仅对 L_p 是复函数空间的情形给出证明,另一情形的证明是类似的.

式 ㉝ 对所有 λ 都成立当且仅当式 ㉝ 对 $\lambda = \pm t$ 及 $\lambda = \pm jt$(其中 t 为任意非负数而 $j = \sqrt{-1}$)都成立. 由定理 5.7,这分别得到等价条件

$$\sum \lambda_i \operatorname{Re} \int_X [\pm h \mid f_i \mid^{p-1} \overline{\operatorname{sign} f_i}] \mathrm{d}\mu \leq 0$$

及

$$\sum \lambda_i \operatorname{Re} \int_X [\pm j h \mid f_i \mid^{p-1} \overline{\operatorname{sign} f_i}] \mathrm{d}\mu \leq 0$$

这四个式子一起又等价于式 ㉑. 证毕.

现在我们来证明本节的主要结果.

定理 5.9 若 K 是 $L_p(1 < p < \infty)$ 的准凸子集,$F \subset L_p, h_0 \in K$,则 h_0 是 F 在 K 中的联合最佳 L_p 逼近的充要条件是

$$\sum \lambda_i \operatorname{Re} \int_X (h - h_0) \mid f_i - h_0 \mid^{p-1} \overline{\operatorname{sign}(f_i - h_0)}] \mathrm{d}\mu \leq 0$$

㉒

对所有 $h \in K$ 都成立.

证 充分性. 对任何 $h \in K$,应用定理 5.7,得出不等式

$$\sum \lambda_i \parallel f_i - h_0 \parallel^p \leq \sum \lambda_i \parallel (f_i - h_0) - \lambda(h - h_0) \parallel^p$$

对所有 $\lambda \geq 0$ 都成立,特别取 $\lambda = 1$,将得出

$$\sum \lambda_i \parallel f_i - h_0 \parallel^p \leq \sum \lambda_i \parallel f_i - h \parallel^p$$

对任何 $h \in K$ 都成立.

附录 V 联合最佳 L_p 逼近

必要性. 若 ㉞ 对某一个 $h \in K$ 不成立, 即

$$\sum \lambda_i \mathrm{Re} \int_X (h - h_0) | f_i - h_0 |^{p-1} \overline{\mathrm{sign}(f_i - h_0)}] \mathrm{d}\mu > 0$$

由引理 5.6, 存在充分小的正数 λ, 特别可以取 $\lambda \leq 1$, 使

$$\sum \lambda_i \| (f_i - h_0) - \lambda (h - h_0) \|^p < \sum \lambda_i \| f_i - h_0 \|^p$$

或写成

$$\sum \lambda_i \| f_i - [\lambda h + (1 - \lambda) h_0] \|^p < \sum \lambda_i \| f_i - h_0 \|^p$$

考虑到 K 之准凸性及泛函 $\Phi_p(h) = \sum \lambda_i \| f_i - h \|^p$ 之连续性, 与定理 5.4 的证明类似, 可知存在 $h_2 \in K$, 使

$$\sum \lambda_i \| f_i - h_2 \|^p < \sum \lambda_i \| f_i - h_0 \|^p$$

从而与 h_0 是 F 在 K 中的联合最佳 L_p 逼近的假设发生矛盾.   证毕.

由定理 5.9 可以推出常用的两个特殊形式的定理.

定理 5.10 设 K 是 $L_p (1 < p < \infty)$ 的子空间, $F \subset L_p, h_0 \in K$. 则 h_0 是 F 在 K 中的联合最佳 L_p 逼近的充要条件是

$$\sum \lambda_i \int_X h | f_i - h_0 |^{p-1} \overline{\mathrm{sign}(f_i - h_0)} \mathrm{d}\mu = 0 \quad ㉝$$

对所有 $h \in K$ 都成立.

证 充分性. 若对所有 $h \in K$ 满足 ㉝, 这样, 对 $h - h_0 \in K$ 也应满足 ㉝, 从而也就满足 ㉞. 因而按定理 5.9, h_0 是联合最佳逼近.

必要性. 由 h_0 是联合最佳逼近, 考虑到子空间也是凸集, 从而 ㉞ 应对所有 $h \in K$ 都成立. 由于 K 是子空间, 因而有

Tschebyscheff 逼近定理

$$h_0 \pm h \in K, h_0 \pm jh \in K \quad (j = \sqrt{-1})$$

将这四个函数分别代替 ⑫ 中的 h,得到这四个式子,将推得式 ⑬.

证毕.

如果 K 是 L_p 中的凸锥,即满足条件

$$K + K \subset K, \lambda K \subset K \quad (\lambda \geq 0)$$

显然凸锥也是凸集,这时成立下面的特征定理.

定理 5.11 设 K 是 $L_p(1 < p < \infty)$ 中的凸锥,$F \subset L_p, h_0 \in K$,则 h_0 是 F 在 K 中的联合最佳 L_p 逼近的充要条件是满足

$$\sum \lambda_i \mathrm{Re} \int_X h \mid f_i - h_0 \mid^{p-1} \overline{\mathrm{sign}(f_i - h_0)}] \mathrm{d}\mu \leq 0$$

$$(\ast)$$

且对所有 $h \in K$ 都成立

$$\sum \lambda_i \mathrm{Re} \int_X h_0 \mid f_i - h_0 \mid^{p-1} \overline{\mathrm{sign}(f_i - h_0)}] \mathrm{d}\mu = 0$$

$$(\ast\ast)$$

证 充分性是显然的,因为满足条件($\ast$)与($\ast\ast$)者必满足 ⑫,从而由定理 4.4 知,h_0 是联合最佳逼近.

必要性. 若 h_0 是联合最佳逼近,由定理 4.4 知,式 ⑫ 应对所有 $h \in K$ 都成立. 特别取 $h = 0$ 与 $h = 2h_0$,由 K 是凸锥,它们都应当属于 K. 从而满足式 ⑫. 由此得到($\ast\ast$). 进而由 ⑫ 与($\ast\ast$)就推得($\ast$).

当 $\lambda_1 = 1, \lambda_2 = \lambda_3 = \cdots = 0$ 时,在第 3 节和第 4 节所得诸定理之特殊形式,分别就是在[7]和[8]中给出的主要结果.

参考文献

[1] REMES E. Sur la détermination des polynomes d'approximation de degré donné[J]. comm. Soc. Math. Kharkoff, 1934,10(4):41-63.

[2] LAURENT P J, PHAM D T. Global approximation of a compact set by elements of a convex set in a normed space[J]. Numer. Math. , 1970,15(2): 137-150.

[3] CHALMERS B L. Uniqueness of best approximation of a function and its derivatives, J. Approx. Theory, 1973,7(3):213-225.

[4] 关肇直. 泛函分析讲义[M]. 北京:高等教育出版社,1958.

[5] KRIPKE B R, RIVLIN T J. Approximation in the metric of $L_1(X,\mu)$[J]. Trans. Amer. Math. Soc. , 1965,119(1):101-122.

[6] CARROLL M P, MCLAUGHLIN H W. L_1 approximation of vector-valued functions[J]. Approx. Theory, 1973,7(2):122-131.

[7] 史应光. 在空间 $L(X,\Sigma,\mu)$ 中的最佳逼近的特征[J],计算数学,1980,2(3):197-203.

[8] HOLMES R B. A coures on optimization and best approximation[M]. Berlin:Springer-Verlag, 1972.

[9] 李文清. 泛函分析[M]. 北京:科学出版社,1960.

多元函数的三角多项式逼近

附录 Ⅵ

1. 引　论

设 D 表示 xOy 平面上的矩形区域: $0 \leqslant x \leqslant 2\pi, 0 \leqslant y \leqslant 2\pi$,我们所考虑的函数 $f(x,y)$ 都是在 D 上确定的周期函数,关于每一变量的周期都是 2π.

假如 $f(x,y)$ 在 D 上有 p 级的连续偏导数,我们就用 $f \in C^p(D)$ 来表示. 当 $p = 0$ 时, $C^0(D)$ 简记作 $C(D)$,表示在 D 上连续的函数类. 设 $f \in C(D)$,我们用

$$\omega(\rho) \equiv \omega(\rho;f) = \max_{(x_1-x_2)^2+(y_1-y_2)^2 \leqslant \rho^2} |f(x_1,y_1) - f(x_2,y_2)|$$

附录 Ⅵ　多元函数的三角多项式逼近

表示 f 的连续模. 当 $f \in C^p(D)$ 时,我们令

$$\omega_p(\rho) \equiv \omega_p(\rho;f) = \max_{\alpha,\beta \geq 0, \alpha+\beta=p} \omega_{\alpha,\beta}(\rho)$$

其中 $\omega_{\alpha,\beta}(\rho)$ 是函数

$$\frac{\partial^p f(x,y)}{\partial x^\alpha \partial y^\beta} \quad (\alpha,\beta \geq 0, \alpha+\beta = p \geq 1)$$

的连续模. $\omega_0(\rho)$ 即规定为 $\omega(\rho)$.

对于任一用复的形式表示的三角多项式

$$T(x,y) = \sum C_{m,n} e^{i(mx+ny)}$$

我们称 $R = \max_{m,n}(m^2+n^2)^{\frac{1}{2}}$ 为三角多项式 $T(x,y)$ 的阶. 并用 $T_R(x,y)$ 表示阶不大于 R 的三角多项式. 本文主要目的是要作出具体的三角多项式 $T_R(x,y)$ 来逼近已经给定的函数 $f(x,y)$,使它们的偏差

$$\max_{0 \leq x,y \leq 2\pi} |f(x,y) - T_R(x,y)|$$

当 R 充分大时的阶达到相当于单元函数的最小偏差的阶.

对于给定的 $f(x,y) \in C(D)$,假定它的 Fourier 级数是

$$f(x,y) \sim \sum_{m,n=-\infty}^{\infty} C_{m,n} e^{i(mx+ny)} \qquad ㉞$$

我们令

$$A_v(x,y) = \sum_{m^2+n^2=v} C_{m,n} e^{i(mx+ny)} \qquad ㉟$$

作级数 $\sum A_v(x,y)$ 的 $\delta(\geq 0)$ 次 Riesz 平均

$$S_R^\delta(x,y;f) = \sum_{v \geq R^2} \left(1 - \frac{v}{R^2}\right)^\delta A_v(x,y) \quad (v = m^2+n^2)$$

㊱

则 $S_R^\delta(x,y;f)$ 是一个阶不大于 R 的三角多项式.

Chandrasekharan 与 Minakshisundaram 曾证明下述结果：

若 $f(x,y)$ 在 D 上一致地满足下面的条件

$$\frac{1}{2\pi}\int_0^{2\pi} f(x+t\cos\theta, y+t\sin\theta)\,\mathrm{d}\theta - f(x,y) = O(t^\alpha)\ (\alpha > 0) \tag{347}$$

则在 D 上，下面的关系式一致地成立

$$S_R^\delta(x,y;f) - f(x,y) = \begin{cases} O(R^{-\alpha}), \left(\alpha + \frac{1}{2} < \delta\right) \\ O(R^{-\alpha}\ln R), \left(\alpha + \frac{1}{2} = \delta\right) \\ O(R^{-\delta+\frac{1}{2}}), \left(\frac{1}{2} < \delta < \alpha + \frac{1}{2}\right) \end{cases} \tag{348}$$

从函数构造论的观点来看，当 $0 < \alpha < 1$ 时，条件(347)的一致成立与下述条件等价

$$\omega(\rho) = O(\rho^\alpha) \tag{349}$$

即 $f(x,y) \in \mathrm{Lip}\ \alpha$. 事实上若对每一 $R > 0$，存在三角多项式 $T_R(x,y)$ 使 $|f(x,y) - T_R(x,y)| \leqslant \frac{A}{R^\alpha}$，$A$ 是常数，则可以利用 Бернштеин 定理的方法，推出 $f(x,y) \in \mathrm{Lip}\ \alpha(0 < \alpha < 1)$，所以当条件(347)一致成立时，可取 $T_R(x,y) = S_R^\delta(x,y;f)\left(\delta > \alpha + \frac{1}{2}\right)$，则由(348)的第一式可推出 $f(x,y) \in \mathrm{Lip}\ \alpha$.

我们直接考虑 $f(x,y)$ 的连续模，可得到更一般的结果如下：

定理 6.1 (i) 设 $f \in C(D)$，$\omega(\rho)$ 为其连续模，则

附录 Ⅵ　多元函数的三角多项式逼近

当 $\delta > \dfrac{1}{2}$，我们有下面的关系式在 D 上一致成立

$$S_R^\delta(x,y;f) - f(x,y) = O\left[\omega\left(\dfrac{1}{R}\right)\right] \qquad ㊂$$

（ii）设 $f \in C^1(D)$，$\omega_1(\rho)$ 的规定如前，则当 $\delta > \dfrac{1}{2}$，我们有下面的关系式在 D 上一致成立

$$S_R^\delta(x,y;f) - f(x,y) = O\left[\dfrac{1}{R}\omega_1\left(\dfrac{1}{R}\right)\right] \qquad ㊶$$

特别当 $f(x,y) \in \mathrm{Lip}\,\alpha(0 < \alpha < 1)$ 时，上述定理的第一部分包括了 Chandrasekharan 与 Minakshisundaram 的结果. 值得指出的是在我们的结论中，$\delta > \dfrac{1}{2}$ 与 α 无关.

上述结果有 $\delta > \dfrac{1}{2}$ 的限制，对于一般的 $\delta \geq 0$，我们可以考虑 $S_R^\delta(x,y;f)$ 在圆周上的平均

$$\mu_t[S_R^\delta(x,y;f)] = \dfrac{1}{2\pi}\int_0^{2\pi} S_R^\delta(x+t\cos\theta, y+t\sin\theta)\,\mathrm{d}\theta$$

下述定理相当于单元函数的 Бернштейн 定理，同时也是 Chandrasekharan 与 Minakshisundaram 另一结果的改进.

定理 6.2　（i）若 $f \in \mathrm{Lip}\,\alpha(0 < \alpha \leq 1)$，则当 $\delta \geq 0$ 时，我们有下面的关系式在 D 上一致成立

$$\mu_{\frac{\lambda_0}{R}}[S_R^\delta(x,y)] - f(x,y) = O(R^{-\alpha}) \qquad ㊷$$

（ii）若 $\dfrac{\partial f}{\partial x}, \dfrac{\partial f}{\partial y} \in \mathrm{Lip}\,\alpha(0 < \alpha \leq 1)$，则当 $\delta \geq 0$ 时下面的关系式在 D 上一致成立

Tschebyscheff 逼近定理

$$\mu_R^{\lambda_0}[S_R^{\delta}(x,y)] - f(x,y) = O(R^{-1-\alpha}) \qquad ㉝$$

关系式 ㉜ 与 ㉝ 中的 λ_0 是零级第一类 Bessel 函数 $J_0(x)$ 的一个正根.

以上我们所考虑的两种方法,逼近的阶都不能高于 R^{-2}. 事实上若 $S_R^{\delta}(x,y;f) - f(x,y) = o(R^{-2})$ 在 D 上一致成立,则两边各乘以 $e^{-i(mx+ny)}$,并在 D 上积分,即得 $C_{m,n} = 0$ 对一切 $m^2 + n^2 \neq 0$ 的 m, n 都成立,故 $f(x, y)$ 恒为常数. 同样可知若 $\mu_R^{\lambda_0}[S_R^{\delta}(x,y)] - f(x,y) = o\left(\dfrac{1}{R^2}\right)$ 在 D 上一致成立时,则 $f(x,y)$ 也恒为常数.

因此当 $f(x,y) \in C^p(D) (p > 1)$ 时,用上面的两种逼近方法所得的阶都不能改进. 很自然地我们应当引进下面的三角多项式

$$S_R^{(k)}(x,y;f) = \sum_{v \leq R^2}\left(1 - \frac{v^{\frac{k}{2}}}{R^k}\right) A_v(x,y) \qquad ㉞$$

其中 k 是正整数. 我们的结果如下:

定理 6.3 设 $f(x,y) \in C^p(D), \omega_p(\rho)$ 的规定同前.

(i) 若 $p < k - 1$,则下面关系式在 D 上一致成立

$$S_R^{(k)}(x,y;f) - f(x,y) = O\left[\frac{1}{R^p}\omega_p\left(\frac{1}{R}\right)\right] \qquad ㉟$$

(ii) 若 $p = k - 1$,则当 k 是偶数时,关系式(12)仍旧成立;当 k 是奇数时,则下面的关系式在 D 上一致成立

$$S_R^{(k)}(x,y;f) - f(x,y) = O\left[\frac{\ln R}{R^p}\omega_p\left(\frac{1}{R}\right)\right] \qquad ㊱$$

以上的结论只要适当地配合变量的空间维数,就可以推广到多元函数的情形. 事实上假定 $f(x_1, \cdots, x_m)$

是 m 维空间的周期函数,关于每一变量的周期都是 2π. 假定它的 Fourier 级数是

$$f(P) \equiv f(x_1,x_2,\cdots,x_m) \sim \sum C_{n_1,\cdots,n_m}\mathrm{e}^{\mathrm{i}(n_1x_1+\cdots+n_mx_m)}$$

令

$$A_v(P)=\sum_{n_1^2+n_2^2+\cdots+n_m^2=v} C_{n_1,\cdots,n_m}\mathrm{e}^{\mathrm{i}(n_1x_1+\cdots+n_mx_m)}$$

作三角多项式

$$S_R^{(k)}(P;f)=\sum_{v\leqslant R^2}\left(1-\frac{v^{\frac{k}{2}}}{R^k}\right)^{\sigma_m}A_v(P)$$

其中 $\sigma_m=\left[\frac{m-1}{2}\right]+1$ 仅与维数 m 有关,k 是正整数.

于是相当于定理 6.3,我们可得下面的结果:

定理 6.4　设 $f(P)$ 有 p 级连续偏导数.

(i) 若 $p<k-1$,则下面的关系式一致成立

$$S_R^{(k)}(P;f)-f(P)=O\left[\frac{1}{R^p}\omega_p\left(\frac{1}{R}\right)\right] \tag{357}$$

(ii) 若 $p=k-1$,则当 k 是偶数时,(357)仍一致成立;当 k 是奇数时,则下面的关系式一致成立

$$S_R^{(k)}(P;f)-f(P)=O\left[\frac{\ln R}{R^p}\omega_p\left(\frac{1}{R}\right)\right] \tag{358}$$

以上 $\omega_p(\rho)$ 仍表示 $f(P)$ 的所有 p 级偏导数的连续模的最大者.

定理 6.5 与定理 6.6 可以看作 Zygmund 定理在多元函数的推广. 定理 6.6 的证明与定理 6.5 的证明相仿,本文不预备详细叙述了.

在上述逼近方法中,若 k 固定,则我们所得到的阶容易看出是不能超过 R^{-k} 的,但对任一固定的函数类 $C^p(p>0)$,若取 k 充分大,则我们所作的三角多项式

(354) 对于这函数类来说所得到的逼近的阶,事实上已达到最小偏差的阶了.

2. 定理 6.2 的证明

首先,由 Bochner 公式知[3]

$$S_R^{\delta}(x,y;f)-f(x,y)=$$

$$2^{\delta}\Gamma(\delta+1)R\int_0^{\infty}\{\varphi_{xy}(t)-f(x,y)\}\frac{J_{\delta+1}(Rt)}{(Rt)^{\delta}}\mathrm{d}t \tag{359}$$

其中

$$\varphi_{xy}(t)=\frac{1}{2\pi}\int_0^{2\pi}f(x+t\cos\theta,y+t\sin\theta)\mathrm{d}\theta \tag{360}$$

今令

$$H_{\delta}(u)=2^{\delta}\Gamma(\delta+1)\frac{J_{\delta+1}(u)}{u^{\delta}}\equiv$$

$$A(\delta)\frac{J_{\delta+1}(u)}{u^{\delta}}$$

则(6) 可写成

$$S_R^{\delta}(x,y;f)-f(x,y)=$$

$$\int_0^{\infty}\left\{\varphi_{xy}\left(\frac{u}{R}\right)-f(x,y)\right\}H_{\delta}(u)\mathrm{d}u \tag{361}$$

由于 $J_{\mu}(u)=O(u^{-\frac{1}{2}})(u\to\infty)$,因此

$$H_{\delta}(u)=O(u^{-\frac{1}{2}-\delta})(u\to\infty)$$

故当 $\delta>\frac{1}{2}$ 时,$H_{\delta}(u)\in L(0,\infty)$,则可设

$$A_0=\int_0^{\infty}|H_{\delta}(u)|\mathrm{d}u$$

又令 $H_{\delta}^{(1)}(u)=\int_u^{\infty}H_{\delta}(t)\mathrm{d}t$,则因

附录Ⅵ 多元函数的三角多项式逼近

$$H_\delta^{(1)}(u) = A(\delta)\int_u^\infty \frac{J_{\delta+1}(t)}{t^\delta}dt = A(\delta)\frac{J_\delta(u)}{u^\delta} =$$
$$O(u^{-\delta-\frac{1}{2}})(u\to\infty)$$

故 $H_\delta^{(1)}(u) \in L(0,\infty)$，同样可设

$$A_1 = \int_0^\infty |H_\delta^{(1)}(u)|\,du$$

又因

$$H_\delta^{(2)}(u) = \int_u^\infty H_\delta^{(1)}(t)\,dt =$$
$$A(\delta)\frac{J_{\delta+1}(u)}{u^\delta} + B(\delta)\int_u^\infty \frac{J_{\delta+1}(t)}{t^{\delta+1}}dt =$$
$$O(u^{-\delta-\frac{1}{2}})(u\to\infty)$$

则又可设

$$A_2 = \int_0^\infty |H_\delta^{(2)}(u)|\,du$$

回到㉛，我们可知若 $f(x,y)$ 有界，则
$$S_R^\delta(x,y;f) - f(x,y) \le 2M_0 A_0$$

而 $\qquad M_0 = \sup|f(x,y)|$ ㊷

若 $f(x,y)$ 有有界的一级偏导数，则利用分部积分知

$$|S_R^\delta(x,y;f) - f(x,y)| =$$
$$\left|\frac{1}{R}\int_0^\infty \varphi'_{xy}\left(\frac{u}{R}\right)\cdot H_\delta^{(1)}(u)\,du\right| \le$$
$$2M_1 A_1 \cdot \frac{1}{R} \qquad\qquad ㊳$$

其中 $\qquad M_1 = \max\{\sup|f'_x|, \sup|f'_y|\}$

若 $f(x,y)$ 有有界的二级偏导数，则再利用一次分部积分，注意到 $\varphi'_{xy}(0) = 0$，即有

Tschebyscheff 逼近定理

$$|S_R^\delta(x,y;f) - f(x,y)| =$$

$$\left| \frac{1}{R^2} \int_0^\infty \varphi''_{xy}\left(\frac{u}{R}\right) \cdot H_\delta^{(2)}(u) \mathrm{d}u \right| \leq$$

$$4M_2 A_2 \cdot \frac{1}{R^2} \qquad \text{⟨364⟩}$$

其中

$$M_2 = \max\{\sup|f''_{xx}|, \sup|f''_{xy}|, \sup|f''_{yy}|\}$$

今设 $G(x,y)$ 是满足以下方程的某一函数

$$\frac{\partial^2 G}{\partial x \partial y} = \frac{\partial^2 G}{\partial y \partial x} = f(x,y)$$

而作下述函数（可以看作一元函数的 Стеклов 函数的推广）

$$f_\rho(x,y) =$$

$$\frac{G(x+h, y+h) - G(x+h, y) - G(x, y+h) + G(x,y)}{h^2}$$

$$\left(h = \frac{\rho}{\sqrt{2}}\right)$$

又令

$$\eta_\rho(x,y)' = f(x,y) = f_\rho(x,y)$$

现在先来证定理 1 中的 (i), 由于

$$|\eta_\rho(x,y)| = |f(x,y) - f(x+\theta_1 h, y+\theta_2 h)| \leq \omega(\rho)$$

故由 ⟨362⟩ 得到

$$|S_R^\delta(x,y;\eta_\rho) - \eta_\rho(x,y)| \leq 2\omega(\rho) A_0 \qquad \text{⟨365⟩}$$

又因

$$\left| \frac{\partial f_\rho}{\partial x} \right| = \left| \frac{G'_x(x+h, y+h) - G'_x(x+h, y) - G'_x(x, y+h) + G'_x(x,y)}{h^2} \right| =$$

$$\left| \frac{f(x+h, y+\theta_3 h) - f(x, y+\theta_3 h)}{h} \right| \leq \frac{\omega(\rho)}{h} = \frac{\sqrt{2}\omega(\rho)}{\rho}$$

附录 Ⅵ　多元函数的三角多项式逼近

同样有

$$\left|\frac{\partial f_\rho}{\partial y}\right| \leq \frac{\sqrt{2}\,\omega(\rho)}{\rho}$$

因此由 ㊅ 即知

$$|S_R^\delta(x,y;f_\rho) - f_\rho(x,y)| \leq$$
$$2\sqrt{2}\,\frac{\omega(\rho)}{\rho}A_1 \cdot \frac{1}{R} \qquad ㊌$$

合并 ㊋ 及 ㊌,取 $\rho = \dfrac{1}{R}$,注意 $f(x,y) = f_\rho(x,y) + \eta_\rho(x,y)$,即有

$$|S_R^\delta(x,y;f) - f(x,y)| \leq 2(A_0 + \sqrt{2}\,A_1)\omega\!\left(\frac{1}{R}\right)$$

此即证明了(ⅰ).

现来证(ⅱ). 设 $f(x,y) \in \mathbf{C}^1$,则因

$$\frac{\partial \eta_\rho}{\partial x} = \frac{\partial f}{\partial x} - \frac{\partial f_\rho}{\partial x} = \frac{\partial f}{\partial x} - \frac{\partial f(x+\theta'_3 h, y+\theta_3 h)}{\partial x}$$

故有

$$\sup\left|\frac{\partial \eta_\rho}{\partial x}\right| \leq \omega_1(\rho)$$

同样地有

$$\sup\left|\frac{\partial \eta_\rho}{\partial y}\right| \leq \omega_1(\rho)$$

则利用了 ㊍ 即得

$$|S_R^\delta(x,y;\eta_\rho) - \eta_\rho(x,y)| \leq 2A_1 \frac{1}{R}\omega_1\!\left(\frac{1}{R}\right) \qquad ㊎$$

又因

$$\frac{\partial^2 f_\rho}{\partial x^2} = \frac{f'_x(x+h, y+\theta_4 h) - f'_x(x, y+\theta_4 h)}{h}$$

故有

Tschebyscheff 逼近定理

$$\sup\left|\frac{\partial^2 f_\rho}{\partial x^2}\right| \leq \frac{\omega_1(\rho)}{h} = \frac{\sqrt{2}\,\omega_1(\rho)}{\rho}$$

同样地有

$$\sup\left|\frac{\partial^2 f_\rho}{\partial y^2}\right| \leq \frac{\sqrt{2}\,\omega_1(\rho)}{\rho}$$

$$\sup\left|\frac{\partial^2 f_\rho}{\partial x \partial y}\right| \leq \frac{\sqrt{2}\,\omega_1(\rho)}{\rho}$$

则由 ㊷ 即得

$$|S_R^\delta(x,y;f_\rho) - f_\rho(x,y)| \leq 4\sqrt{2}\,\frac{\omega_1(\rho)}{\rho}A_2 \cdot \frac{1}{R^2} \quad ㊸$$

合并 ㊷ 及 ㊸，取 $\rho = \dfrac{1}{R}$，即得

$$|S_R^\delta(x,y;f) - f(x,y)| \leq 2(A_1 + 2\sqrt{2}A_2)\frac{1}{R}\omega_1\!\left(\frac{1}{R}\right)$$

也即证明了(ii)．

3. 定理 6.3 的证明

我们将用下述的记号

$$S(u) = \sum_{v \leq u} A_v(x,y)$$

$$S^0(u) = \sum_{v \leq u^2} A_v(x,y)$$

$$S^1(u) = \frac{1}{u^2}\sum_{v \leq u^2}(u^2 - v)A_v(x,y)$$

则

$$\mu_{\frac{\lambda_0}{R}}(S_R^\delta) = \frac{1}{2\pi}\int_0^{2\pi} S_R^\delta\!\left(x + \frac{\lambda_0}{R}\cos\theta,\; y + \frac{\lambda_0}{R}\sin\theta\right)\mathrm{d}\theta =$$

$$\sum_{v \leq R^2}\left(1 - \frac{v}{R^2}\right)^\delta A_v(x,y) J_0\!\left(\frac{\lambda_0}{R}\sqrt{v}\right) =$$

附录 VI 多元函数的三角多项式逼近

$$-\frac{1}{R^{2\delta}}\int_0^{R^2} S(u)\,\mathrm{d}\left\{(R^2-u)^\delta J_0\left(\frac{\lambda_0}{R}\sqrt{u}\right)\right\}=$$

$$-\frac{1}{R^{2\delta}}\int_0^{R} S^0(u)\,\mathrm{d}\left\{(R^2-u^2)^\delta J_0\left(\frac{\lambda_0}{R}\sqrt{u}\right)\right\}=$$

$$\frac{2\delta}{R^{2\delta}}\int_0^R S^0(u)u(R^2-u^2)^\delta J_0\left(\frac{\lambda_0}{R}u\right)\mathrm{d}u +$$

$$\frac{\lambda_0}{R^{2\delta+1}}\int_0^R S^0(u)(R^2-u^2)^\delta J_1\left(\frac{\lambda_0}{R}u\right)\mathrm{d}u =$$

$$I_1 + I_2$$

先证(i),设 $f \in \mathrm{Lip}\,\alpha$,为简单计可设 $f(x,y)=0$,则由定理 6.1 的结果可知

$$S^1(u)=O(u^{-\alpha}) \qquad ⑩$$

现要证 $\mu_{\frac{\lambda_0}{R}}(S_R^\delta)=O(R^{-\alpha})$.

首先来估计 I_1,利用关系式

$$u^2 S^1(u)=2\int_0^u v^{k-1}S^0(v)\,\mathrm{d}v \qquad ⑳$$

对 I_1 进行一次分部积分可得

$$I_1=\frac{\delta}{R^{2\delta}}u^2 S^1(u)(R^2-u^2)^{\delta-1}J_0\left(\frac{\lambda_0 u}{R}\right)\bigg|_0^R +$$

$$\frac{2\delta(\delta-1)}{R^{2\delta}}\int_0^R u^3(R^2-u^2)^{\delta-2}J_0\left(\frac{\lambda_0 u}{R}\right)S^1(u)\mathrm{d}u +$$

$$\frac{\delta\lambda_0}{R^{2\delta+1}}\int_0^R u^2(R^2-u^2)^{\delta-1}J_1\left(\frac{\lambda_0 u}{R}\right)S^1(u)\mathrm{d}u$$

注意到 $J_0(\lambda_0)=0$,即知上式第一项等于 0,后两项分别以 I_{11} 及 I_{12} 记之. 又令

$$I_{11}=\frac{2\delta(\delta-1)}{R^{2\delta}}\left\{\int_0^{R^{\frac{1}{2}}}+\int_{R^{\frac{1}{2}}}^R\right\}u^3(R^2-u^2)^{\delta-2}J_0\left(\frac{\lambda_0 u}{R}\right)S^1(u)\mathrm{d}u =$$

$$I_{11}^{(1)}+I_{11}^{(2)}$$

则利用 (369) 即知

$$I_{11}^{(1)}=O\left(R^{-2\delta}\int_0^{R^{\frac{1}{2}}}u^3R^{2\delta-4}\mathrm{d}u\right)=O(R^{-2})$$

$$I_{11}^{(2)}=$$

$$O\left(R^{-2\delta}\int_{R^{\frac{1}{2}}}^{R}u^{3-\alpha}(R^2-u^2)^{\delta-1}\cdot\frac{1}{R^2}\cdot\frac{\left|J_0\left(\frac{\lambda_0u}{R}\right)\right|}{1-\frac{u}{R}}\mathrm{d}u\right)=$$

$$O\left(R^{-2\delta+2-\alpha-2}\int_{R^{\frac{1}{2}}}^{R}u(R^2-u^2)^{\delta-1}\mathrm{d}u\right)=O(R^{-\alpha})$$

故我们得到

$$I_{11}=O(R^{-\alpha}) \tag{371}$$

同样地令

$$I_{12}=\frac{\delta\lambda_0}{R^{2\delta+1}}\left\{\int_0^{R^{\frac{1}{3}}}+\int_{R^{\frac{1}{3}}}^{R}\right\}u^3(R^2-u^2)^{\delta-1}J_1\left(\frac{\lambda_0u}{R}\right)S^1(u)\mathrm{d}u=I_{12}^{(1)}+I_{12}^{(2)}$$

利用 (369) 即知

$$I_{12}^{(1)}=O\left(R^{-2\delta-1}\int_0^{-\frac{1}{3}}u^2R^{2\delta-2}\mathrm{d}u\right)=O(R^{-2})$$

$$I_{11}^{(2)}=O\left(R^{-2\delta-1}\int_{R^{\frac{1}{3}}}^{R}u^{1-\alpha}u(R^2-u^2)^{\delta-1}\mathrm{d}u\right)=O(R^{-\alpha})$$

故我们得到

$$I_{12}=O(R^{-\alpha}) \tag{372}$$

合并 (371) 及 (372) 即得

$$I_1=O(R^{-\alpha}) \tag{373}$$

其次来估计 I_2，利用 (369) 进行一次分部积分并注意

附录 VI 多元函数的三角多项式逼近

$$\frac{\mathrm{d}}{\mathrm{d}x}\left(\frac{J_1(x)}{x}\right) = -\frac{J_2(x)}{x}$$

即可得到

$$I_2 = \frac{\lambda_0}{2R^{2\delta+1}}(R^2-u^2)^\delta J_1\left(\frac{\lambda_0 u}{R}\right) uS^1(u)\bigg|_0^R +$$

$$\frac{\lambda_0 \delta}{R^{2\delta+1}}\int_0^R u^2 S^1(R^2-u^2)^{\delta-1}J_1\left(\frac{\lambda_0 u}{R}\right)\mathrm{d}u +$$

$$\frac{\lambda_0^2}{2R^{2\delta+2}}\int_0^R (R^2-u^2)^\delta J_2\left(\frac{\lambda_0 u}{R}\right) uS^1(u)\mathrm{d}u$$

上式第一项为零,而第二项即为 I_{12},最后一项记作 I_{21}, 而令

$$I_{21} = \frac{\lambda_0^2}{2R^{2\delta+2}}\left\{\int_0^A + \int_A^R\right\} u(R^2-u^2)^\delta J_2\left(\frac{\lambda_0 u}{R}\right) S^1(u)\mathrm{d}u =$$
$$I_{21}^{(1)} + I_{21}^{(2)}$$

则易知

$$I_{21}^{(1)} = O\left(R^{-2\delta-2}\int_0^A R^{2\delta}u\mathrm{d}u\right) = O(R^{-2})$$

$$I_{21}^{(2)} = O\left(R^{-2\delta-2}\int_A^R R^{2\delta}u^{1-\alpha}\mathrm{d}u\right) = O(R^{-\alpha})$$

故有

$$I_{21} = O(R^{-\alpha})$$

因此也就有

$$I_2 = O(R^{-\alpha}) \qquad \text{③⑦④}$$

合并 ③⑦③ 及 ③⑦④,最后即得到

$$\mu_{\frac{\lambda_0}{R}}(S_R^\delta) = O(R^{-\alpha})$$

此即证明了(i).

现在设 $\frac{\partial f}{\partial x}, \frac{\partial f}{\partial y} \in \mathrm{Lip}\,\alpha$,由定理 6.1 的结果,知

Tschebyscheff 逼近定理

$$S^1(u) = O(u^{-1-\alpha})$$

此时对 $I_{11}^{(1)}$ 的估计仍成立,而对 $I_{11}^{(2)}$ 我们有

$$I_{11}^{(2)} = O\left(R^{-2\delta}\int_{R^{\frac{1}{2}}}^{R} u^{2-\alpha}(R^2 - u^2)^{\delta-1}\frac{1}{R^2}\mathrm{d}u\right) =$$
$$O(R^{-1-\alpha})$$

因此有
$$I_{11} = O(R^{-1-\alpha})$$

又对 $I_{11}^{(2)}$,我们有

$$I_{12}^{(2)} = O\left(R^{-2\delta-1}\int_{R^{\frac{1}{3}}}^{R} u^{1-\alpha}(R^2 - u^2)^{\delta-1}\cdot\frac{u}{R}\mathrm{d}u\right) =$$
$$O(R^{-1-\alpha})$$

因此也有
$$I_{12} = O(R^{-1-\alpha})$$

最后对 $I_{21}^{(2)}$ 可得

$$I_{21}^{(2)} = O\left(R^{-2\delta-2}\int_{A}^{R} R^{2\delta} u^{-\alpha}\cdot\frac{u^2}{R^2}\mathrm{d}u\right) =$$
$$O(R^{-1-\alpha})$$

故也有
$$I_{21} = O(R^{-1-\alpha})$$

合并以上的结论即得到
$$\mu_{\frac{\lambda_0}{R}}(S_R^\delta) = O(R^{-1-\alpha})$$

此即证明了(ii).

4. 定理 6.3 的另一证明

首先我们有
$$S_R^{(k)}(x,y;f) = \int_0^\infty \varphi_{xy}\left(\frac{t}{R}\right) H_k(t)\mathrm{d}t \qquad ㊵$$

其中 φ_{xy} 由 ㊱ 定义,而

附录 Ⅵ 多元函数的三角多项式逼近

$$H_k(t) = t\int_0^1 u(1-u^k)J_0(ut)\,du$$

在 ㊵ 中令 $f \equiv 1$,即有

$$\int_0^\infty H_k(t)\,dt = 1 \qquad ㊶$$

因此有

$$S_R^{(k)}(x,y;f) - f(x,y) =$$
$$\int_0^\infty \left\{\varphi_{xy}\left(\frac{t}{R}\right) - f(x,y)\right\} H_k(t)\,dt \qquad ㊷$$

现在设 k 是偶数,即 $k = 2k_1$,则利用 Bessel 函数的关系式

$$xJ'_n(x) + nJ_n(x) = xJ_{n-1}(x) \qquad ㊸$$

不难得到

$$H_k(t) = \sum_{k=1}^{k_1}(-1)^{i+1}\frac{2^i k_1(k_1-1)\cdots(k_1-i+1)}{t^i}J_{i+1}(t)$$

$$㊹$$

注意到 $J_\mu(t) = O(t^{-\frac{1}{2}})\ (t \to \infty)$,因此知

$$H_k(t) = O(t^{-\frac{3}{2}}) \quad (t \to \infty)$$

则 $H_k(t) \in L(0,\infty)$,令

$$H_k^{(1)}(u) = \int_u^\infty H_k(t)\,dt$$

利用表达式 ㊹ 而进行分部积分即知

$$H_k^{(1)}(u) = O(u^{-\frac{3}{2}}) \quad (u \to \infty)$$

用类似的方法递推下去可知,若我们令

$$H_k^{(j)}(u) = \int_u^\infty H_k^{(j-1)}(t)\,dt$$

而 $$H_k^{(0)}(t) = H_k(t)$$

则 $H_k^{(j)}(u) \in L(0,\infty)$,且有

Tschebyscheff 逼近定理

$$H_k^{(j)}(u) = O\ (u^{-\frac{3}{2}})\quad (u \to \infty; j = 0,1,2,\cdots)$$

故可令

$$A_j = \int_u^\infty |H_k^{(j)}(u)|\ \mathrm{d}u \quad (j = 0,1,2,\cdots)$$

由 ㊟ 知 $H_k^{(j)}(0) = 1$.

今在 ㊟ 中特别取 $f(x,y) = \mathrm{e}^{\mathrm{i}(x+y)}$，则注意此时

$$S_R^{(k)}(x,y;\mathrm{e}^{\mathrm{i}(x+y)}) = \left(1 - \frac{2^{\frac{k}{2}}}{R^k}\right)\mathrm{e}^{\mathrm{i}(x+y)}$$

$$\varphi_{xy}(t) = J_0(\sqrt{2}t)\mathrm{e}^{\mathrm{i}(x+y)}$$

再令 $x = 0, y = 0$，即得

$$1 - \frac{2^{\frac{k}{2}}}{R^k} = \int_0^\infty J_0\left(\frac{\sqrt{2}t}{R}\right) H_k(t)\ \mathrm{d}t$$

对上式右边进行二次分部积分，因 $H_k^{(j)}(0) = 1$ 及 $J_0'(0) = 0$，则可得到

$$1 - \frac{2^{\frac{k}{2}}}{R^k} = 1 - \frac{2}{R^2}\int_0^\infty H_k^{(2)}(t) J_0''\left(\frac{\sqrt{2}t}{R}\right)\ \mathrm{d}t$$

也即

$$\int_0^\infty H_k^{(2)}(t) J_0''\left(\frac{\sqrt{2}t}{R}\right)\mathrm{d}t = \frac{2^{\frac{k}{2}}}{R^k} \cdot \frac{R^2}{2} \qquad ㊳$$

若 $k > 2$，则令 $R \to \infty$ 即得

$$\int_0^\infty H_k^{(2)}(t)\mathrm{d}t = H_k^{(3)}(0) = 0$$

若 $k > 4$，则对 ㊳ 左边再用分部积分即可推知

$$H_k^{(5)}(0) = 0$$

总之，只需注意

$$J_0^{(2i+1)}(0) = 0$$
$$J_0^{(2i)}(0) = (-1)^i \frac{(2i-1)(2i-3)\cdots 1}{2^i \cdot i!}$$

附录 Ⅵ 多元函数的三角多项式逼近

即可推知
$$H_k^{(3)}(0) = H_k^{(5)}(0) = \cdots = H_k^{(k-1)}(0) = 0$$
利用上述结果以及 $\varphi_{xy}^{(2i+1)}(0) = 0$,我们对 ㊍ 进行 j 次分部积分,而 $j \leqslant k$,则即有
$$S_R^{(k)}(x,y;f) - f(x,y) = \frac{1}{R^j} \int_0^\infty \varphi_{xy}^{(j)}\left(\frac{t}{R}\right) H_k^{(j)}(t)\,\mathrm{d}t \quad ㊿$$
因此若已知
$$\left|\frac{\partial^j f}{\partial x^\alpha \partial y^\beta}\right| \leqslant M \quad (\alpha + \beta = j)$$
则即有
$$| S_R^{(k)}(x,y;f) - f(x,y) | \leqslant 2^j M A_j \frac{1}{R^j} \quad ㊁$$

今设 $f \in C^p$,而 $p \leqslant k-1$,与证明定理 1 时一样,我们引出函数 $G(x,y)$,$f_\rho(x,y)$ 及 $\eta_\rho(x,y)$,则易知
$$\left|\frac{\partial^p \eta_\rho}{\partial x^\alpha \partial y^\beta}\right| \leqslant \omega_p(\rho) \quad (\alpha + \beta = p)$$
及
$$\left|\frac{\partial^{p+1} f_\rho}{\partial x^\alpha \partial y^\beta}\right| \leqslant \frac{\sqrt{2}}{\rho}\omega_p(\rho) \quad (\alpha + \beta = p+1)$$
因 $p+1 \leqslant k$,故利用 ㊁ 即有
$$| S_R^{(k)}(x,y;\eta_\rho) - \eta_\rho(x,y) \leqslant 2^p A_p \omega_p(\rho)\frac{1}{R^p}$$
$$| S_R^{(k)}(x,y;f_\rho) - f_\rho(x,y) \leqslant 2^{p+1}\frac{\sqrt{2}}{\rho}\omega_p(\rho)A_{p+1}\frac{1}{R^{p+1}}$$
合并上两式,取 $\rho = \frac{1}{R}$ 即知
$$| S_R^{(k)}(x,y;f) - f(x,y) \leqslant$$
$$2^p(A_p + 2\sqrt{2}A_{p+1})\frac{1}{R^p}\omega_p\left(\frac{1}{R}\right) \quad ㊃$$

Tschebyscheff 逼近定理

故当 k 是偶数时定理已得证.

现在设 k 是奇数,同样利用关系式 ⑰⑧ 可得

$$H_k(t) =$$

$$\sum_{i=0}^{k}(-1)^i\frac{k(k-2)(k-4)\cdots(k-2i)}{t^{i+1}}J_{i+2}(t) +$$

$$(-1)^{\frac{k-1}{2}}\frac{[k(k-2)(k-4)\cdots 1]^2(k+2)}{t^{k+1}} \cdot$$

$$\int_0^t \frac{J_{k+2}(u)}{u}\mathrm{d}u$$

故此时当 $j < k$ 时,仍有

$$H_k^{(j)}(u) = O(u^{-\frac{3}{2}}) \quad (u \to \infty)$$

而当 $j = k$ 时有

$$H_k^{(k)}(u) = B_k \cdot \frac{1}{u} + O(u^{-\frac{3}{2}}) \quad (u \to \infty) \qquad ㊴$$

其中

$$B_k = (-1)^{\frac{k-1}{2}}\frac{[k(k-2)\cdots 1]^2 \cdot (k+2)}{k!} \cdot$$

$$\int_0^\infty \frac{J_{k+2}(t)}{t}\mathrm{d}t$$

又同样有

$$H_k^{(3)}(0) = H_k^{(5)}(0) = \cdots = H_k^{(k)}(0) = 0$$

因此当 $j < k$ 时 ㊂ 仍成立. 与偶数的情形完全类似地即可推知当 $f(x,y) \in C^p$ 而 $p < k-1$ 时仍有 ㊃ 成立.

式 ㊂ 中当 $j = k$ 时,也即若

$$\left|\frac{\partial^k f}{\partial x^\alpha \partial y^\beta}\right| \leq M \quad (\alpha + \beta = k)$$

则有

$$S_R^{(k)}(x,y;f) - f(x,y) =$$

附录 Ⅵ　多元函数的三角多项式逼近

$$\frac{1}{R^k}\int_0^\infty \varphi_{xy}^{(k)}\left(\frac{t}{R}\right) H_k^{(k)}(t)\,\mathrm{d}t =$$

$$\frac{1}{R^k}\left\{\int_0^1 + \int_1^\infty\right\} \varphi_{xy}^{(k)}\left(\frac{t}{R}\right) H_k^{(k)}(t)\,\mathrm{d}t =$$

$$O(R^{-k}) + \frac{1}{R^k}\int_1^\infty \varphi_{xy}^{(k)}\left(\frac{t}{R}\right) H_k^{(k)}(t)\,\mathrm{d}t$$

注意到 ㊞ 即得

$$S_R^{(k)}(x,y;f) - f(x,y) =$$

$$\frac{B_k}{R^k}\int_{\frac{1}{R}}^\infty \varphi_{xy}^{(k)}(u)\,\frac{1}{u}\mathrm{d}u + O(R^{-k}) =$$

$$O\left(\frac{\ln R}{R^k}\right)$$

利用上式与以前的推理一样,即知若 $f \in C^p$ 而 $p = k - 1$,则应用式 ㊟ 成立. 定理 6.3 的证明已毕.

值得指出的是当 $p = k - 1$ 时, 若 $\omega_p(\rho) = O(\rho^\alpha)\left(0 \leqslant \alpha < \frac{1}{2}\right)$,则直接利用等式

$$S_R^{(k)}(x,y;f) - f(x,y) =$$

$$\frac{1}{R^p}\int_0^\infty \varphi_{xy}^{(p)}\left(\frac{t}{R}\right) H_k^{(p)}(t)\,\mathrm{d}t$$

并注意到

$$\int_0^\infty H_k^{(p)}(t)\,\mathrm{d}t = H_k^{(p+1)}(0) = H_k^{(k)}(0) = 0$$

即得

$$|S_R^{(k)}(x,y;f) - f(x,y)| \leqslant O\left(\frac{1}{R^{p+\alpha}}\right) \qquad ㊟$$

这是由于当 $\alpha < \frac{1}{2}$ 时,积分

$$\int_0^\infty t^\alpha |H_k^{(p)}(t)|\,\mathrm{d}t$$

Tschebyscheff 逼近定理

是收敛的缘故. 关系式 ㊟ 表示对于 $\omega_p(\rho) = O(\rho^\alpha)\left(0 < \alpha < \dfrac{1}{2}\right)$ 的特殊情形来说

$$S_R^{(k)}(x,y;f) - f(x,y) = O\left[\dfrac{1}{R^p}\omega_p\left(\dfrac{1}{R}\right)\right] \quad (p = k-1)$$

当 k 是奇数时还是成立的.

多元周期函数的非整数次积分与三角多项式逼近

附录 VII

1. 多元周期函数的非整数次积分

设 $f(P)$ 是在整个 m 维空间 Ω 上连续的函数,并且假定积分

$$\int_\Omega f(Q) r_{PQ}^{\alpha-m} dQ \quad (0 < \alpha < m)$$

绝对收敛,其中

$$r_{PQ} = [(x_1 - \xi_1)^2 + \cdots + (x_m - \xi_m)^2]^{\frac{1}{2}}$$

表示任一固定点 $P(x_1, \cdots, x_m)$ 到点 $Q(\xi_1, \cdots, \xi_m)$ 的距离. 对于这样的函数 $f(P)$, M. Riesz 用下面的关系式来定义 $f(P)$ 的 α 次积分

$$I_\alpha[f(p)] = \frac{1}{H_m(\alpha)} \int_\Omega f(Q) r_{PQ}^{\alpha-m} dQ \quad \text{⑯}$$

Tschebyscheff 逼近定理

其中

$$H_m(\alpha) = e^{\frac{\pi}{2}i\alpha} \frac{2^\alpha \pi^{\frac{m-1}{2}} \Gamma\left(\frac{\alpha}{2}\right)}{\Gamma\left(\frac{m-\alpha}{2}\right)} =$$

$$e^{\frac{\pi}{2}i\alpha} \int_\Omega e^{i\xi_1} (\xi_1^2 + \cdots + \xi_m^2)^{\frac{\alpha-m}{2}} d\xi_1 \cdots d\xi_m \quad ① \quad ㊛$$

Riesz 的定义不适用于一般的周期函数，我们的目的是要对周期函数来建立非整次积分的概念，这可以看作是单元情形 Weyl 积分的推广. 我们下面所讨论的函数 $f(P)$ 都假定满足下面两个条件：

(1) $f(P)$ 对每一变量都以 1 为周期，且

$$\int_E f(P) dP = 0 \quad ㊝$$

其中 E 是方形区域 $0 \le x_i \le 1 (i = 1, 2, \cdots, m)$.

(2) $f(P)$ 在 E 上是连续的.

为了要对这样的周期函数 $f(P)$ 来建立非整数次积分的概念，我们需要先证以下几个引理：

引理 7.1 设

$$g(P) = \begin{cases} \dfrac{1}{H_m(\alpha)} r_{0P}^{-\beta}, (P \ne 0) \\ 0, (P = 0) \end{cases}$$

其中 $m - 1 < \beta < m$，而 $\alpha = m - \beta$. 又设

$$G_N(P) = G_{n_1, \cdots, n_m}(x_1, \cdots, x_m) =$$

$$\sum_{v_1 = -n_1}^{n_1} \cdots \sum_{v_m = -n_m}^{n_m} g(x_1 + v_1, \cdots, x_m + v_m)$$

---

① 我们采用的 $H_m(\alpha)$ 和 Riesz 所定义的差一因子 $e^{\frac{\pi}{2}i\alpha}$.

附录 Ⅶ 多元周期函数的非整数次积分与三角多项式逼近

$$(0 \leqslant x_i \leqslant 1; i = 1, 2, \cdots, m)$$

$$K_N = K_{n_1, \cdots, n_m} = \int_E G_N(P) \, dP$$

则函数

$$H(P) = \lim_{n_1, \cdots, n_m \to \infty} \{ G_N(P) - K_N \}$$

可积,且 $H(P)$ 的 Fourier 级数具有下面的形式

$$H(P) \sim \sum_{-\infty}^{\infty}{}' \left\{ \int_\Omega g(Q) e^{-2\pi i NQ} dQ \right\} e^{2\pi i NP}$$

这里 NP 表示 $n_1 x_1 + \cdots + n_m x_m$,$\sum'$ 表示和号中 $n_1 = n_2 = \cdots = n_m = 0$ 的项不出现,而无穷积分 $\int_\Omega g(Q) e^{-2\pi i NQ} dQ$ 是依 Cauchy 意义收敛的.

证明 我们只需验证序列

$$\{ G_N(P) - K_N \}$$

在 E 上的一致收敛性. 事实上,如果这一点成立,即由 K_N 的定义可知

$$\int_E H(P) \, dP = \lim_{N \to \infty} \int_E \{ G_N(P) - K_N \} \, dP = 0$$

其次由于

$$\int_E e^{-2\pi i NP} dP = 0$$

可知

$$\int_E H(P) e^{-2\pi i NP} dP = \lim_{N \to \infty} \int_E G_N(P) e^{-2\pi i NP} dP$$

再根据 $G_N(P)$ 的定义我们得到

$$\int_E H(P) e^{-2\pi i NP} dP =$$

$$\lim_{N \to \infty} \int_E \left\{ \sum_{-n_1}^{n_1} \cdots \sum_{-n_m}^{n_m} g(x_1 + v_1, \cdots, x_m + v_m) \right\} e^{-2\pi i NP} dP =$$

$$\lim_{n_1,\cdots,n_m\to\infty}\int_{-n_1}^{n_1+1}\cdots\int_{-n_m}^{n_m+1}g(x_1,\cdots,x_m)\mathrm{e}^{-2\pi\mathrm{i}(n_1x_1+\cdots+n_mx_m)}\mathrm{d}x_1\cdots\mathrm{d}x_m=$$

$$\int_\Omega g(P)\mathrm{e}^{-2\pi\mathrm{i}NP}\mathrm{d}P$$

这样 $H(P)$ 的 Fourier 系数已确定,即得引理 7.1 的结果.

今来验证序列 (389) 的一致收敛性,由 K_N 的定义知

$$G_N(P)-K_N=\int_E\{G_N(P)-G_N(Q)\}\mathrm{d}Q=$$

$$\sum_{-n_1}^{n_1}\cdots\sum_{-n_m}^{n_m}\int_0^1\cdots\int_0^1\{g(x_1+v_1,\cdots,x_m+v_m)-$$

$$g(y_1+v_1,\cdots,y_m+v_m)\}\mathrm{d}y_1\cdots\mathrm{d}y_m$$

令

$$C_{v_1,\cdots,v_m}(P)=\int_0^1\cdots\int_0^1\{g(x_1+v_1,\cdots,x_m+v_m)-$$

$$g(y_1+v_1,\cdots,y_m+v_m)\}\mathrm{d}y_1\cdots\mathrm{d}y_m$$

则我们的问题是要证明多重级数

$$\sum C_{v_1,\cdots,v_m}(P)$$

在 $P\in E$ 上是一致收敛的.

不难看出当 $v=v_1^2+\cdots+v_m^2$ 充分大时

$$|g(x_1+v_1,\cdots,x_m+v_m)-$$

$$g(y_1+v_1,\cdots,y_m+v_m)|<Kv^{-\frac{1}{2}(\beta+1)}$$

其中K是与$(x_1,\cdots,x_m)\in E$及$(y_1,\cdots,y_m)\in E$无关的常数. 因而

$$|C_{v_1,\cdots,v_m}(P)|<K(v_1^2+\cdots+v_m^2)^{-\frac{1}{2}(\beta+1)}$$

今用 $A_v(P)$ 来记下面的和

$$A_v(P)=\sum_{v_1^2+\cdots+v_m^2=v}|C_{v_1,\cdots,v_m}(P)|$$

附录 Ⅶ 多元周期函数的非整数次积分与三角多项式逼近

则 v 充分大时

$$A_v(P) < Kv^{-\frac{1}{2}(\beta+1)} B(v)$$

其中 $B(v)$ 表示在 $x_1^2 + \cdots + x_m^2 = v$ 的圆周上的格子点的数目. 则

$$S(n) = \sum_{v=1}^{n} B(v)$$

正好是圆 $x_1^2 + \cdots + x_m^2 \leq n$ 上格子点的数目, 我们已知

$$S(n) = O(n^{\frac{m}{2}})$$

因此有

$$\sum_{v=1}^{M} A_v(P) = O\left[\sum_{v=1}^{M} v^{-\frac{1}{2}(\beta+1)} B(v)\right] =$$

$$O\left[S(M) M^{-\frac{1}{2}(\beta+1)} + \sum_{v=1}^{M-1} S(v) v^{-\frac{1}{2}(\beta+1)-1}\right]$$

注意到 $m - 1 < \beta$, 即知下式关于 $P \in E$ 一致地成立

$$\sum_{v=1}^{M} A_v(P) = O(1) \quad (M \to \infty)$$

这也就证明了级数

$$\sum |C_{v_1, \cdots, v_m}(P)|$$

在 E 上的一致收敛性.

引理 7.2 设 $f(P)$ 是我们所考虑的周期函数满足前面规定的两个条件, 又设 $f(P)$ 的 Fourier 级数如下

$$f(P) \sim {\sum}' C_{n_1, \cdots, n_m} e^{2\pi i(n_1 x_1 + \cdots + n_m x_m)}$$

而令

$$A_v(P) = \sum_{n_1^2 + \cdots + n_m^2 = v} C_{n_1, \cdots, n_m} e^{2\pi i(n_1 x_1 + \cdots + n_m x_m)}$$

则下述等式成立

Tschebyscheff 逼近定理

$$\lim_{R\to\infty}\sum_{v\leqslant R^2}\left(1-\frac{v}{R^2}\right)^{\delta}\frac{A_v(P)}{(2\pi)^{\alpha}v^{\frac{1}{2}\alpha}e^{\frac{\pi}{2}i\alpha}} =$$

$$\int_E f(Q)H(Q-P)\mathrm{d}Q \qquad \text{㊚}$$

其中 $\delta > \dfrac{m-1}{2}, 0 < \alpha = m - \beta < 1$，而 $H(P)$ 即为引理 7.1 中所确定的极限函数.

证明 由引理 7.1 知 $H(P)$ 的 Fourier 系数是

$$\int_\Omega g(Q)\mathrm{e}^{-2\pi iNQ}\mathrm{d}Q = \frac{1}{H_m(\alpha)}\int_\Omega \mathrm{e}^{-2\pi iNQ}r_{0Q}^{\alpha-m}\mathrm{d}Q$$

而经过正交变换可得

$$\int_\Omega \mathrm{e}^{-2\pi iNQ}r_{0Q}^{\alpha-m}\mathrm{d}Q = (n_1^2+\cdots+n_m^2)^{-\frac{\alpha}{2}}(2\pi)^{-\alpha}\cdot$$

$$\int_\Omega \mathrm{e}^{i\xi_1}(\xi_1^2+\cdots+\xi_m^2)^{\frac{\alpha-m}{2}}\mathrm{d}\xi_1\cdots\mathrm{d}\xi_m$$

再根据 ㊇ 即得

$$\int_\Omega g(Q)\mathrm{e}^{-2\pi iNQ}\mathrm{d}Q = (n_1^2+\cdots+n_m^2)^{-\frac{\alpha}{2}}(2\pi)^{-\alpha}\mathrm{e}^{-\frac{\pi}{2}i\alpha}$$

因此如果令

$$G(P) = \int_E f(Q)H(Q-P)\mathrm{d}Q$$

则 $G(P)$ 的 Fourier 级数为

$$G(P) \sim \sum{}' \frac{C_{n_1,\cdots,n_m}}{(2\pi)^{\alpha}v^{\frac{1}{2}\alpha}\mathrm{e}^{\frac{\pi}{2}i\alpha}}\mathrm{e}^{2\pi iNP}$$

$$(v = n_1^2+\cdots+n_m^2)$$

因 $f(P)$ 连续,则 $G(P)$ 也连续. 因此利用多元函数的 Fourier 级数方面的已知结果,即知 $G(P)$ 的 Fourier 级数可以用 $\delta > \dfrac{m-1}{2}$ 级的 Riesz 平均求和,且其和即为 $G(P)$. 这也就证明了我们的结论.

定义　设 $\delta_m = \left[\frac{m-1}{2}\right] + 1, \alpha \geqslant 0$,我们称

$$I_\alpha^*[f(P)] = \lim_{R\to\infty} \sum_{v\leqslant R^2}{}' \left(1 - \frac{v}{R^2}\right)^{\delta_m} \frac{A_v(P)}{(2\pi)^\alpha v^{\frac{1}{2}\alpha} e^{\frac{\pi}{2}i\alpha}}$$

为 $f(P)$ 的 α 次积分,其中 $f(P)$ 是满足我们条件的周期函数,$A_v(P)$ 为 $f(P)$ 的Fourier级数的圆形部分和之一般项.

我们有下述的定理:

定理7.3　对于任一 $\alpha > 0$,$I_a^*[f(P)]$ 存在,且当 $0 < \alpha < 1$ 时

$$I_\alpha^*[f(P)] = \frac{I}{H_m(\alpha)}\int_\Omega f(Q) r_{PQ}^{\alpha-m} dQ = I_\alpha[f(P)] \tag{391}$$

证明　当 $0 < \alpha < 1$ 时,$I_\alpha^*[f(P)]$ 的存在以及(391)的成立即为引理 2 的直接推论. 事实上,由于 $f(P)$ 满足条件(388),因而(390)的右端可以写成

$$\lim_{n_1,\cdots,n_m\to\infty} \int_{-n_1}^{n_i+1}\cdots\int_{-n_m}^{n_m+1} f(x_1+y_1,\cdots,x_m+y_m)\cdot g(y_1,\cdots,y_m) dy_1\cdots dy_m = \frac{1}{H_m(\alpha)}\int_\Omega f(P+Q) r_{0Q}^{\alpha-m} dQ$$

所以由(390)即得(391).

至于对一般 $\alpha > 0$ 的情形要证明 $I_\alpha^*[f(P)]$ 的存在,只需注意到以下事实:即从极限

$$\lim_{R\to\infty} \sum_{v\leqslant R^2}{}' \left(1 - \frac{v}{R^2}\right)^{\delta_m} A_v(P) \tag{392}$$

的存在可以推出极限

$$\lim_{R\to\infty} \sum_{v\leqslant R^2}{}' \left(1 - \frac{v}{R^2}\right)^{\delta_m} \frac{A_v(P)}{(2\pi)^\alpha v^{\frac{1}{2}\alpha} e^{\frac{\pi}{2}i\alpha}} \tag{393}$$

的存在. 这是由于一方面 δ_m 是正整数,另一方面 δ_m 级的 Riesz 求和与 (C,δ_m) 求和是等价的,所以反复地应用 Abel 变换(即和差变换)即可证得这事实. 进一步由于 $f(P)$ 的连续性,则 ㉒ 一致收敛,因此可推知 ㉓ 也是一致收敛的.

2. 非整数次积分的性质

我们限于讨论 $m = 2$ 的情形.

首先来证明下面的引理:

引理 7.4 设 $f(x,y)$ 为对每一变量以 1 为周期的连续函数,且满足下列条件

$$\int_0^1 \int_0^1 f(x,y)\,\mathrm{d}x\mathrm{d}y = 0 \qquad ㉔$$

$$\int_0^a \frac{\omega(\rho)}{\rho}\mathrm{d}\rho < \infty \quad (a > 0) \qquad ㉕$$

其中 $\omega(\rho)$ 为 $f(x,y)$ 的连续模

$$\omega(\rho) = \max_{(x_1-x_2)^2+(y_1-y_2)^2 \leqslant \rho^2} | f(x_1,y_1) - f(x_2,y_2) |$$

则存在一连续周期函数 $G(P) \equiv G(x,y)$,使在整个 xOy 平面上满足 Poisson 方程

$$\Delta G(P) = f(P) \quad (\Delta\ 表示\ Laplace\ 运算子) \qquad ㉖$$

证明 设

$$f(x,y) \sim \sum_{m,n=-\infty}^{\infty} C_{mn} e^{2\pi i(mx+ny)}$$

而其圆形和的一般项记作

$$A_v(x,y) = \sum_{m^2+n^2=v} C_{mn} e^{2\pi i(mx+ny)}$$

由于 $f(x,y)$ 连续,则 $f(x,y) \in L^p(p>1)$,因此级数

$$\sum_{v=1}^{\infty} \frac{A_v(x,y)}{v}$$

附录 Ⅶ 多元周期函数的非整数次积分与三角多项式逼近

绝对且一致收敛.

今令
$$G(x,y) = -\sum_{v=1}^{\infty} \frac{A_v(x,y)}{(2\pi)^2 v} \qquad ㊳$$
我们来证明 $G(x,y)$ 即满足我们的要求.

显然 $G(x,y)$ 是对每一变量以 1 为周期的连续函数,今设 E_1 为任一包含区域 $E:0 \leq x \leq 1, 0 \leq y \leq 1$ 在其内部的区域,我们只需证明 E_1 内满足方程 ㊳ 就够了. 作函数
$$u(P) \equiv u(x,y) = -\frac{1}{2\pi}\int_{E_1} f(Q)\ln\frac{1}{r}\mathrm{d}S_Q$$
其中 $r = |PQ|$. 由条件 ㊳,我们即知在 E_1 内有
$$\Delta u(P) = \widetilde{\Delta} u(P) = f(P) \qquad ㊳$$
这里
$$\widetilde{\Delta} u(P) = \lim_{t\to 0}\frac{4}{t^2}\{\mu(u;P;t) - u(P)\}$$
而
$$\mu(u;P;t) = \frac{1}{2\pi}\int_0^{2\pi} u(x+t\cos\theta, y+t\sin\theta)\mathrm{d}\theta$$

又因 $f(x,y)$ 连续,故级数 $\sum_{1}^{\infty} A_v(x,y)$ 可以 $(C,1)$ 求和. 因此由 ㊳ 定义的函数 $G(x,y)$ 在整个平面上满足关系
$$\widetilde{\Delta} G(x,y) = f(x,y) \qquad ㊳$$
只需注意现在由于条件 (1) 知 $C_{00} = 0$.

联合 ㊳ 及 ㊳ 即得在 E_1 内
$$\widetilde{\Delta}(G - u) = 0$$
因此 $G - u$ 是 E_1 内的调和函数,也即

Tschebyscheff 逼近定理

$$\Delta(G-u)=0$$

最后再注意到 E_1 内

$$\Delta u = f(P)$$

所以在 E_1 内就有

$$\Delta G = f(P)$$

引理于是证毕.

首先来证以下定理：

定理 7.5 (i) 对于任意的 $\alpha \geqslant 0, \beta \geqslant 0$ 有

$$I^*_{\alpha+\beta}[f(P)] = I^*_\alpha\{I^*_\beta[f(P)]\} \qquad ⑩$$

(ii) 如果 $I^*_\alpha[f(P)](\alpha \geqslant 0)$ 的连续模 $\omega(\rho)$ 满足条件 ㊴，则下式成立

$$\Delta I^*_{\alpha+2}[f(P)] = I^*_\alpha[f(P)] \qquad ⑪$$

证明 设

$$f(x,y) \sim \sum_{m,n=-\infty}^{\infty}{}' C_{mn} e^{2\pi i(mx+ny)}$$

$C_{00}=0$ 是由于 $f(P)$ 满足条件 ㊞.

根据定义知

$$I^*_\beta[f(x,y)] = \lim_{R\to\infty}\left(1-\frac{v}{R^2}\right)\frac{A_v(x,y)}{(2\pi)^\beta v^{\frac{\beta}{2}} e^{\frac{\pi}{2}i\beta}}$$

其中 $A_v(x,y)$ 是 $f(x,y)$ 的 Fourier 级数圆形部分和的一般项. 令

$$C'_{mn} = \frac{C_{mn}}{e^{\frac{\pi}{2}i\beta} v^{\frac{\beta}{2}}(2\pi)^\beta} = O(v^{-\frac{\beta}{2}}) \quad (v=m^2+n^2)$$

因此三角级数

$$\sum_{m,n=-\infty}^{\infty} C'_{mn} e^{2\pi i(mx+ny)}$$

属于 U 类,也即级数

附录 Ⅶ 多元周期函数的非整数次积分与三角多项式逼近

$$\sum_{v=1}^{\infty} \frac{A'(x,y)}{v}$$

一致收敛,这里

$$A'(x,y) = \sum_{m^2+n^2=v} C'_{mn} e^{2\pi i(mx+ny)} = \frac{A_v(x,y)}{e^{\frac{\pi}{2}i\beta} v^{\frac{\beta}{2}} (2\pi)^\beta}$$

现在再根据三角级数的唯一性即知

$$I_\beta^*[f(x,y)] \sim \sum_{m,n=-\infty}^{\infty}{}' C'_{mn} e^{2\pi i(mx+ny)}$$

同样地当然就有

$$I_\alpha^*\{I_\beta^*[f(x,y)]\} \sim \sum_{m,n=-\infty}^{\infty}{}' C''_{mn} e^{2\pi i(mx+ny)}$$

而其中

$$C''_{mn} = \frac{C_{mn}}{e^{\frac{\pi}{2}i\alpha} v^{\frac{\alpha}{2}} (2\pi)^\alpha} = \frac{C_{mn}}{e^{\frac{\pi}{2}i(\alpha+\beta)} v^{\frac{\alpha+\beta}{2}} (2\pi)^{\alpha+\beta}}$$

又在上面讨论的以 $\alpha+\beta$ 代替 β 就得到

$$I_{\alpha+\beta}^*[f(x,y)] \sim \sum_{m,n=-\infty}^{\infty}{}' C''_{mn} e^{2\pi i(mx+ny)}$$

最后根据 Fourier 级数的唯一性以及函数 $I_\alpha^*\{I_\beta^*[f(x,y)]\}$ 与 $I_{\alpha+\beta}^*[f(x,y)]$ 都是连续的即推知 ⑩ 成立.

其次来证 ⑩. 由上面讨论可知

$$I_\alpha^*[f(x,y)] \sim \sum_{m,n=-\infty}^{\infty} \frac{C_{mn}}{e^{\frac{\pi}{2}i\alpha} v^{\frac{\alpha}{2}} (2\pi)^\alpha} e^{2\pi i(mx+ny)}$$
$$(v = m^2 + n^2) \quad ⑫$$

$$I_{\alpha+2}^*[f(x,y)] \sim$$
$$\sum_{m,n=-\infty}^{\infty}{}' \left(-\frac{1}{(2\pi)^2 v}\right) \frac{C_{mn}}{e^{\frac{\pi}{2}i\alpha} v^{\frac{\alpha}{2}} (2\pi)^\alpha} e^{2\pi i(mx+ny)} \quad ⑬$$

根据定理的假设,由引理 7.4 知存在连续周期函数 $G(x,y)$,使

Tschebyscheff 逼近定理

$$\Delta G(x,y) = I_\alpha^* [f(x,y)]$$

其中 $G(x,y)$ 由 ㊗ 定义(当然此时 ㊗ 中的 $A_v(x,y)$ 代表 $I_\alpha^*[f(x,y)]$ 的 Fourier 级数圆形部分和之一般项).
由此易知

$$G(x,y) \sim \sum_{m,n=-\infty}^{\infty}{}' \left(-\frac{1}{(2\pi)^2 v}\right) C'_{mn} e^{2\pi i(mx+ny)}$$

其中 C'_{mn} 为 $I_\alpha^*[f(x,y)]$ 的 Fourier 系数. 注意到 ㊿㊿,利用 Fourier 级数的唯一性得到

$$G(x,y) = I_{\alpha+2}^*[f(x,y)]$$

因此 ㊶ 成立,定理 7.5 完全证明了.

现在我们来引进非整数次 Laplace 运算的概念.

定义 如果下式右端的运算有意义,我们即定义它为关于 $f(P)$ 的非整数次 Laplace 运算 Δ^β

$$\Delta^\beta f(P) = \Delta I_{2(1-\beta)}^*[f(P)] \quad (0 < \beta < 1)$$

下面两个定理说明了运算 I_β^* 与 Δ^β 间的互逆关系.

定理 7.6 如果

$$f(x,y) = I_{2\beta}^*[g(x,y)] \quad (0 < \beta < 1)$$

且 $g(x,y)$ 的连续模 $\omega(\rho)$ 满足条件 ㊕,则 $\Delta^\beta f(x,y)$ 存在且

$$\Delta^\beta f(x,y) = g(x,y)$$

证明 由 ㊀ 知

$$I_{2(1-\beta)}^*[f(x,y)] = I_{2(1-\beta)}^*\{I_{2\beta}^*[g(x,y)]\} = I_2^*[g(x,y)]$$

再由 ㊶($\alpha = 0$)即得我们的结论.

定理 7.7 如果 $\Delta^\beta f(x,y) = g(x,y)$ 存在且连续,则

附录 Ⅶ 多元周期函数的非整数次积分与三角多项式逼近

$$I_{2\beta}^{*}[g(x,y)] = f(x,y) \quad (0 < \beta < 1)$$

证明 设

$$f(x,y) \sim \sum_{m,n=-\infty}^{\infty}{}' C_{mn} e^{2\pi i(mx+ny)}$$

$$G(x,y) = I_{2(1-\beta)}^{*}[f(x,y)]$$

由定理 7.5 的证明过程知

$$G(x,y) \sim \sum_{m,n=-\infty}^{\infty}{}' \frac{C_{mn}}{(2\pi)^{2(1-\beta)} v^{(1-\beta)} e^{\pi i(1-\beta)}} e^{2\pi i(mx+ny)}$$

因 $\Delta G(x,y) = g(x,y)$,故

$$g(x,y) \sim \sum_{m,n=-\infty}^{\infty}{}' (2\pi)^{2\beta} v^{\beta} e^{i\alpha\beta} C_{mn} e^{2\pi i(mx+ny)}$$

再由非整数次积分的定义即得

$$I_{2\beta}^{*}[g(x,y)] = \lim_{R\to\infty} \sum_{v \leqslant R^2} \left(1 - \frac{v}{R^2}\right) A_v(x,y)$$

今因 $f(x,y)$ 连续,故它的 Fourier 级数也即三角级数

$$\sum_{m,n=-\infty}^{\infty}{}' C_{mn} e^{2\pi i(mx+ny)}$$

属于 U 类. 根据三角级数的唯一性知上级数即为 $I_{2\beta}^{*}[g(x,y)]$ 的 Fourier 级数,最后根据 Fourier 级数的唯一性即得

$$I_{2\beta}^{*}[g(x,y)] = f(x,y)$$

为了进一步讨论非整数次积分的性质,我们先要引进下面两个函数类

我们称 $f(x,y) \in \text{Lip } \alpha$,如果关于 (x,y) 一致地有
$$f(x+h, y+k) - f(x,y) = O(\rho^{\alpha})$$
$$(\rho = \sqrt{h^2 + k^2} \to 0, 0 < \alpha \leqslant 1)$$

我们称 $f(x,y) \in \text{Lip}^{*} \alpha$,如果关于点 $P(x,y)$ 一致地有

Tschebyscheff 逼近定理

$$\frac{1}{2\pi t}\int_{C(P;t)} f(R)\,dR - f(P) = O(t^\alpha)$$

$$(t \to 0, 0 < \alpha \leqslant 2)$$

其中 $C(P,t)$ 表示以 P 为中心, t 为半径的圆周.

我们已知当 $0 < \alpha < 1$ 时, $f(P) \in \mathrm{Lip}\,\alpha$ 与 $f(P) \in \mathrm{Lip}^*\alpha$ 是等价的.

今来证以下的定理, 关于 $f(x,y)$ 我们仍假定对每一变量以 1 为周期, 且满足 ㊴.

定理 7.8 设 $f(x,y) \in \mathrm{Lip}\,\alpha(0 < \alpha \leqslant 1)$, 又设 $0 < \beta < 1$, 则

$$I_\beta^*[f(x,y)] \in \mathrm{Lip}^*(\alpha + \beta)$$

特别当 $\alpha + \beta < 1$ 时, 则由 $f(x,y) \in \mathrm{Lip}\,\alpha(0 < \alpha < 1)$ 可得

$$I_\beta^*[f(x,y)] \in \mathrm{Lip}(\alpha + \beta)$$

上述定理的最后一部分, N. du Plessis 对 Riesz 的非整数次积分 $I_\beta[f(P)]$ 在较强的假设下有同样的结果.

证明 以 Ω 表整个 xOy 平面. 由于 $f(P) \in \mathrm{Lip}\,\alpha$, 因此满足引理 7.4 中的条件 ㊵. 则根据引理 7.4, 存在连续周期函数 $G(P)$ 在 Ω 上满足 Poisson 方程 ㊶.

令 P 固定, 作函数

$$\Phi(Q) = G(Q) - G(P)$$

应用 Green 公式到环形区域 $S(0;\varepsilon,R)$

$$\varepsilon^2 \leqslant x^2 + y^2 \leqslant R^2$$

我们得到

$$\int_{S(0;\varepsilon,R)} f(P+Q) r_{0Q}^{\beta-2}\,dQ =$$

$$\int_{S(0;\varepsilon,R)} \Delta_Q \Phi(P+Q) r_{0Q}^{\beta-2}\,dQ =$$

附录 Ⅶ 多元周期函数的非整数次积分与三角多项式逼近

$$\int_{S(0;\varepsilon,R)} \Phi(P+Q)\Delta r_{0Q}^{\beta-2}\mathrm{d}Q +$$

$$\int_{C(R)} \left[\frac{\partial}{\partial r}\Phi(P+Q)\right] r_{0Q}^{\beta-2}\mathrm{d}Q -$$

$$\int_{C(R)} \left[\frac{\partial}{\partial r}r_{0Q}^{\beta-2}\right] \Phi(P+Q)\mathrm{d}Q -$$

$$\int_{C(\varepsilon)} \left[\frac{\partial}{\partial r}\Phi(P+Q)\right] r_{0Q}^{\beta-2}\mathrm{d}Q +$$

$$\int_{C(\varepsilon)} \left[\frac{\partial}{\partial r}r_{0Q}^{\beta-2}\right] \Phi(P+Q)\mathrm{d}Q$$

其中 $C(R)$ 及 $C(\varepsilon)$ 分别表示以原点为中心,以 R 及 ε 为半径的圆周.

今注意

$$\int_{C(\varepsilon)} \frac{\partial}{\partial r}\Phi(P+Q)r_{0Q}^{\beta-2}\mathrm{d}Q = \varepsilon^{\beta-2}\int_{S(\varepsilon)} f(P+Q)\mathrm{d}Q$$

其中 $S(\varepsilon)$ 表示以原点为中心,ε 为半径的圆域. 因此

$$\int_{C(\varepsilon)} \frac{\partial}{\partial r}\Phi(P+Q)r_{0Q}^{\beta-2}\mathrm{d}Q = O(\varepsilon^{\beta}) \to 0 \quad (\varepsilon \to 0)$$

另外,注意到函数 $\Phi(Q)$ 以及 $\dfrac{\partial \Phi}{\partial x},\dfrac{\partial \Phi}{\partial y}$ 的有界性,我们可以得到

$$\int_{C(R)} \frac{\partial}{\partial r}\Phi(P+Q)r_{0Q}^{\beta-2}\mathrm{d}Q =$$

$$R^{\beta-1}\int_{C(1)} \frac{\partial}{\partial r}\Phi(P+RQ)\mathrm{d}Q =$$

$$O(R^{\beta-1}) \to 0 \quad (R \to \infty)$$

$$\int_{C(\varepsilon)} \frac{\partial}{\partial r}r_{0Q}^{\beta-2}(G(P+Q) - G(P))\mathrm{d}Q =$$

$$(\beta-2)\varepsilon^{\beta-2}\int_{C(1)} [G(P+\varepsilon Q) - G(P)]\mathrm{d}Q =$$

Tschebyscheff 逼近定理

$$O(\varepsilon^{\beta-2} \cdot \varepsilon^2 f(P+Q^*)) =$$
$$O(\varepsilon^{\beta}) \to 0 \quad (\varepsilon \to \infty)$$
$$\int_{C(R)} \frac{\partial}{\partial r} r_{0Q}^{\beta-2} \Phi(P+Q) \mathrm{d}Q =$$
$$(\beta-2) R^{\beta-2} \int_{C(1)} \Phi(P+QR) \mathrm{d}Q =$$
$$O(R^{\beta-2}) \to 0 \quad (R \to \infty)$$

因 $0 < \beta < 1$,则由定理 7.3 我们得到下式
$$g_1(P) = I_{\beta}^*[f(P)] =$$
$$\frac{(\beta-2)^2}{H_2(\beta)} \int_{\Omega} \{G(P+Q) - G(P)\} r_{0Q}^{\beta-4} \mathrm{d}Q$$

今考虑
$$\frac{1}{2\pi t} \int_{C(P,t)} g_1(R) \mathrm{d}R - g_1(R) =$$
$$\frac{(\beta-2)^2}{H_2(\beta)} \{ \int_{\Omega} r_{0Q}^{\beta-4} \mathrm{d}Q [\frac{1}{2\pi t} \int_{C(O,t)} (G(P+Q+R) -$$
$$G(P+R) - G(P+Q) + G(P)) \mathrm{d}R] \} =$$
$$\frac{(\beta-2)^2}{H_2(\beta)} \{ \int_{S(t)} + \int_{S'(t)} \} = \frac{(\beta-2)^2}{H_2(\beta)} \{I_1 + I_2\}$$

其中 $S'(t) = \Omega - S(t)$.

先考虑 I_1,我们有
$$I_1 = \frac{1}{2\pi t} \int_{C(t)} \{ \int_0^t r^{\beta-4} \cdot 2\pi r [(\frac{1}{2\pi r} \int_{C(r)} G(P+Q+R) \mathrm{d}Q -$$
$$G(P+R)) - (\frac{1}{2\pi t} \int_{C(r)} G(P+Q) \mathrm{d}Q - G(P))] \mathrm{d}r \} \mathrm{d}R =$$
$$\frac{1}{2\pi t} \int_{C(t)} \mathrm{d}R \int_0^t r^{\beta-4} \cdot 2\pi r^3 \{f(P+R+\theta Q) - f(P+\theta Q)\} \mathrm{d}r =$$
$$O(t^{\alpha} \int_0^t r^{\beta-1} \mathrm{d}r) = O(t^{\alpha+\beta})$$

附录Ⅶ 多元周期函数的非整数次积分与三角多项式逼近

对于 I_2 我们同样地有

$$I_2 = \int_{S'(t)} r_{0Q}^{\beta-4} \left\{ \left[\frac{1}{2\pi t} \int_{C(t)} G(P+Q+R) dR - G(P+Q) \right] - \left[\frac{1}{2\pi t} \int_{C(t)} G(P+R) dR - G(P) \right] \right\} dQ =$$

$$O\left(t^2 \int_0^\infty r^{\alpha+\beta-3} dr \right) = O(t^{\alpha+\beta})$$

由此最后得到

$$\frac{1}{2\pi t} \int_{C(P,t)} g_1(R) dR - g_1(P) = O(t^{\alpha+\beta})$$

这也就表示 $g_1(P) = I_\beta^* [f(P)] \in \text{Lip}^*(\alpha+\beta)$，定理证毕.

由此定理即可推知当 $f(x,y) \in \text{Lip} \gamma (\gamma > 0)$ 时，⑩ 成立.

又由此定理及定理 7.7 直接推出

定理 7.9 如果 $\Delta^\beta f(x,y) = g(x,y) \in \text{Lip} \alpha$ $(0 < \alpha \leq 1)$，且 $0 < \beta < \frac{1}{2}$，则

$$f(x,y) \in \text{Lip}^*(\alpha+2\beta)$$

作为定理 7.9 的逆定理，我们来证明下述定理：

定理 7.10 设 $f(P) \equiv f(x,y) \in \text{Lip}^* \alpha$，且 $1 < 2\beta < \alpha \leq 2$，则

$$\Delta^\beta f(x,y) \in \text{Lip}^*(\alpha-2\beta)$$

证明 令 $\delta = 2(1-\beta)$，则由假设的条件，$0 < \delta < 1$，因此由定理 7.3 知

$$I_\delta^* [f(P)] = \frac{1}{H_2(\delta)} \int_\Omega f(P+Q) r_{0Q}^{\delta-2} dQ$$

与证明定理 7.8 一样，我们利用引理 7.4 中的 $G(P)$ 可以得到

$$I_\delta^*[f(P)] = \frac{(\delta-2)^2}{H_2(\delta)} \int_\Omega \{G(P+Q) - G(P)\} r_{0Q}^{\delta-4} dQ$$

因此

$$g(P) \equiv \Delta^\beta f(P) = \Delta I_\delta^*[f(P)] =$$
$$\frac{(\delta-2)^2}{H_2(\delta)} \int_\Omega \{f(P+Q) - f(P)\} r_{0Q}^{\delta-4} dQ$$

利用上述表达式即有

$$\frac{1}{2\pi t} \int_{C(P,t)} g(R) dR - g(P) =$$
$$\frac{(\delta-2)^2}{H_2(\delta)} \int_\Omega r_{0Q}^{\delta-4} \{\frac{1}{2\pi t} \int_{C(t)} [f(P+Q+R) -$$
$$f(P+R) - f(P+Q) + f(P)] dR\} dQ =$$
$$\frac{(\delta-2)^2}{H_2(\delta)} [\int_{S(t)} + \int_{S'(t)}] = \frac{(\delta-2)^3}{H_2(\delta)} [I_1 + I_2]$$

但利用 $f(P) \in \text{Lip}^* \alpha$ 可以得到

$$I_1 = \frac{1}{2\pi t} \int_{C(t)} dR \cdot O(\int_0^t r^{\delta-3+\alpha} dr) =$$
$$O(t^{\alpha+\delta-2}) = O(t^{\alpha-2\beta})$$
$$I_2 = O(t^\alpha \int_t^\infty r^{\delta-3} dr) = O(t^{\alpha+\delta-2}) = O(t^{\alpha-2\beta})$$

这就表示 $g(P) = \Delta^\beta f(P) \in \text{Lip}^*(\alpha-2\beta)$. 定理证得了.

3. 三角多项式逼近

下面我们要来考虑多元周期函数的三角多项式逼近,主要讨论 $m=2$ 的情形.

在这一节中,我们设 $f(x,y)$ 为对每一变量以 2π 为周期的函数,作为逼近 $f(x,y)$ 的工具是下列形状的三角形多项式

附录 Ⅶ 多元周期函数的非整数次积分与三角多项式逼近

$$S_R^{\delta,k}(x,y;f) = \sum_{v \leqslant R^2} \left(1 - \frac{v^{\frac{k}{2}}}{R^k}\right)^\delta A_v(x,y)$$

其中 $A_v(x,y)$ 是 $f(x,y)$ 的 Fourier 级数圆形部分和的一般项,δ 与 k 都是正数,将在下面定理中加以规定.

我们以 C^{2p} 表示具有直到 $2p$ 级连续偏导数的函数集合,又以 Δ^p 表示连续进行 p 次 Laplace 运算的运算子. 我们来证明下面的结果:

定理 7.11 设 $f(x,y) \in C^{2p}$,且 $\Delta^p f(x,y)$ 的连续模为 $\omega_p(\rho)$,则当 $k > 2p$ 及 $\delta > 2p + \dfrac{5}{2}$ 时,下式关于 (x,y) 一致地成立

$$f(x,y) - S_R^{\delta,k}(x,y;f) = O\left[\frac{1}{R^{2p}}\omega_p\left(\frac{1}{R}\right)\right]$$

证明

$$S_R^{\delta,k}(x,y;f) - f(x,y) =$$
$$\int_0^\infty \left\{\varphi_{xy}\left(\frac{u}{R}\right) - f(x,y)\right\} H_{\delta,k}(u)\,\mathrm{d}u$$

其中

$$\varphi_{xy}(t) = \frac{1}{2\pi}\int_0^{2\pi} f(x + t\cos\theta, y + t\sin\theta)\,\mathrm{d}\theta$$

$$H_{\delta,k}(u) = u\int_0^1 t(1 - t^k)^\delta J_0(ut)\,\mathrm{d}t$$

而 $J_0(x)$ 是零级的第一类 Bessel 函数.

注意到

$$H_{\delta,k}(u) = \frac{1}{u}\int_0^u v\left(1 - \frac{v^k}{u^k}\right) J_0(v)\,\mathrm{d}v =$$
$$\frac{k\delta}{u^{k+1}}\int_0^u J_1(v) v^k \left(1 - \frac{v^k}{u^k}\right)^{\delta-1}\,\mathrm{d}v =$$

Tschebyscheff 逼近定理

$$\frac{(2-k)k\delta}{u^{k+1}}\int_0^u J_2(v)v^{k-1}\left(1-\frac{v^k}{u^k}\right)^{\delta-1}dv +$$

$$\frac{k^2(\delta-1)\delta}{u^{2k+1}}\int_0^u J_2(v)v^{2k-1}\left(1-\frac{v^k}{u^k}\right)^{\delta-1}dv =$$

$$O(u^{-\delta-\frac{1}{2}}) \quad (u\to\infty)$$

又 $\delta > 2p + \dfrac{5}{2}$,因此

$$\int_0^\infty u^i |H_{\delta,k}(u)| du < \infty \quad (i=0,1,\cdots,2p+2) \quad ⑤$$

今在 ④ 中,特别以 $f(x,y) = e^{i(x+y)}$ 代入,并令 $x = y = 0$,利用 Taylor 公式,并注意 $\Delta^j e^{i(x+y)}|_{(0,0)} = (-1)^j 2^j$ 即得

$$\left(1 - \frac{2^{\frac{k}{2}}}{R^k}\right)^\delta - 1 = \sum_{j=1}^{m-1}\frac{(-1)^j}{j!R^{2j}}\int_0^\infty u^{2j}H_{\delta,k}(u)du +$$

$$\frac{(-1)^m}{m!R^{2m}}\int_0^\infty e^{i(x^*+y^*)}u^{2m}H_{\delta,k}(u)du$$

其中 $m \leq p+1$,而 (x^*, y^*) 是以原点为中心,$\dfrac{u}{R}$ 为半径的圆内某一点. 在上式中先取 $m = 2$ 并令 $R \to \infty$,注意到 $k > 2$ 及 ⑤ 即得

$$\int_0^\infty u^2 H_{\delta,k} du = 0$$

然后再逐次地取 $m = 3, 4, \cdots, p+1$,而令 $R \to \infty$. 由于假设 $k > 2p$ 以及 ⑤,我们得到

$$\int_0^\infty u^{2i}H_{\delta,k}(u)du = 0 \quad (i = 1,2,\cdots,p) \quad ⑥$$

今因 $f(x,y) \in C^{2p}$,则利用 Taylor 公式到 ④ 的右边并注意 ⑥ 即可得到下式

$$S_R^{\delta,k}(x,y;f) - f(x,y) =$$

附录 Ⅶ　多元周期函数的非整数次积分与三角多项式逼近

$$\frac{1}{2^p p!} \cdot \frac{1}{R^{2p}} \int_0^\infty \Delta^p f\left(x + \theta_1 \frac{u}{R}, y + \theta_2 \frac{u}{R}\right) u^{2p} H_{\delta,k}(u) \mathrm{d}u =$$

$$\frac{1}{2^p p!\, R^{2p}} \int_0^\infty \left\{\Delta^p f\left(x + \theta_1 \frac{u}{R}, y + \theta_2 \frac{u}{R}\right) - \Delta^p f(x,y)\right\} \cdot$$

$$u^{2p} H_{\delta,k}(u) \mathrm{d}u$$

最后由于 ⑤ 即得关于 (x,y) 一致地有

$$|S_R^{\delta,k}(x,y;f) - f(x,y)| \leqslant$$

$$\frac{1}{2^p p!\, R^{2p}} \omega_p\left(\frac{1}{R}\right) \int_0^\infty u^{2p+1} |H_{\delta,k}(u)| \mathrm{d}u =$$

$$O\left[\frac{1}{R^{2p}} \omega_p\left(\frac{1}{R}\right)\right]$$

这也就证明了我们的定理.

对于非整数次的 Laplace 运算,我们有下面的结果:

定理 7.12　设 $f(x,y)$ 为对每一变量以 2π 为周期的连续函数且满足条件

$$\int_0^{2\pi} \int_0^{2\pi} f(x,y) \mathrm{d}x \mathrm{d}y = 0$$

又如果 $\Delta^\beta f(x,y)$ 存在且连续,其连续模为 $\omega(\rho)$.

（i）若 $0 < \beta < \frac{1}{2}$,则当 $k > 2, \delta > 2 + \frac{5}{2}$ 时,式 ⑦ 关于 (x,y) 一致成立

$$f(x,y) - S_R^{\delta,k}(x,y;f) = O\left[\frac{1}{R^{2\beta}} \omega\left(\frac{1}{R}\right)\right] \qquad ⑦$$

（ii）若 $\frac{1}{2} \leqslant \beta < 1$,则当 $k > 4, \delta > 4 + \frac{5}{2}$ 时,⑦ 仍关于 (x,y) 一致成立.

证明　先设 $0 < \beta < \frac{1}{2}$. 我们令

Tschebyscheff 逼近定理

$$\Delta^\beta f(x,y) = g(x,y)$$

则由于定理 7.7，下式成立

$$f(x,y) = I_{2\beta}^*[g(x,y)]$$

但 $0 < 2\beta < 1$，故由定理 7.3 可得

$$f(P) = \frac{1}{H_2(2\beta)}\int_\Omega g(Q) r_{0Q}^{2\beta-2} \mathrm{d}Q$$

其中 Ω 表示整个 xOy 平面.

今设 C_{mn} 与 C'_{mn} 分别为函数 $f(P)$ 与 $g(P)$ 的 Fourier 系数，由定理 7.7 的证明中知

$$C'_{mn} = C_{mn} v^\beta e^{i\pi\beta} \quad (v = m^2 + n^2)$$

利用上述关系，易得下式

$$f(P) - \sum_{v \leqslant R^2}\left(1 - \frac{v^{\frac{k}{2}}}{R^k}\right)^\delta A_v(P) =$$

$$\frac{1}{H_2(2\beta)}\int_\Omega \left\{g(Q) - \sum_{v \leqslant R^2}\left(1 - \frac{v^{\frac{k}{2}}}{R^k}\right)^\delta A'_v(Q)\right\} r_{PQ}^{2\beta-2} \mathrm{d}Q =$$

$$\frac{1}{H_2(2\beta)}\left\{\int_{S(P,\frac{1}{R})} + \int_{S'(P,\frac{1}{R})}\right\} =$$

$$\frac{1}{H_2(2\beta)}[I_1 + I_2]$$

其中 $A'_v(Q)$ 是 $g(Q)$ 的 Fourier 级数圆形部分和的一般项，而 $S\left(P,\frac{1}{R}\right)$ 与 $S'\left(P,\frac{1}{R}\right)$ 分别表示以 P 为中心，$\frac{1}{R}$ 为半径的圆域的内部与外部.

直接由 ⑩ 容易知道定理 7.11 当 $p = 0$ 时还是成立的，这样就得到了 I_1 的估计

$$I_1 = O\left[\omega\left(\frac{1}{R}\right)\int_0^{\frac{1}{R}} r^{2\beta-1} \mathrm{d}r\right] = O\left[\frac{1}{R^{2\beta}}\omega\left(\frac{1}{R}\right)\right]$$

其次来估计 I_2. 令

附录 Ⅶ 多元周期函数的非整数次积分与三角多项式逼近

$$G(x,y) = I^*_{2(1-\beta)}[f(x,y)] \qquad ㊽$$

则 $G(x,y)$ 是连续周期函数满足关系

$$\Delta G(x,y) = g(x,y) \qquad ㊾$$

又 $G(x,y)$ 的 Fourier 系数是 $-\dfrac{C'_{mn}}{v}$ $(v = m^2 + n^2)$. 今对 I_2 来用 Green 公式得到

$$I_2 = \int_{S'(P,\frac{1}{R})} \left\{ G(Q) - \sum_{v \leq R^2} \left(1 - \frac{v^{\frac{k}{2}}}{R^k}\right)^\delta A_v^*(Q) \right\} \Delta r_{PQ}^{2\beta-2}\, \mathrm{d}Q -$$

$$\int_{C(P,\frac{1}{R})} r_{PQ}^{2\beta-2} \frac{\partial}{\partial r}\left\{ G(Q) - \sum_{v \leq R^2}\left(1 - \frac{v^{\frac{k}{2}}}{R^k}\right)^\delta A_v^*(Q) \right\} \mathrm{d}Q +$$

$$\int_{C(P,\frac{1}{R})} \left\{ G(Q) - \sum_{v \leq R^2}\left(1 - \frac{v^{\frac{k}{2}}}{R^k}\right)^\delta A_v^*(Q) \right\} \frac{\partial}{\partial r} r_{PQ}^{2\beta-2}\, \mathrm{d}Q =$$

$$I_{21} + I_{22} + I_{23}$$

其中 $C\left(P,\dfrac{1}{R}\right)$ 表示 $S\left(P,\dfrac{1}{R}\right)$ 的边界，而 $A_v^*(Q)$ 表示 $G(Q)$ 的 Fourier 级数圆形部分和的一般项.

由于 ㊾ 及 $k > 2$ 与 $\delta > 2 + \dfrac{5}{2}$，我们可以利用定理 7.11 在 $p = 1$ 的情形即得

$$I_{21} = O\left[\frac{1}{R^2}\omega\left(\frac{1}{R}\right)\int_{\frac{1}{R}}^{\infty} r^{2\beta-3}\mathrm{d}r\right] = O\left[\frac{1}{R^{2\beta}}\omega\left(\frac{1}{R}\right)\right]$$

另外注意到

$$I_{22} = \left(\frac{1}{R}\right)^{2\beta-2}\int_{S(P,\frac{1}{R})}\left\{g(Q) - \sum_{v \leq R^2}(1 - \frac{v^{\frac{k}{2}}}{R^k})^\delta A'_v(Q)\right\}\mathrm{d}Q = O\left[\frac{1}{R^{2\beta}}\omega\left(\frac{1}{R}\right)\right]$$

$$I_{23} = O\left[\frac{1}{R^2}\omega\left(\frac{1}{R}\right)\left(\frac{1}{R}\right)^{2\beta-2}\right] = O\left[\frac{1}{R^{2\beta}}\omega\left(\frac{1}{R}\right)\right]$$

Tschebyscheff 逼近定理

因此

$$I_2 = O\left[\frac{1}{R^{2\beta}}\omega\left(\frac{1}{R}\right)\right]$$

合并关于 I_1 与 I_2 的估计即得 ⑩.

现在来考虑 $\frac{1}{2} \leqslant \beta < 1$ 的情形. 令

$$g_1(x,y) = I_\beta^*[g(x,y)]$$

则由定理 7.7 及定理 7.5 的(i) 可知

$$f(x,y) = I_\beta^*[g_1(x,y)] \qquad ⑩$$

设 $G(x,y)$ 仍由 ⑩ 所定义,则注意到 $G(x,y)$ 的 Fourier 系数与 $g(x,y)$ 的 Fourier 系数之间的关系,可得

$$I_2^*[g(x,y)] = G(x,y)$$

又因

$$\Delta I_{2(1-\frac{\beta}{2})}^*[g_1(x,y)] = \Delta I_{2-\beta}^*\{I_\beta^*[g(x,y)]\} =$$
$$\Delta I_2^*[g(x,y)] =$$
$$\Delta G(x,y) = g(x,y)$$

所以

$$\Delta^{\frac{\beta}{2}} g_1(x,y) = g(x,y) \qquad ⑪$$

现在假定

$$g_1(x,y) \sim \sum_{m,n=-\infty}^{\infty}{}' C''_{mn} e^{i(mx+ny)}$$

则

$$C''_{mn} = C'_{mn} v^{-\frac{\beta}{2}} e^{-\frac{\pi}{2}i\beta} = C_{mn} v^{\frac{\beta}{2}} e^{\frac{\pi}{2}i\beta} \qquad ⑫$$

于是从 ⑩ 及 ⑫ 可得

$$f(x,y) - \sum_{v \leqslant R^2}\left(1 - \frac{v^{\frac{k}{2}}}{R^k}\right)^\delta A_v(x,y) =$$

附录 Ⅶ　多元周期函数的非整数次积分与三角多项式逼近

$$\frac{1}{H_2(\beta)}\int_\Omega \left\{ g_1(Q) - \sum_{v \leqslant R^2}\left(1 - \frac{v^{\frac{k}{2}}}{R^k}\right)^\delta A''_v(Q) \right\} r_{PQ}^{\beta-2} dQ =$$

$$\frac{1}{H_2(\beta)}\left\{ \int_{S(P,\frac{1}{R})} + \int_{S'(P,\frac{1}{R})} \right\} =$$

$$\frac{1}{H_2(\beta)}[I_1 + I_2]$$

其中 $A''_v(Q)$ 是 $g_1(Q)$ 的 Fourier 级数圆形部分和的一般项，$S\left(P,\frac{1}{R}\right)$ 与 $S'\left(P,\frac{1}{R}\right)$ 的规定同前．

今因 $0 < \frac{\beta}{2} < \frac{1}{2}$，于是根据 ⑪ 可以利用已经证得的 (i) 的结果，知道下式关于 Q 一致成立

$$g_1(Q) - \sum_{v \leqslant R^2}\left(1 - \frac{v^{\frac{k}{2}}}{R^k}\right)^\delta A''_v(Q) = O\left[\frac{1}{R^\beta}\omega\left(\frac{1}{R}\right)\right] \quad ⑬$$

因而

$$I_1 = O\left[\frac{1}{R^\beta}\omega\left(\frac{1}{R}\right)\int_0^{\frac{1}{R}} r^{\beta-1} dr\right] = O\left[\frac{1}{R^{2\beta}}\omega\left(\frac{1}{R}\right)\right]$$

又由 (10) 可以推出 $g_1(P) \in \text{Lip}\,\beta$，因此 $g_1(P)$ 满足条件 ㊟，于是若令

$$G_1(x,y) = I_2^*[g_1(x,y)]$$

则根据定理 7.5 的 ⑩ ($\alpha = 0$ 的情形) 可知

$$\Delta G_1(x,y) = g_1(x,y)$$

又 $G_1(x,y)$ 的 Fourier 系数即为 $-\frac{C''_{mn}}{v}$．因此对 I_2 用 Green 公式可得

$$I_2 = \int_{S'(P,\frac{1}{R})} \{G_1(Q) - S_R^{\delta,k}(Q;G_1)\} \Delta r_{PQ}^{\beta-2} dQ -$$

$$\int_{C(P,\frac{1}{R})} r_{PQ}^{\beta-2}\frac{\partial}{\partial r}\{G_1(Q) - S_R^{\delta,k}(Q;G_1)\} dQ +$$

Tschebyscheff 逼近定理

$$\int_{C(P,\frac{1}{R})} \{G_1(Q) - S_R^{\delta,k}(Q;G_1)\} \frac{\partial}{\partial r} r_{PQ}^{\beta-2} dQ$$

我们如果能够证明下式关于 Q 一致成立

$$G_1(Q) - S_R^{\delta,k}(Q;G_1) = O\left[\frac{1}{R^{2+\beta}}\omega\left(\frac{1}{R}\right)\right] \quad ⑭$$

则与上面情形(i)时估计 I_2 一样即可得到

$$I_2 = O\left[\frac{1}{R^{2\beta}}\omega\left(\frac{1}{R}\right)\right]$$

这样也就证明了 ⑦ 当 $\frac{1}{2} \leq \beta < 1$ 时仍成立,定理即已完全证明.

要验证 ⑭ 只要注意到

$$G_1(P) = I_2^*[g_1(P)] = I_{2+\beta}^*[g(P)] = I_\beta^*[G(P)]$$

以及 $G_1(P)$ 的 Fourier 系数与 $G(P)$ 的 Fourier 系数间的关系,我们就可得到

$$G_1(P) - S_R^{\delta,k}(P;G_1) = I_\beta^*[G(P) - S_R^{\delta,k}(P;G)] =$$

$$\frac{1}{H_2(\beta)}\int_\Omega \{G(Q) - S_R^{\delta,k}(Q;G)\} r_{PQ}^{\beta-2} dQ$$

重复上面所用的估计方法,把以上积分分成在 $S\left(P,\frac{1}{R}\right)$ 与 $S'\left(P,\frac{1}{R}\right)$ 上的两部分,对于第一部分利用定理 7.11 当 $p=1$ 时的结果,而对于第二部分利用一次 Green 公式再利用定理 7.11 当 $p=2$ 时的结果,最后即可得到 ⑭ 关于 P 一致地成立,这时我们需假定 $k > 4$ 及 $\delta > 4 + \frac{5}{2}$.

利用上述逼近定理,从函数构造论的观点来看可得下面的推论:

附录 Ⅶ 多元周期函数的非整数次积分与三角多项式逼近

1. 如果 $\Delta^\beta f(x,y)(0 < \beta < 1)$ 存在且连续,则

(i) $f(x,y) \in \mathrm{Lip}^* 2\beta \left(0 < \beta \leqslant \dfrac{1}{2}\right)$.

(ii) $f(x,y) \in C^1$,且 $\dfrac{\partial f}{\partial x}, \dfrac{\partial f}{\partial y} \in \mathrm{Lip}^*(2\beta - 1)\left(\dfrac{1}{2} < \beta < 1\right)$.

2. 如果 $\Delta^\beta f(x,y) \in \mathrm{Lip}\,\alpha\,(0 < \alpha \leqslant 1, 0 < \beta < 1)$,则

(i) $f(x,y) \in \mathrm{Lip}^*(\alpha + 2\beta)(0 < \alpha + 2\beta \leqslant 1)$.

(ii) $f(x,y) \in C^1$,且 $\dfrac{\partial f}{\partial x}, \dfrac{\partial f}{\partial y} \in \mathrm{Lip}^*(\alpha + 2\beta - 1)$ $(1 < \alpha + 2\beta \leqslant 2)$.

(iii) $f(x,y) \in C^2$,且 $\dfrac{\partial^2 f}{\partial x^2}, \dfrac{\partial^2 f}{\partial x \partial y}, \dfrac{\partial^2 f}{\partial y^2} \in \mathrm{Lip}^*(\alpha + 2\beta - 2)(2 < \alpha + 2\beta < 3)$.

上述推论中有一部分已在定理 7.9 中得到.

要证明上述推论,我们只需利用下面结果(并注意 $\mathrm{Lip}\,\alpha$ 与 $\mathrm{Lip}^*\alpha$ 当 $0 < \alpha < 1$ 时是等价的):

如果对任一 $R > 0$,存在三角多项式

$$T_R(x,y) = \sum_{m^2+n^2 \leqslant R^2} C_{mn} \mathrm{e}^{\mathrm{i}(mx+ny)}$$

使以下不等式成立

$$|f(x,y) - T_R(x,y)| \leqslant \frac{A}{R^{p+\alpha}}$$

其中 A 是绝对常数, $p \geqslant 0$ 是整数, $0 < \alpha \leqslant 1$,则

(i) 当 $0 < \alpha < 1$ 时,可以推出 $f \in C^p$,且

$$\frac{\partial^p f}{\partial x^\alpha \partial y^\beta} \in \mathrm{Lip}\,\alpha \quad (\alpha \geqslant 0, \beta \geqslant 0, \alpha + \beta = p)$$

Tschebyscheff 逼近定理

(ii) 当 $\alpha = 1$ 时,可以推出 $f \in C^p$,且

$$\frac{\partial^p f}{\partial x^\alpha \partial y^\beta} \in \text{Lip}^* 1 \quad (\alpha \geq 0, \beta \geq 0, \alpha + \beta = p)$$

上述结果可以看成是单元情形 Бернштейн 定理及 Zygmund 定理的推广,其证明可以与这两个定理的证明类似地作出,我们这里不写出来了.

在具有基的 Banach 空间中的最佳逼近问题

附录 VIII

定义 设 G 是 Banach 空间 X 的一子集. 设 $x \in X$, 如果元素 $y_0 \in G$ 使得

$$\|x - y_0\| = \inf_{y \in G} \|x - y\| \qquad ⑮$$

那么称 y_0 是 x 关于 G 的最佳逼近元素, 特别地当 $\dim G = n < \infty$ 时, G 中的元素叫作多项式; 当 G 是有限余维的闭子空间时, G 中的元素叫作多项式余元.

尽管在一些具体情况下最佳逼近元素的存在性和唯一性都有保证(例如区间 $[a, b]$ 上连续函数用次数小于或等于 n 的多项式逼近), 但最佳逼近

Tschebyscheff 逼近定理

元的计算则十分困难,这是因为映射 $\Pi_G : x \to y_0$(其中 $y_0 = \Pi_G(x)$ 是 x 的最佳逼近元素)一般说来不是线性的. 但对于具有基的 Banach 空间中关于有限维空间和有限余维空间我们可以通过引进一个等价范数使得映射 Π_G 成为线性映射而且使得最佳元素的计算也变得容易些. 对于新范数逼近阶数与对原来范数相同.

2. 定义 设 X 为具有基 $\{x_n\}_{n=1}^{\infty}$ 的 Banach 空间. X 上的范数叫作关于基 $\{x_n\}_{n=1}^{\infty}$ 的 T-范数(简称 T-范数),如果

(a) 对于每个 $x \in X$ 和 $n = 1, 2, \cdots$,存在 x 的唯一的多项式最佳逼近元素 $y_0 = \Pi_{P_{(n)}}(x) \in P_{(n)} = [x_1, x_2, \cdots, x_n]$;

(b) 这个多项式与 x 关于基 $\{x_n\}_{n=1}^{\infty}$ 的表示式的 n 项部分和一致,等.

$$\Pi_{P_{(n)}}(x) = s_n(x) \quad (x \in X, n = 1, 2, \cdots) \quad ⑯$$

我们将用 ((((表示 T-范数.

定义 具有基 $\{x_n\}_{n=1}^{\infty}$ 的 Banach 空间 X 上的范数叫作关于 $\{x_n\}_{n=1}^{\infty}$ 的 K-范数,如果

(a) 对于每个 $x \in X$ 和 $n = 1, 2, \cdots$ 存在 x 的唯一的多项式余元最佳逼近 $y_0 \in \Pi_{P^{(n)}}(x) \in P^{(n)} = [x_{n+1}, x_{n+2}, \cdots]$;

(b) 这个多项式余元与 x 关于 $\{x_n\}_{n=1}^{\infty}$ 的表示式的 n-余项一致,等.

$$\Pi_{P^{(n)}}(x) = r_n(x) = x - s_n(x) \quad (x \in X, n = 1, 2, \cdots)$$

⑰

我们将用))   )) 表示 K-范数.

定义 具有基 $\{x_n\}_{n=1}^{\infty}$ 的 Banach 空间 X 上的范数

附录 Ⅷ 在具有基的 Banach 空间中的最佳逼近问题

叫作关于 $\{x_n\}_{n=1}^{\infty}$ 的 TK- 范数,如果这个范数关于基 $\{x_n\}_{n=1}^{\infty}$ 既是 T- 范数又是 K- 范数.

我们将用 $((\ ))$ 表示 TK- 范数.

现在我们来研究这些范数的特征.

定理 设 X 是具有基 $\{x_n\}_{n=1}^{\infty}$ 的 Banach 空间.则

(a) X 中范数是 T- 范数当且仅当对于每个数列 $\{\alpha_n\}_{n=l-1}^{\infty} \subset \Phi$(系数域)$,\alpha_{l-1} \neq 0$ 且级数 $\sum_{i=l}^{\infty} \alpha_i x_i$ 收敛有

$$\|\sum_{i=l}^{\infty} \alpha_i x_i\| < \|\sum_{i=l-l}^{\infty} \alpha_i x_i\| \qquad ⑱$$

(b) X 中范数是 K- 范数当且仅当对于任意数列 $\{\alpha_1,\alpha_2,\cdots,\alpha_{n+1}\} \subset \Phi$(系数域)且 $\alpha_{n+1} \neq 0$ 有

$$\|\sum_{i=1}^{n} \alpha_i x_i\| < \|\sum_{i=1}^{n+1} \alpha_i x_i\| \qquad ⑲$$

(此时称基 $\{x_n\}_{n=1}^{\infty}$ 是严格单调的).

(c) 如果 X 中范数是 TK- 范数,那么对于任意数列 $\alpha_{l-1},\alpha_l,\cdots,\alpha_n,\alpha_{n+1} \in \Phi$ 且 $|\alpha_{l-1}|+|\alpha_{n+1}| \neq 0$ 有

$$\|\sum_{i=l}^{n} \alpha_i x_i\| < \|\sum_{i=l-1}^{n+1} \alpha_i x_i\| \qquad ⑳$$

证明 (a) 假设 X 中范数是 T- 范数,设级数 $\sum_{i=1}^{\infty} \alpha_i x_i$ 是收敛的.则 $\sum_{i=l-1}^{\infty} \alpha_i x_i$ 在 $P_{(l-1)} = [x_1,x_2,\cdots,x_{l-1}]$ 中有唯一的最佳逼近元素,即

$$\Pi_{P_{(l-1)}}(\sum_{i=l-1}^{\infty} \alpha_i x_i) = s_{l-1}(\sum_{i=l-1}^{\infty} \alpha_i x_i) = \alpha_{l-1} x_{l-1}$$

从而,因为 $0 \in P_{(l-1)}$ 有

$$\|\sum_{i=l}^{\infty} \alpha_i x_i\| = \|\sum_{i=l-l}^{\infty} \alpha_i x_i - \Pi_{P_{(l-1)}}(\sum_{i=l-1}^{\infty} \alpha_i x_i)\| <$$

Tschebyscheff 逼近定理

$$\| \sum_{i=l-1}^{\infty} \alpha_i x_i - 0 \|$$

即得式 ⑱.

反之,假设 ⑱ 成立. 则对每个 $x = \sum_{i=l}^{\infty} \alpha_i x_i \in X$ 和 $p = \sum_{i=1}^{n} \beta_i x_i \in P_{(n)}$ 且 $p \neq s_n(x)$(其中 $1 \leq n < \infty$),有

$$\| x - s_n(x) \| = \| \sum_{i=n+1}^{\infty} \alpha_i x_i \| <$$
$$\| \sum_{i=n+1}^{\infty} \alpha_i x_i - \sum_{i=1}^{n} (\beta_i - \alpha_i) x_i \| =$$
$$\| x - p \|$$

因此 X 中的范数为 T-范数.

(b) 假设 X 中的范数为 K-范数,设 $\alpha_1, \alpha_2, \cdots, \alpha_{n+1}$ 是数列且 $\alpha_{n+1} \neq 0$. 则 $\sum_{i=1}^{n+1} \alpha_i x_i$ 在 $P^{(n)} = [x_{n+1}, x_{n+2}, \cdots]$ 中有唯一的最佳逼近元素,即

$$\Pi P^{(n)} (\sum_{i=l}^{n+1} \alpha_i x_i) = \sum_{i=1}^{n+1} \alpha_i x_i - s_n (\sum_{i=1}^{n+1} \alpha_i x_i) = \alpha_{n+1} x_{n+1}$$

由于 $0 \in P^{(n)}$,所以

$$\| \sum_{i=1}^{n} \alpha_i x_i \| = \| \sum_{i=1}^{n+1} \alpha_i x_i - \Pi P^{(n)} (\sum_{i=1}^{n+1} \alpha_i x_i) \| <$$
$$\| \sum_{i=1}^{n+1} \alpha_i x_i - 0 \|$$

式 ⑲ 成立.

反之,假设式 ⑲ 成立. 设 $x = \sum_{i=1}^{\infty} \alpha_i x_i \in X, p = \sum_{i=n+1}^{\infty} \beta_i x_i \in P^{(n)}$ 是任意的且 $p = r_n(x) = \sum_{i=n+1}^{\infty} \alpha_i x_i$. 则存在

附录 Ⅷ 在具有基的 Banach 空间中的最佳逼近问题

一最小下标,譬如说 $n+m$,使得 $\beta_{n+m} \neq \alpha_{n+m}$. 从而应用式 ⑲ 得到

$$\|x - r_n(x)\| = \|\sum_{i=1}^{n} \alpha_i x_i\| =$$

$$\|\sum_{i=1}^{n} \alpha_i x_i - \sum_{i=n+1}^{n+m-1}(\beta_i - \alpha_i) x_i\| \leqslant$$

$$\|\sum_{i=1}^{n} \alpha_i x_i - \sum_{i=n+1}^{n+m}(\beta_i - \alpha_i) x_i\| \leqslant$$

$$\|\sum_{i=1}^{n} \alpha_i x_i - \sum_{i=n+1}^{n+m+1}(\beta_i - \alpha_i) x_i\| \leqslant \cdots \leqslant$$

$$\|\sum_{i=1}^{n} \alpha_i x_i - \sum_{i=n+1}^{\infty}(\beta_i - \alpha_i) x_i\| =$$

$$\|x - p\|$$

因此 X 中范数是 K-范数.

(c) 是 (a),(b) 的必要性的直接结果.

T-范数和 K-范数是不相同的. 由下面两个例子可以看出.

存在不是 K-范数的 T-范数.

例如在 c_0 上定义数

$$|||x||| = \max_{1 \leqslant n < \infty}\left(\frac{1}{n}\sum_{i=1}^{n}\alpha_i + \sup_{n+1 \leqslant j < \infty}|a_j|\right) \quad \text{㉑}$$

$$(x = \{a_i\}_{i=1}^{\infty} \in c_0)$$

那么 $|||\cdot|||$ 是 c_0 上的范数,$|||\cdot|||$ 等价于 c_0 的原来范数,且 $|||\cdot|||$ 关于 c_0 的单位向量基 $\{e_n\}_{n=1}^{\infty}$ 是 T-范数而不是 K-范数.

事实上,由下面定理 8.2 知 $|||\cdot|||$ 是等价于 c_0 原范数的 T-范数. 另一方面有

$$|||e_1 + e_2||| = \max\{1+1, \frac{1}{2}(1+1), \frac{1}{3}(1+1), \cdots\} = 2$$

$$((\boldsymbol{e}_1+\boldsymbol{e}_2+\boldsymbol{e}_3((=\max\{1+1,\frac{1}{2}(1+1)+1,$$

$$\frac{1}{3}(1+1+1),$$

$$\frac{1}{4}(1+1+1),\cdots\}=2$$

所以由式 (419) 知((((不是 K- 范数.

存在不是 T- 范数的 K- 范数. 例如,对于每个 $n\geqslant 2$ 令 $\Pi_{1,n}$ 表示集合 $\{2,3,\cdots,n-1,n+1,n+2,\cdots\}$ 上排列的全体,则数

$$))x))=\sup_{2\leqslant n<\infty}\sup_{\sigma\leqslant\Pi_{1,n}}\left(\frac{|a_1|}{n\cdot 2^n}+\sum_{i=2}^{\infty}\frac{a_{\sigma(i)}}{2^i}\right) \quad (422)$$

$$(x=\{a_i\}_{i=1}^{\infty}\in c_0)$$

定义 c_0 上与 c_0 原来范数等价的一个范数,范数)))) 关于 c_0 的单位向量基是 K- 范数但不是 T- 范数.

事实上,显然)))) 是 c_0 上的范数. 现在来证明这个范数与 c_0 的原来范数 $\|x\|=\sup_{1\leqslant n<\infty}|a_i|$ 是等价的. 因为对于一切 $n\geqslant 2$ 和 $\sigma\in\Pi_{1,n}$ 有

$$\frac{|a_1|}{n\cdot 2^n}\leqslant\frac{1}{8}|a_1|,\sum_{i=2,i\neq n}^{\infty}\frac{a_{\sigma(i)}}{2^i}\leqslant\frac{1}{2}\sup_{2\leqslant j<\infty}|a_j|$$

所以有

$$))x))\leqslant\frac{5}{8}\|x\| \quad (x\in c_0) \quad (423)$$

另一方面有

$$|a_1|\leqslant 8\left(\frac{|a_1|}{2\cdot 2^2}+\sum_{i=3}^{\infty}\frac{|a_i|}{2^i}\right)\leqslant 8))x))$$

$$(x=\{a_i\}_{i=1}^{\infty}\in c_0)$$

$$|a_2|\leqslant 4\left(\frac{|a_1|}{3\cdot 2^3}+\frac{|a_2|}{2^2}+\sum_{i=4}^{\infty}\frac{|a_i|}{2^i}\right)\leqslant 4))x))$$

附录 Ⅷ 在具有基的 Banach 空间中的最佳逼近问题

$$(x = \{a_i\}_{i=1}^{\infty} \in c_0)$$

且对于 $j > 2$ 选择一正整数 $n > 2, n \neq j$,选取一个排列 $\sigma_j \in \Pi_{1,n}$,且 $\sigma_j(2) = j$,得到

$$|a_j| \leqslant 4\left(\frac{|a_1|}{n \cdot 2^n} + \sum_{i=2, i \neq n}^{\infty} \frac{|a_{\sigma_j(i)}|}{2^i}\right) \leqslant 4)))x))$$

$$(x = \{a_i\}_{i=1}^{\infty} \in c_0)$$

所以

$$\frac{1}{8}\|x\| \leqslant)))x))\quad (x \in c_0) \qquad ㉔$$

㉔ 与 ㉓ 一起知)))) 与 ‖ ‖ 等价.

现在我们证明:对于任意一对有限指标集 d_1 和 d_2 且 $d_1 \subset d_2$ 和对每一有限数列 $\{a_i\}_{i \in d_2}$ 且 $\sum_{i \in d_2 \setminus d_1} |\alpha_i| \neq 0$ 有

$$))\sum_{i \in d_1} \alpha_i e_i)) <))\sum_{i \in d_2} \alpha_i e_i)) \qquad ㉕$$

特别地)))) 满足式 ⑳,从而)))) 是 K- 范数.

事实上,由于 d_1 是有限集,根据范数)))) 的定义式 ㉒ 知)) $\sum_{i \in d_1} \alpha_i e_i$)) 一定在某个 $n_0 \geqslant 2$ 和某个排列 $\sigma_0 \in \Pi_{1,n}$ 上达到. 又因 $\sum_{i \in d_2 \setminus d_1} |\alpha_i| \neq 0$,当同样的 n_0 和 σ_0 用于级数 $\sum_{i \in d_2} a_i e_i$ 时得到的)) $\sum_{i \in d_2} a_i e_i$)) 必然大于)) $\sum_{i \in d_1} a_i e_i$)),从而式 ㉕ 得证.

现在我们来证明

$$))\sum_{m=1}^{\infty} \frac{1}{m} e_m)) \geqslant))\sum_{m=2}^{\infty} \frac{1}{m} e_m)) \qquad ㉖$$

从而,根据定理 8.1(a) 知)))) 不是 T- 范数.

Tschebyscheff 逼近定理

事实上，显然有

$$\|\sum_{m=1}^{\infty} \frac{1}{m} e_m\| \geq \|\sum_{m=2}^{\infty} \frac{1}{m} e_m\| \qquad ㊷$$

固定 $n \geq 2$. 如果对于一对正整数 $i,j \in \{2,3,\cdots,n-1, n+1,\cdots\}$ 和一排列 $\sigma \in \Pi_{1,n}$ 有 $\dfrac{1}{\sigma(i)} < \dfrac{1}{\sigma(i+j)}$，那么对于定义为

$$\sigma'(i) = \sigma(i+j), \sigma'(i+j) = \sigma(i),$$
$$\sigma'(k) = \sigma(k) \quad (k \neq i, k \neq j)$$

的排列 $\sigma' \in \Pi_{1,n}$，有

$$\sum_{m=2, m\neq n}^{\infty} \frac{1}{\sigma'(m) 2^m} > \sum_{m=2, m\neq n}^{\infty} \frac{1}{\sigma(m) 2^m}$$

（因为当 $\alpha > \beta \geq 0$ 时有 $\dfrac{\alpha}{2^i} + \dfrac{\beta}{2^{i+j}} > \dfrac{\beta}{2^i} + \dfrac{\alpha}{2^{i+j}}$）. 因此，对于每个 $n \geq 2$ 和 $\sigma \in \Pi_{1,n}$，有

$$\frac{1}{n 2^n} + \sum_{m=2, m\neq n}^{\infty} \frac{1}{\sigma(m) 2^m} \leq \frac{1}{n 2^n} + \sum_{m=2, m\neq n}^{\infty} \frac{1}{m 2^m} =$$

$$\sum_{m=2}^{\infty} \frac{1}{m 2^m} =$$

$$\sup_{2 \leq n < \infty} \sum_{m=2, m\neq n}^{\infty} \frac{1}{m 2^m} =$$

$$\sup_{2 \leq n < \infty} \sup_{\tau \in \Pi_{1,n}} \sum_{m=2, m\neq n}^{\infty} \frac{1}{\tau(m) 2^m} =$$

$$\|\sum_{m=2}^{\infty} \frac{1}{m} e_m\|$$

与式 ㊷ 一起可推出式 ㊻. 从而 $\|\cdot\|$ 是 K-范数.

定理 8.2 设 X 是具有基 $\{x_n\}_{n=1}^{\infty}$ 的 Banach 空间. $\{x_n^*\}_{n=1}^{\infty}$ 是相应于 $\{x_n\}_{n=1}^{\infty}$ 的系数泛函序列. 则

附录 Ⅷ 在具有基的 Banach 空间中的最佳逼近问题

(a) 对于每个 $x \in X$,定义

$$\|x\|_1 = \max_{1 \leq n < \infty} \left\{ \frac{1}{n} \sum_{i=1}^{n} \|x_i^*(x) x_i\| + \sum_{i=n+1}^{\infty} \|x_i^*(x) x_i\| \right\} \qquad ⑫⑧$$

$$\|x\|_2 = \sum_{i=1}^{\infty} \frac{1}{2^i} \|x_i^*(x) x_i\| + \max_{1 \leq n < \infty} \left\| \sum_{i=n}^{n} x_i^*(x) x_i \right\| \qquad ⑫⑨$$

$\|\cdot\|_1$ 及 $\|\cdot\|_2$ 都是 X 上关于 $\{x_n\}_{n=1}^{\infty}$ 的 T-范数且与原来范数 $\|\cdot\|$ 等价.

(b) 对于每个 $x \in X$ 定义

$$\|x\| = \sum_{i=1}^{\infty} \frac{1}{2^i} \|x_i^*(x) x_i\| + \sup_{1 \leq n < \infty} \left\| \sum_{i=1}^{n} x_i^*(x) x_i \right\| \qquad ⑬⓪$$

$\|\cdot\|$ 是等价于原范数 $\|\cdot\|$ 的关于 $\{x_n\}_{n=1}^{\infty}$ 的 K-范数.

(c) 对于每个 $x \in X$ 定义

$$\|(x)\| = \sum_{i=1}^{\infty} \frac{1}{2^i} \|x_1^*(x) x_i\| + \sup_{1 \leq n, m < \infty} \left\| \sum_{i=n}^{m} x_i^*(x) x_i \right\| \qquad ⑬①$$

$\|(\cdot)\|$ 是 X 上关于 $\{x_n\}_{n=1}^{\infty}$ 的 TK-范数且等价于原来的范数 $\|\cdot\|$.

证明 (a) 首先注意由于 $\{x_n\}_{n=1}^{\infty}$ 是基,那么式 ⑫⑧ 中右端圆括号中的两项当 $n \to \infty$ 时均趋向于 0,所以式 ⑫⑧ 的最大值必然在某个自然数 n 时达到.

现在我们证明式 ⑫⑧ 定义 X 上一范数. 显然

333

Tschebyscheff 逼近定理

$((x((_1 \geq 0$ 且 $((0((_1 = 0$. 如果我们有 $((x((_1 = 0$,那么由

$$((x((_1 \geq \|x_1^*(x)x_1\| + \|\sum_{i=2}^{\infty} x_i^*(x)x_i\| \geq$$

$$\|\sum_{i=1}^{\infty} x_i^*(x)x_i\| = \|x\| \quad (x \in X) \quad ⑫$$

有 $x = 0$. 由于对于某个适当的 n_0 有

$$((x+y((_1 = \frac{1}{n_0} \sum_{i=1}^{n_0} \|x_i^*(x+y)x_i\| +$$

$$\|\sum_{i=n_0+1}^{\infty} x_i^*(x+y)x_i\|$$

立即推出三角不等式成立,即

$$((x+y((_1 \leq ((x((_1 + ((y((_1$$

由式 ⑱ 立即可得 $((\alpha x((_1 = |\alpha| ((x((_1$.

现在证明 $((\quad ((_1$ 与原范数 $\| \quad \|$ 的等价性.

由于

$$((x((_1 = \frac{1}{n_0} \sum_{i=1}^{n_0} \|x_i^*(x)x_i\| + \|\sum_{i=n_0+1}^{\infty} x_i^*(x)x_i\| \leq$$

$$\max_{1 \leq i < \infty} \|x_i^*(x)x_i\| + \|x - \sum_{i=1}^{n_0} x_i^*(x)x_i\| \leq$$

$$\max_{1 \leq i < n_0} (\|\sum_{j=1}^{i} x_j^*(x)x_j\| + \|\sum_{j=1}^{i-1} x_j^*(x)x_j\|) +$$

$$\|x\| + \|\sum_{i=1}^{n_0} x_i^*(x)x_i\| \leq$$

$$3 \sup_{1 \leq n < \infty} \|\sum_{i=1}^{n} x_i^*(x)x_i\| + \|x\| \leq$$

$$(3K+1)\|x\|$$

其中 K 是 $\{x_n\}_{n=1}^{\infty}$ 的基常数,上面不等式与式 ⑫ 一起说明 $((\quad ((_1$ 等价于 $\| \quad \|$.

附录 Ⅷ　在具有基的 Banach 空间中的最佳逼近问题

最后让我们证明 $((\quad ((_1$ 是 T- 范数. 设 $\{\alpha_i\}_{i=l-1}^{\infty}$ 是 $\alpha_{l-1} \neq 0$ 且使 $\sum_{i=1}^{\infty} \alpha_i x_i$ 收敛的数列. 那么对于某一适当的 $n_0, l \leq n_0 < \infty$ 有

$$((\sum_{i=l}^{\infty} \alpha_i x_i ((_1 = \frac{1}{n_0} \sum_{i=l}^{n_0} \| \alpha_i x_i \| + \| \sum_{i=n_0+1}^{\infty} \alpha_i x_i \| <$$

$$\frac{1}{n_0} \sum_{i=l-1}^{n_0} \| \alpha_i x_i \| + \| \sum_{i=n_0+1}^{\infty} \alpha_i x_i \| \leq$$

$$\max_{l-1 \leq n < \infty} \left(\frac{1}{n} \sum_{i=l-1}^{n} \| \alpha_i x_i \| + \| \sum_{i=n+1}^{\infty} \alpha_i x_i \| \right) =$$

$$((\sum_{i=l-1}^{\infty} \alpha_i x_i ((_1$$

所以由定理 8.1(a) 知 $((\quad ((_1$ 是 T- 范数.

现在来考虑式 ㊉ 所定义的 $((\quad ((_2$. 因为 $\{x_n\}_{n=1}^{\infty}$ 是基, 所以式 ㊉ 右端的 max 必然在某个 n 达到. $((\quad ((_2$ 显然是 X 上的范数. $((\quad ((_2$ 是等价于 $\| \quad \|$ 的, 因为对于每个 $x \in X$ 有

$$\| x \| = \| \sum_{i=1}^{\infty} x_i^*(x) x_i \| \leq ((x((_2 \leq$$

$$\max_{1 \leq i < \infty} \| x_i^*(x) x_i \| + \max_{1 \leq n < \infty} \| \sum_{i=n}^{\infty} x_i^*(x) x_i \| \leq$$

$$\max_{1 \leq i < \infty} (\| \sum_{j=i}^{\infty} x_j^*(x) x_j \| + \| \sum_{j=i+1}^{\infty} x_j^*(x) x_j \|) +$$

$$\max_{1 \leq n < \infty} \| \sum_{i=n}^{\infty} x_i^*(x) x_i \| \leq$$

$$3 \max_{1 \leq n < \infty} \| \sum_{i=n}^{\infty} x_i^*(x) x_i \| \leq$$

$$3 \| x \| + \sup_{1 \leq n < \infty} \| \sum_{i=1}^{n} x_i^*(x) x_i \| \leq$$

Tschebyscheff 逼近定理

$3(1+K)\|x\|$ (K 是 $\{x_n\}_{n=1}^{\infty}$ 的基常数)

最后证明(($\cdot$ ((${}_2$ 是 T-范数. 设 $\{\alpha_i\}_{i=l-1}^{\infty}$, $\alpha_{l-1} \neq 0$ 是使 $\sum_{i=l}^{\infty} \alpha_i x_i$ 收敛的数列. 那么

$$((\sum_{i=l}^{\infty} \alpha_i x_i ((_2 = \sum_{i=l}^{\infty} \frac{1}{2^i} \|\alpha_i x_i\| + \max_{l \leqslant n < \infty} \|\sum_{i=l}^{\infty} \alpha_i x_i\| <$$

$$\sum_{i=l-1}^{\infty} \frac{1}{2^i} \|\alpha_i x_i\| + \max_{l-1 \leqslant n < \infty} \|\sum_{i=n}^{\infty} \alpha_i x_i\| =$$

$$((\sum_{i=l-1}^{\infty} \alpha_i x_i ((_2$$

所以由定理 8.1(a) 知(($\cdot$ ((${}_2$ 是 T-范数.

(b) 显然式 ㊹ 所定义的)) $\cdot$)) 是 X 上的范数. 下面证明)) $\cdot$)) 与 $\|\cdot\|$ 等价, 因为对于每个 $x \in X$ 有

$$\|x\| \leqslant))x)) \leqslant \max_{1 \leqslant i < \infty} \|x_i^*(x) x_i\| +$$

$$\sup_{1 \leqslant n < \infty} \|\sum_{i=1}^{n} x_i^*(x) x_i\| \leqslant$$

$$\max_{1 \leqslant i < \infty} (\|\sum_{j=1}^{i} x_j^*(x) x_j\| + \|\sum_{j=1}^{i-1} x_j^*(x) x_j\|) +$$

$$\sup_{1 \leqslant n < \infty} \|\sum_{i=1}^{n} x_i^*(x) x_i\| \leqslant$$

$3K\|x\|$ (K 为 $\{x_n\}_{n=1}^{\infty}$ 的基常数).

最后, 对于任意数列 $\alpha_1, \alpha_2, \cdots, \alpha_{n+1}$ 且 $\alpha_{n+1} \neq 0$ 有

$$)) \sum_{i=1}^{n} \alpha_i x_i)) = \sum_{i=1}^{n} \frac{1}{2^i} \|\alpha_i x_i\| + \max_{1 \leqslant k \leqslant n} \|\sum_{i=1}^{k} \alpha_i x_i\| <$$

$$\sum_{i=1}^{n+1} \frac{1}{2^i} \|\alpha_i x_i\| + \max_{1 \leqslant k \leqslant n+1} \|\sum_{i=1}^{k} \alpha_i x_i\| =$$

$$)) \sum_{i=1}^{n+1} \alpha_i x_i))$$

附录 Ⅷ 在具有基的 Banach 空间中的最佳逼近问题

根据定理 8.1(b) 知)）)) 是 X 上的 K- 范数.

（c）与证明(((($_2$ 和 (b) 相类似地证明.

推论 8.3 在定理 8.2 的要件下,对于每个元素 $x \in X$,定义

$$((x)) = \sup_{1 \leqslant n < \infty} \{((\sum_{i=1}^{n} x_i^*(x)x_i((+)) \sum_{i=n+1}^{\infty} x_i^*(x)x_i))\} \quad ⑬$$

其中((((是 T- 范数,)）)) 是 K- 范数,则(( )) 是 X 上等价于原范数 $\| \ \|$ 的 TK- 范数.

注 对于双正交系 $\{x_n, x_n^*\}_{n=1}^{\infty}$ 当 $[x_n]_{n=1}^{\infty} = X$ 时可与定理 8.2 中类似引进一个 T- 范数. 对于定理 8.2 的某种意义下的逆定理成立:

如果对于双正交系 $\{x_n, x_n^*\}_{n=1}^{\infty}$ 有一个与原范数等价的 T- 范数且 $[x_n]_{n=1}^{\infty} = X$,则 $\{x_n\}_{n=1}^{\infty}$ 是 X 的基.

推论 8.4 设 X 是具有基 $\{x_n\}_{n=1}^{\infty}$ 的 Banach 空间. 令

$$b_n(x) = \inf_{y \in P_{(n)}} \| x - y \| \quad (x \in X, n = 1, 2, \cdots) \quad ⑭$$

则存在一仅依赖于基 $\{x_n\}_{n=1}^{\infty}$ 的常数 $M, 0 < M < 1$ 使得

$$M \| x - s_n(x) \| \leqslant b_n(x) \leqslant \| x - s_n(x) \|$$
$$(x \in X, n = 1, 2, \cdots) \quad ⑮$$

其中 s_n 是关于 $\{x_n\}_{n=1}^{\infty}$ 的部分和算子,$P_{(n)} = [x_1, \cdots, x_n]$.

证明 设 $x \in X, n$ 是任意的正整数. 设 $y_0 \in P_{(n)}$ 是 x 的最佳逼近元素,因为 $\dim P_{(n)} < \infty$,故这样的元素 y_0 是存在的. 那么 $b_n(x) = \| x - y_0 \|$. 根据定理 8.2,设((((是 X 上等价于原来范数 $\| \ \|$ 的 T- 范数. 设

Tschebyscheff 逼近定理

常数 M_1 和 M_2 使得对于一切 $x \in X$ 有
$$M_1 \|x\| \leq \|\|x\|\| \leq M_2 \|x\|$$
则
$$b_n(x) = \|x - y_0\| \geq \frac{1}{M_2} \|\|x - y_0\|\| \geq$$
$$\frac{1}{M_2} \|\|x - s_n(x)\|\| \geq$$
$$\frac{M_1}{M_2} \|x - s_n(x)\|$$

这样得到式 ㊵ 的第一个不等式,$M = M_1/M_2$,㊵ 的第二个不等式由 ㊴ 得到. 所以式 ㊵ 成立.

这个推论说明 $\{b_n(x)\}_{n=1}^{\infty}$ 与 $\{\|x - s_n(x)\|\}_{n=1}^{\infty}$ 具有相同的"阶",如果我们要计算 $\{b_n(x)\}_{n=1}^{\infty}$ 收敛于 0 的速度时,可用计算数列 $\{\|x - s_n(x)\|\}_{n=1}^{\infty}$ 来代替.

注 推论 8.4 在某种意义下的逆定理也成立. 如果对于双正交系 $\{x_n, x_n^*\}_{n=1}^{\infty}$ 且 $[x_n]_{n=1}^{\infty} = X$ 有式 ㊵ 的第一个不等式,那么
$$\|s_n(x)\| \leq \|x\| + \|x - s_n(x)\| \leq$$
$$\|x\| + \frac{1}{M} |b_n(x)| \leq$$
$$\left(1 - \frac{1}{M}\right) \|x\| \quad (x \in X, n = 1, 2, \cdots)$$

可得 $\{x_n\}_{n=1}^{\infty}$ 是 X 的基. 此外,对于序列 $\{x_n\}_{n=1}^{\infty} \subset X$, $x_n \neq 0 (n = 1, 2, \cdots)$ 且 $[x_n]_{n=1}^{\infty} = X$ 引入算子
$$Q_n\left(\sum_{i=1}^{n} \alpha_i x_i\right) = \begin{cases} \sum_{i=1}^{n} \alpha_i x_i, & n = 1, 2, \cdots, k \\ \sum_{i=1}^{k} \alpha_i x_i, & n = k+1, k+2, \cdots \end{cases}$$

附录 Ⅷ 在具有基的 Banach 空间中的最佳逼近问题

用 Q_n 代替 s_n 并用同样的方法讨论知如果 $\{x_n\}_{n=1}^{\infty} \subset X$, $[x_n]_{n=1}^{\infty} = X$ 且 $x_n \neq 0 (n=1,2,\cdots)$, 存在一常数 M, $0 < M \leqslant 1$ 使得

$$\|y - Q_n(y)\| \leqslant \frac{1}{M} b_n(y) \quad (y \in \bigcup_{k=1}^{\infty} P_{(k)}, n = 1,2,\cdots)$$

则 $\{x_n\}_{n=1}^{\infty}$ 是 X 的基.

此外,还可在具有基的 Banach 空间中引入弱 T-范数和弱 K-范数等.

对于具有无条件基 Banach 空间,我们可以与对于具有基的 Banach 空间相似地给出 T-范数和 K-范数和 TK-范数等的定义并讨论它们之间的关系. 但此时它们之间的关系与在具有基的 Banach 空间的情况很不相同.

定义 具有无条件基 $\{x_n\}_{n=1}^{\infty}$ 的 Banach 空间中的范数叫作关于 $\{x_n\}_{n=1}^{\infty}$ 的 NT-范数, 如果

(a) 对于每个 $x \in X$ 和 $d = \{i_1, i_2, \cdots, i_n\} \in \mathscr{D}$ ($\mathscr{D} = \{\{i_1, i_2, \cdots, i_n\} \subset \mathbf{N}; 1 \leqslant n < \infty \}$), x 有唯一的多项式最佳逼近元素

$$y_0 = \Pi_{P_{(d)}}(x) \in P_{(d)} = [x_{i_1}, x_{i_2}, \cdots, x_{i_n}]$$

(b) 这个多项式与 x 关于基 $\{x_n\}_{n=1}^{\infty}$ 表示式的 d-部分和一致

$$\Pi_{P_{(d)}}(x) = s_d(x) = \sum_{j=1}^{n} x_{i_j}^{*}(x) x_{i_j}$$
$$(x \in X; d = \{i_1, i_2, \cdots, i_n\} \in \mathscr{D})$$

用 $((\quad ((_u$ 表示 X 的 NT-范数.

定义 具有无条件基 $\{x_n\}_{n=1}^{\infty}$ 的 Banach 空间的范数叫作关于 $\{x_n\}_{n=1}^{\infty}$ 的 NK-范数, 如果

Tschebyscheff 逼近定理

(a) 对于每个 $x \in X$ 和 $d = \{i_1, i_2, \cdots, i_n\} \in \mathscr{D}$, x 有唯一的多项式余元佳逼近 $y_0 = \Pi_{P(d)}(x) \in P^{(d)} = [x_j]_{j \in \mathbf{N} \setminus d}$;

(b) 这个多项式余元与 x 关于基 $\{x_n\}_{n=1}^{\infty}$ 的表示式的 d-余基一致

$$\Pi_{P_{(d)}}(x) = r_d(x) = x - s_d(x)$$
$$(x \in X, d = \{i_1, i_2, \cdots, i_n\} \in \mathscr{D})$$

用 $)$ $)_u$ 表示 NK-范数.

定义 具有无条件基 $\{x_n\}_{n=1}^{\infty}$ 的 Banach 空间 X 中的范数叫关于 $\{x_n\}_{n=1}^{\infty}$ 的 NTK-范数,如果它对于这个基既是 NT-范数又是 NK-范数.

用 $(($ $))_u$ 表示 NTK-范数.

例如可分 Hilbert 空间 H 的自然范数是关于 H 的任何正交基的 NTK-范数.

定理 8.5 设 X 为具有无条件基的 Banach 空间. 则

(a) X 中范数是 NT-范数当且仅当对于每对 $d_1, d_2 \in \mathscr{D}, d_1 \subset d_2$ 和对于每个使级数 $\sum_{i \in \mathbf{N} \setminus d_1} \alpha_i x_i$, $\sum_{i \in \mathbf{N} \setminus d_2} \alpha_i x_i$ 收敛且 $\sum_{i \in d_2 \setminus d_1} |\alpha_i| \neq 0$ 的序列 $\{\alpha_i\}_{i \in \mathbf{N} \setminus d_1}$ 有

$$\|\sum_{i \in \mathbf{N} \setminus d_2} \alpha_i x_i\| < \|\sum_{i \in \mathbf{N} \setminus d_1} \alpha_i x_i\| \qquad ㊻$$

(b) X 中范数是 NK-范数当且仅当对于每一对 $d_1, d_2 \in \mathscr{D}, d_1 \subset d_2$ 和每个有限数列 $\{\alpha_i\}_{i \in d_2}$ 且 $\sum_{i \in d_2 \setminus d_1} |\alpha_i| \neq 0$ 有

$$\|\sum_{i \in d_1} \alpha_i x_i\| < \|\sum_{i \in d_2} \alpha_i x_i\| \qquad ㊼$$

附录 Ⅷ　在具有基的 Banach 空间中的最佳逼近问题

证明　(a) 假设 X 中范数是一 NT-范数. 设 $d_2 \in D$ 且 $\sum_{i \in \mathbf{N} \backslash d_2} \alpha_i x_i$ 收敛,设 $d_1 \subset d_2$ 是任意的. 则 $\sum_{i \in \mathbf{N} \backslash d_1} \alpha_i x_i$ 在 $P_{(d_2)} = [x_i]_{i \in d_2}$ 中有唯一最佳逼近元,即

$$\Pi_{P_{(d_2)}}\left(\sum_{i \in \mathbf{N} \backslash d_2} \alpha_i x_i\right) = s_{d_2}\left(\sum_{i \in \mathbf{N} \backslash d_1} \alpha_i x_i\right) = \sum_{i \in d_2 \backslash d_1} \alpha_i x_i$$

由于 $0 \in P_{(d_2)}$,从而

$$\left\|\sum_{i \in \mathbf{N} \backslash d_2} \alpha_i x_i\right\| = \left\|\sum_{i \in \mathbf{N} \backslash d_1} \alpha_i x_i - \Pi_{P_{(d_2)}}\left(\sum_{i \in \mathbf{N} \backslash d_1} \alpha_i x_i\right)\right\| <$$
$$\left\|\sum_{i \in \mathbf{N} \backslash d_1} \alpha_i x_i - 0\right\|$$

式 ㊹ 成立.

反之,假设式 ㊹ 成立. 那么对于每个 $x = \sum_{i=1}^{\infty} \alpha_i x_i \in X, d \in \mathscr{D}$ 和 $y = \sum_{i \in d} \beta_i x_i \in P_{(d)}$ 且 $y \neq s_d(x)$(在式 ㊹ 中用 $d_2 = d, d_1 = \varnothing$) 有

$$\|x - s_d(x)\| = \left\|\sum_{i \in \mathbf{N} \backslash d} \alpha_i x_i\right\| <$$
$$\left\|\sum_{i \in \mathbf{N} \backslash d} \alpha_i x_i - \sum_{i \in d}(\beta_i - \alpha_i) x_i\right\| =$$
$$\|x - y\|$$

因此 X 中这一范数为 NT-范数.

(b) 假设 X 中范数是 NK-范数. 设 $\{\alpha_k\}_{i \in d_2}$ 是一有限数列,设 $d_1 \subset d_2$ 且 $\sum_{i \in d_2 \backslash d_1} |\alpha_i| \neq 0$. 则 $\sum_{i \in d_2} \alpha_i x_i$ 在 $P^{(d)} = [x_i]_{i \in \mathbf{N} \backslash d_1}$ 中有唯一最佳逼近元,即

$$\Pi_{P^{(d_1)}}\left(\sum_{i \in d_2} \alpha_i x_i\right) = \sum_{i \in d_2} \alpha_i x_i - s_{d_1}\left(\sum_{i \in d_2} \alpha_i x_i\right) = \sum_{i \in d_2 \backslash d_1} \alpha_i x_i$$

由于 $0 \in P^{(d_1)}$,所以

$$\left\|\sum_{i \in d_1} \alpha_i x_i\right\| = \left\|\sum_{i \in d_2} \alpha_i x_i - \Pi_{P^{(d_1)}}\left(\sum_{i \in d_2} \alpha_i x_i\right)\right\| <$$

Tschebyscheff 逼近定理

$$\|\sum_{i \in d_2} \alpha_i x_i\|$$

式 ㊲ 成立.

反之,假设式 ㊲ 成立,设 $x = \sum_{i=1}^{\infty} \alpha_i x_i, d \in \mathscr{D}$ 和 $y = \sum_{i \in \mathbf{N} \setminus d} \beta_i x_i \in P^{(d)}$ 且 $y \neq r_d(x) = \sum_{i \in d} \alpha_i x_i$ 是任意的. 那么,在 $\mathbf{N} \setminus d$ 中有一个最小的指标 i_0 使得 $\beta_{i_0} \neq \alpha_{i_0}$. 所以,多次应用式 ㊲ 之后得

$$\|x - r_d(x)\| = \|\sum_{i \in d} \alpha_i x_i\| =$$
$$\|\sum_{i \in d} \alpha_i x_i + \sum_{i \in \mathbf{N} \setminus d, i < i_0} (\beta_i - \alpha_i) x_i\| <$$
$$\|\sum_{i \in d} \alpha_i x_i + \sum_{i \in \mathbf{N} \setminus d, i < i_0} (\beta_i - \alpha_i) x_i\| \leqslant$$
$$\|\sum_{i \in d} \alpha_i x_i + \sum_{i \in \mathbf{N} \setminus d, i \leqslant i_0+1} (\beta_i - \alpha_i) x_i\| \leqslant \cdots \leqslant$$
$$\|\sum_{i \in d} \alpha_i x_i - \sum_{i \in \mathbf{N} \setminus d} (\beta_i - \alpha_i) x_i\| =$$
$$\|x - y\|$$

因此 X 中这一范数是 NK- 范数.

现在我们来考察 NT- 范数与 NK- 范数之间的关系.

定理 8.6 设 X 是具有无条件基 $\{x_n\}_{n=1}^{\infty}$ 的 Banach 空间. 则每个关于 $\{x_n\}_{n=1}^{\infty}$ 的 NT- 范数是关于 $\{x_n\}_{n=1}^{\infty}$ 的 NK- 范数,从而也是关于 $\{x_n\}_{n=1}^{\infty}$ 的 NTK- 范数.

证明 我们证明式 ㊱ 蕴涵式 ㊲.

设 $d_1, d_2 \in \mathscr{D}, d_1 \subset d_2$ 且 $\{\alpha_i\}_{i \in d_2}$ 使得 $\sum_{i \in d_2 \setminus d_1} |\alpha_i| \neq 0$. 令

附录 Ⅷ 在具有基的 Banach 空间中的最佳逼近问题

$$\beta_i = \begin{cases} \alpha_i, & \text{对于 } i \in d_2 \\ 0, & \text{对于 } i \in \mathbf{N}\setminus d_2 \end{cases} \quad ㊳$$

$$d_1' = \varnothing, \, d_2' = d_2 \setminus d_1 \quad ㊴$$

那么 $d_1', d_2' \in \mathscr{D}, d_1' \subset d_2', \sum\limits_{i \in d_2' \setminus d_1'} |\beta_i| = \sum\limits_{i \in d_2 \setminus d_1} |\alpha_i| \neq 0$ 且级数 $\sum\limits_{i \in \mathbf{N} \setminus d_1'} \beta_i x_i = \sum\limits_{i \in d_2} \alpha_i x_i$ 收敛. 因为这个范数是关于 $\{x_n\}_{n=1}^{\infty}$ 的 NT-范数,由式 ㊱ 所以就有

$$\|\sum_{i \in d_1} \alpha_i x_i\| = \|\sum_{i \in d_2 \setminus d_2'} \alpha_i x_i\| = \|\sum_{i \in \mathbf{N} \setminus d_2'} \beta_i x_i\| <$$
$$\|\sum_{i \in \mathbf{N} \setminus d_1'} \beta_i x_i\| = \|\sum_{i \in d_2} \beta_i x_i\|$$

式 ㊲.

这个定理的逆不成立.

例 存在不是 NT-范数的 NK-范数. 仍以 c_0 为例.)))) 表示由式 ㊔ 所确定的等价范数,那么令 $))x))_n =))x))(x \in c_0)$ 是一个 NK-范数但不是关于 c_0 的单位向量基 $\{e_n\}_{n=1}^{\infty}$ 的 NT-范数.

事实上,由定理 8.1 和式 ㊕ 及 ㊖ 知 $))x))_n$ 是关于 $\{e_n\}_{n=1}^{\infty}$ 的 NK-范数但不是 NT-范数.

定理 8.7 设 X 是具有无条件基 $\{x_n\}_{n=1}^{\infty}$ 的 Banach 空间,设 $\{x_n^*\}_{n=1}^{\infty} \subset X^*$ 是相关于 $\{x_n\}_{n=1}^{\infty}$ 的系数泛函序列,则下式

$$((x))_n = \sum_{i=1}^{\infty} \frac{1}{2^i} \|x_i^*(x) x_i\| +$$
$$\sup_{(i_1, i_2, \cdots, i_n) \in \mathscr{D}} \|\sum_{j=1}^{n} x_{i_j}^*(x) x_{i_j}\| \quad (x \in X) \quad ㊵$$

是 X 上关于 $\{x_n\}_{n=1}^{\infty}$ 的等价于 X 上原来范数的 NTK-范数.

Tschebyscheff 逼近定理

证明 $((x))_n$ 显然是 X 上的范数. $((\ ))_n$ 等价于 X 上原来的范数,因为对于每个 $x \in X$ 有

$$\|x\| = ((x))_n \leq \max_{1 \leq i < \infty} \|x_i^*(x)x_i\| +$$

$$\sup_{(i_1, i_2, \cdots, i_n) \in \mathscr{D}} \| \sum_{j=1}^{n} x_{i_j}^*(x) x_{i_j} \| \leq$$

$$\max_{1 \leq i < \infty} (\| \sum_{j=1}^{i} x_j^*(x) x_j \| + \| \sum_{j=1}^{i-1} x_j^*(x) x_j \|) +$$

$$\sup_{(i_1, i_2, \cdots, i_n) \in \mathscr{D}} \| \sum_{j=1}^{n} x_{i_j}^*(x) x_{i_j} \| \leq$$

$3K\|x\|$ (K 是 $\{x_n\}_{n=1}^{\infty}$ 的无条件基常数)

最后,为了证明 $((\ ))_n$ 是 NTK-范数,根据定理 8.6 只要证明 $((\ ))_u$ 是 NT-范数便可. 设 $d_1, d_2 \in \mathscr{D}$, $d_1 \subset d_2$, 设 $\{\alpha_i\}_{i \in \mathbf{N} \setminus d_2}$, $\sum_{i \in d_2 \setminus d_1} |\alpha_i| \neq 0$ 且使得级数 $\sum_{i \in \mathbf{N} \setminus d_2} \alpha_i x_i$ 收敛. 那么有 $\mathbf{N} \setminus d_1 = (\mathbf{N} \setminus d_2) \cup (d_2 \setminus d_1)$, 从而

$$((\sum_{i \in \mathbf{N} \setminus d_2} \alpha_i x_i))_u =$$

$$\sum_{i \in \mathbf{N} \setminus d_2} \frac{1}{2^i} \|\alpha_i x_i\| + \sup_{(i_1, i_2, \cdots, i_n) \in \mathscr{D} \cap (\mathbf{N} \setminus d_2)} \| \sum_{j=1}^{n} \alpha_{i_j} x_{i_j} \| <$$

$$\sum_{i \in \mathbf{N} \setminus d_1} \frac{1}{2^i} \|\alpha_i x_i\| + \sup_{(i_1, i_2, \cdots, i_n) \in \mathscr{D} \cap (\mathbf{N} \setminus d_1)} \| \sum_{j=1}^{n} \alpha_{i_j} x_{i_j} \| =$$

$$((\sum_{i \in \mathbf{N} \setminus d_1} \alpha_i x_i))_u$$

因此,由定理 8.5(a) 知 $((\ ))_u$ 是 NT-范数.

值得指出的是在定理 8.7 的条件下并且用类似的方法可以证明下式

$$((x))_u = \sum_{i=1}^{\infty} \frac{1}{2^i} \|x_i^*(x)x_i\| +$$

附录 Ⅷ 在具有基的 Banach 空间中的最佳逼近问题

$$\sup_{1\leqslant n<\infty, \varepsilon_i=\pm 1}\|\sum_{i=1}^{n}\varepsilon_i x_i^*(x)x_i\| =$$

$$\sum_{i=1}^{\infty}\frac{1}{2^i}\|x_i^*(x)x_i\| +$$

$$\sup_{x^*\in X^*, \|x^*\|\leqslant 1}\sum_{i=1}^{n}|x_i^*(x)x^*(x)|$$

⑷⒈

是 X 上等价于原范数的 NTK- 范数.

С. Н. Мергелян 定理的推广[①]

附录 IX

　　С. Н. Мергелян 在他的博士论文中[1], 给出复数域逼近论的一逆定理, 即由 $f(z)$ 在区域 D 中的逼近度 $\rho_n(f,D)$ 给出 $f(z)$ 的连续性. 本文把他的结果推广为 De la Vallée Poussin 在实变数逼近论中[2]相应定理的形式.

　　兹先介绍本文中引用的符号:

　　区域 D 是具有连通补集的卡拉切奥多利域. L_R 是 D 的外平准线, 它是把 $\overline{D}$ 的补集保角映照于 $|w|>1$ 的映照下, $|w|=R>1$ 所对应的曲线. Γ 是 D 的境界线, $D(\xi;R)$ 是 Γ 上一点 ξ

---

① 郭竹瑞. 浙江大学.

到 L_R 的距离. 对于 $\xi \in \Gamma, B_\xi$ 是 D 的部分区域,它具有下面的性质: B_ξ 内的任意一点到 ξ 的距离与到 Γ 的距离的比恒小于某一常数 $C_1, \omega(\delta)$ 是 $f(z)$ 在 D 中的连续模,它表示

$$|f(z')-f(z'')|$$

对于所有 $\overline{D}$ 中的点 z', z'' 能够用全在 $\overline{D}$ 中长度不超过 δ 的有长曲线连续接的上确界. $\omega_r(\delta)$ 是 $f(z)$ 的 r 阶微商 $f^{(r)}(z)$ 在 D 中的连续模.

引理 9.1 存在正整数序列

$$n_1 < n_2 < \cdots < n_k < \cdots$$

使

$$\frac{1}{4}D\left(\xi; 1+\frac{1}{n_{k-1}}\right) < D\left(\xi; 1+\frac{1}{n_k}\right) < \frac{1}{2}D\left(\xi; 1+\frac{1}{n_{k-2}}\right) \quad ⑷⑷②$$

证明 对于每一整数 $k > 0$ 可决定正整数 n_k 使

$$D\left(\xi; 1+\frac{1}{n_k}\right) \geqslant \frac{1}{2^k} > D\left(\xi; 1+\frac{1}{n_k+1}\right)$$

于是

$$D\left(\xi; 1+\frac{1}{n_k}\right) \leqslant D\left(\xi; 1+\frac{1}{n_{k-1}+1}\right) < \frac{1}{2^{k-1}} \leqslant$$

$$D\left(\xi; 1+\frac{1}{n_{k-1}}\right) \leqslant$$

$$D\left(\xi; 1+\frac{1}{n_{k-2}+1}\right) < \frac{1}{2^{k-2}} \leqslant$$

$$D\left(\xi; 1+\frac{1}{n_{k-2}}\right)$$

从而推得

Tschebyscheff 逼近定理

$$D\left(\xi;1+\frac{1}{n_k}\right) < \frac{1}{2}D\left(\xi;1+\frac{1}{n_{k-2}}\right)$$

又可证

$$D\left(\xi;1+\frac{1}{n_k}\right) > \frac{1}{4}D\left(\xi;1+\frac{1}{n_{k-1}}\right)$$

所以

$$\frac{1}{4}D\left(\xi;1+\frac{1}{n_{k-1}}\right) < D\left(\xi;1+\frac{1}{n_k}\right) < \\ \frac{1}{2}D\left(\xi;1+\frac{1}{n_{k-2}}\right)$$

引理 9.1 证毕.

引理 9.2 设 n 次多项式 $P_n(z)$ 满足

$$\max_{z\in\bar{D}}|P_n(z)| = M$$

那么

$$\max_{z\in B_\xi}|P_n^{(r)}(z)| \leqslant \frac{LMr!}{\left[D\left(\xi;1+\frac{1}{n}\right)\right]^r}$$

其中 L 是和 n 与 r 无关的常数.

证明见 Мергелян 论文[1]第一章引理 4.

定理 9.3 设 $\Omega(x)$ 是实变数 x 单调不增的函数,又

$$\lim_{x\to\infty}\Omega(x) = 0 \qquad ⑷⑷③$$

$f(z)$ 是定义于 $\bar{D}$ 中的函数,若

$$\rho_n(f,D) < \left\{D\left(\xi;1+\frac{1}{n}\right)\right\}^r\Omega(n) \qquad ⑷⑷④$$

其中 $r \geqslant 0$ 是整数,如果

$$\int^\infty \frac{\Omega(x)}{D\left(\xi;1+\frac{1}{x}\right)}\mathrm{d}D\left(\xi;1+\frac{1}{x}\right) < \infty \qquad ⑷⑷⑤$$

附录 IX　C. H. Мергелян 定理的推广

那么, $f(z)$ 在 $\bar{B}_\xi$ 中具有 r 级连续微商, 且 $f^{(r)}(z)$ 的连续模 $\omega_r(\delta)$ 满足

$$\omega_r(\delta) \leqslant C\left[-\delta\int_{n_1}^{D^{-1}(\frac{\delta}{4})} \frac{\Omega(x)}{\left[D\left(\xi;1+\frac{1}{x}\right)\right]^2} \mathrm{d}D\left(\xi;1+\frac{1}{x}\right) - \right.$$
$$\left. \int_{D^{-1}(\delta)}^{\infty} \frac{\Omega(x)}{D\left(\xi;1+\frac{1}{x}\right)} \mathrm{d}D\left(\xi;1+\frac{1}{x}\right)\right]$$

其中 C, n_1 是常数, $D^{-1}(\delta)$ 是 $D\left(\xi;1+\frac{1}{x}\right)$ 的反函数, 它表示使

$$D\left(\xi;1+\frac{1}{x}\right) = \delta$$

的数值 x.

如果对于任意的正整数 N 都有 $\Omega_N(x)$ 满足[443][445]使

$$\rho_n(f,D) < \left\{D\left(\xi;1+\frac{1}{x}\right)\right\}^N \Omega_N(n)$$

成立, 则 $f(z)$ 在 $\bar{B}_\xi$ 中有无限级微商存在.

证明　设 $P_n(z)$ 是 $f(z)$ 在 D 中的 n 次最佳逼近多项式, 显见

$$f(z) = P_{n_1}(z) + \sum_{k=1}^{\infty}\left[P_{n_{k+1}}(z) - P_{n_k}(z)\right]$$

记

$$R_{n_k}(z) = R_{n_k}(z) - R_{n_{k-1}}(z)$$

则有

$$\max_{z\in\bar{D}}|R_{n_k}(z)| \leqslant \max_{z\in\bar{D}}|P_{n_k}(z) - f(z)| +$$
$$\max_{z\in\bar{D}}|P_{n_{k-1}}(z) - f(z)| \leqslant$$

Tschebyscheff 逼近定理

$$2\rho_{n_{k-1}}(f,D)$$

由引理 9.1, 9.2 及定理 9.3 的假设得到

$$\max_{z\in \bar{B}_\xi} |R_{n_k}^{(r)}(z)| \le \max_{z\in \bar{B}_\xi} |R_{n_{k+1}}^{(r)}(z)| \le$$

$$\frac{2Lr!\,\rho_{n_{k-1}}(f,D)}{\left[D\!\left(\xi;1+\dfrac{1}{n_k}\right)\right]^r} + \frac{2Lr!\,\rho_{n_k}(f,D)}{\left[D\!\left(\xi;1+\dfrac{1}{n_{k+1}}\right)\right]^r} \le$$

$$\frac{2^{2r+1}Lr!\,\rho_{n_{k-1}}(f,D)}{\left[D\!\left(\xi;1+\dfrac{1}{n_{k-1}}\right)\right]^r} + \frac{2^{2r+1}Lr!\,\rho_{n_k}(f,D)}{\left[D\!\left(\xi;1+\dfrac{1}{n_k}\right)\right]^r} \le$$

$$2^{2r+1}Lr!\,\Omega(n_{k-1}) + 2^{2r+1}Lr!\,\Omega(n_k) \le$$

$$C'\Omega(n_{k-1}) \le$$

$$-2C'\frac{\Omega(n_{k-1})}{D\!\left(\xi;1+\dfrac{1}{n_{k-3}}\right)}\left[D\!\left(\xi;1+\dfrac{1}{n_{k-1}}\right) - D\!\left(\xi;1+\dfrac{1}{n_{k-3}}\right)\right] \le$$

$$-C''\int_{n_{k-3}}^{n_{k-1}} \frac{\Omega(x)}{D\!\left(\xi;1+\dfrac{1}{x}\right)}\,\mathrm{d}D\!\left(\xi;1+\dfrac{1}{x}\right)$$

这里 $C''>0$ 是和 n_k 及 x 无关的常数

$$\sum_{K=k}^{\infty} \max_{z\in \bar{B}_\xi} |R_{n_K}^{(r)}(z)| =$$

$$(\max_{z\in \bar{B}_\xi} |R_{n_k}^{(r)}(z)| + \max_{z\in \bar{B}_\xi} |R_{n_{k+1}}^{(r)}(z)|) +$$

$$(\max_{z\in \bar{B}_\xi} |R_{n_{k+2}}^{(r)}(z)| + \max_{z\in \bar{B}_\xi} |R_{n_{k+3}}^{(r)}(z)| + \cdots \le$$

$$-C''\left[\int_{n_{k-3}}^{n_{k-1}} \frac{\Omega(x)}{D\!\left(\xi;1+\dfrac{1}{x}\right)}\,\mathrm{d}D\!\left(\xi;1+\dfrac{1}{x}\right) + \right.$$

(446)

$$\int_{n_{k-1}}^{n_{k+1}} \frac{\Omega(x)}{D\left(\xi;1+\frac{1}{x}\right)} \mathrm{d}D\left(\xi;1+\frac{1}{x}\right) + \cdots \Bigg] =$$

$$- C'' \int_{n_{k-3}}^{\infty} \frac{\Omega(x)}{D\left(\xi;1+\frac{1}{x}\right)} \mathrm{d}D\left(\xi;1+\frac{1}{x}\right)$$

由定理 9.3 的假设 ㊺ 知上面不等式中最后一积分是收敛的.

多项式级数

$$P_{n_1}^{(r)}(z) + \sum_{k=1}^{\infty} \left[P_{n_{k+1}}^{(r)}(z) - P_{n_k}^{(r)}(z) \right]$$

在 D 中就是 $f^{(r)}(z)$,由上面的证明知道它在 $\bar{B}_\xi$ 中是绝对且一致收敛的,从而 $f^{(r)}(z)$ 在 $\bar{B}_\xi$ 中是连续的.

现在来估计 $f^{(r)}(z)$ 在 $\bar{B}_\xi$ 中的连续模:设 z',z'' 是 $\bar{B}_\xi$ 中任意两点,显见

$$|f^{(r)}(z') - f^{(r)}(z'')| \leqslant$$
$$|P_{n_1}^{(r)}(z') - P_{n_1}^{(r)}(z'')| +$$
$$\sum_{k=2}^{3} |P_{n_k}^{(r)}(z') - P_{n_{k-1}}^{(r)}(z') - P_{n_k}^{(r)}(z'') + P_{n_{k-1}}(z'')| +$$
$$\sum_{k=4}^{m} |P_{n_k}^{(r)}(z') - P_{n_{k-1}}^{(r)}(z') - P_{n_k}^{(r)}(z'') + P_{n_{k-1}}(z'')| +$$
$$\sum_{k=m+1}^{\infty} |P_{n_k}^{(r)}(z') - P_{n_{k-1}}^{(r)}(z') - P_{n_k}(z'') + P_{n_{k-1}}(z'')| \quad ㊼$$

这里的 m 留待下面决定.

式 ㊼ 右边第三项和数中相邻两项的和不超过

$$\max_{z \in \bar{B}_\xi} |P_{n_k}^{(r+1)}(z) - P_{n_{k-1}}^{(r+1)}(z)| \int_{z'}^{z''} \mathrm{d}S +$$
$$\max_{z \in \bar{B}_\xi} |P_{n_{k+1}}^{(r+1)}(z) - P_{n_k}^{(r+1)}(z)| \int_{z'}^{z''} \mathrm{d}S =$$

Tschebyscheff 逼近定理

$$\left[\max_{z\in \bar{B}_\xi}|R_{n_k}^{(r+1)}(z)|+\max_{z\in \bar{B}_\xi}|R_{n_{k+1}}^{(r+1)}(z)|\right]\int_{z'}^{z''}\mathrm{d}S \quad ⑭⑧$$

这里积分路线位于 $\bar{B}_\xi$ 之中.

仿式 ⑭⑥ 并利用引理 9.1 可以估计式 ⑭⑧ 右边第一因子

$$\max_{z\in \bar{B}_\xi}|R_{n_k}^{(r+1)}(z)|+\max_{z\in \bar{B}_\xi}|R_{n_{k+1}}^{(r+1)}(z)|\leqslant$$

$$\frac{C'''\Omega(n_{k-1})}{D\left(\xi;1+\frac{1}{n_{k-1}}\right)}+\frac{C'''\Omega(n_k)}{D\left(\xi;1+\frac{1}{n_k}\right)}\leqslant$$

$$\frac{5C'''\Omega(n_{k-1})}{D\left(\xi;1+\frac{1}{n_{k-1}}\right)}<\frac{80C'''\Omega(n_{k-1})}{D\left(\xi;1+\frac{1}{n_{k-3}}\right)}<$$

$$\frac{-160C'''\Omega(n_{k-1})}{\left[D\left(\xi;1+\frac{1}{n_{k-3}}\right)\right]^2}\left[D\left(\xi;1+\frac{1}{n_{k-1}}\right)-D\left(\xi;1+\frac{1}{n_{k-3}}\right)\right]\leqslant$$

$$-160C'''\int_{n_{k-2}}^{n_{k-1}}\frac{\Omega(x)}{\left[D\left(\xi;1+\frac{1}{x}\right)\right]^2}\mathrm{d}D\left(\xi;1+\frac{1}{x}\right)$$

取 $\int_{z'}^{z''}\mathrm{d}S\leqslant\delta$, 即积分路径为一有长曲线,其长不超过 δ, 于是式 ⑭⑧ 不超过

$$-C_3\delta\int_{n_{k-3}}^{n_{k-1}}\frac{\Omega(x)}{\left[D\left(\xi;1+\frac{1}{x}\right)\right]^2}\mathrm{d}D\left(\xi;1+\frac{1}{x}\right) \quad ⑭⑨$$

这里 $C_3>0$ 为一与 n_k 无关的常数.

现在模仿式 ⑭⑥ 来估计式 ⑭⑦ 右边最后一项和数中相邻两项之和

$$|R_{n_k}^{(r)}(z')-R_{n_k}^{(r)}(z'')|+|R_{n_{k+1}}^{(r)}(z')-R_{n_{k+1}}^{(r)}(z'')|\leqslant$$

$$2C_2\Omega(n_{k-1})+2C_2\Omega(n_k)\leqslant 4C_2\Omega(n_{k-1})<$$

附录 IX　C. H. Мергелян 定理的推广

$$-8C_2 \frac{\Omega(n_{k-1})}{D\left(\xi;1+\frac{1}{n_{k-3}}\right)}\left[D\left(\xi;1+\frac{1}{n_{k-1}}\right)-\right.$$

$$\left.D\left(\xi;1+\frac{1}{n_{k-3}}\right)\right]\leqslant$$

$$-8C_2\int_{n_{k-3}}^{n_{k-1}}\frac{\Omega(x)}{D\left(\xi;1+\frac{1}{x}\right)}\mathrm{d}D\left(\xi;1+\frac{1}{x}\right)$$

于是

$$\mid R^{(r)}_{n_k}(z')-R^{(r)}_{n_k}(z'')\mid+\mid R^{(r)}_{n_{k+1}}(z')-R_{n_{k+1}}(z'')\mid<$$

$$-C_4\int_{n_{k-3}}^{n_{k-1}}\frac{\Omega(x)}{D\left(\xi;1+\frac{1}{x}\right)}\mathrm{d}D\left(\xi;1+\frac{1}{x}\right) \qquad ㊄$$

这里 $C_4>0$ 为一常数.

基于 ㊇㊈㊄ 可得 $f^{(r)}(z)$ 连续模的估计

$$\mid f^{(r)}(z')-f^{(r)}(z'')\mid\leqslant\mid P^{(r)}_{n_1}(z')-P^{(r)}_{n_1}(z'')\mid+$$

$$\delta\sum_{k=1}^{3}\max_{z\in\bar{B}_\xi}\mid R^{(r+1)}_{n_k}(z)\mid-$$

$$C_3\delta\sum_{k=1}^{m-3}\int_{n_k}^{n_{k+1}}\frac{\Omega(x)}{\left[D\left(\xi;1+\frac{1}{x}\right)\right]^2}\mathrm{d}D\left(\xi;1+\frac{1}{x}\right)-$$

$$C_4\sum_{k=m-2}^{\infty}\int_{n_k}^{n_{k+1}}\frac{\Omega(x)}{D\left(\xi;1+\frac{1}{x}\right)}\mathrm{d}D\left(\xi;1+\frac{1}{x}\right)\leqslant$$

$$C_5\delta+C_6\delta-$$

$$C_3\delta\int_{n_1}^{n_{m-2}}\frac{\Omega(x)}{\left[D\left(\xi;1+\frac{1}{x}\right)\right]^2}\mathrm{d}D\left(\xi;1+\frac{1}{x}\right)-$$

$$C_4\int_{n_{m-2}}^{\infty}\frac{\Omega(x)}{D\left(\xi;1+\frac{1}{x}\right)}\mathrm{d}D\left(\xi;1+\frac{1}{x}\right)$$

取 m 使

Tschebyscheff 逼近定理

$$n_{m-2} > D^{-1}(\delta) \geqslant n_{m-2} - 1$$

又只要 $f^{(r)}(z)$ 不为常数,必有 $C_7 > 0$ 存在使得它的连续模 $\omega_r(\delta)$ 满足

$$\omega_r(\delta) > C_7 \delta^{①}$$

于是得

$$\omega_r(\delta) \leqslant C\left[-\delta \int_{n_1}^{D^{-1}\left(\frac{\delta}{4}\right)} \frac{\Omega(x)}{\left[D\left(\xi;1+\frac{1}{x}\right)\right]^2} \mathrm{d}D\left(\xi;1+\frac{1}{x}\right) - \int_{D^{-1}(\delta)}^{\infty} \frac{\Omega(x)}{D\left(\xi;1+\frac{1}{x}\right)} \mathrm{d}D\left(\xi;1+\frac{1}{x}\right)\right]$$

定理的第一部分证毕. 至于 $f(z)$ 具有无限级微商的情况,由上面的证明显然可见.

推论 9.4 如果

$$\lim_{n\to\infty} \frac{\ln \dfrac{1}{\rho_n(f,D)}}{\ln \dfrac{1}{D\left(\xi;1+\dfrac{1}{n}\right)}} = A$$

对于任意的 $\varepsilon > 0$,记 $[A-\varepsilon] = r$,则在 $\bar{B}_\xi$ 中 $f^{(r)}(z) \in \mathrm{Lip}(B_\xi; A-\varepsilon-r)$. 若 $A = \infty$,则 $f(z)$ 在 $\bar{B}_\xi$ 中具有无限极微商.

推论 9.5 如果

$$\rho_n(f,D) \sim \frac{M\left[D\left(\xi;1+\dfrac{1}{n}\right)\right]^r}{\left[\ln \dfrac{1}{D\left(\xi;1+\dfrac{1}{n}\right)}\right]^{1+\alpha}}$$

其中 $r \geqslant 0$ 是整数,$\alpha > 0$,则在 $\bar{B}_\xi$ 中 $f^{(r)}(z)$ 的连续模

---

① 可以仿[1] p.17 的证明.

附录 IX C. H. Мергелян 定理的推广

满足

$$\omega_r(\delta) < \frac{C_8}{\left(\ln\dfrac{1}{\delta}\right)^\alpha}$$

附记 1 推论 9.1 就是 Мергелян 定理 3.1[1].

附记 2 在推论 9.2 的条件下如果应用 Мергелян 定理，只能得到在 $\bar{B}_\xi$ 中

$$f^{(r-1)}(z) \in \mathrm{Lip}(B_\xi; 1-\varepsilon)$$

得不到推论 2 的结果.

定理 9.3 只在 $\bar{B}_\xi$ 中成立，而对于具有连通补集的卡拉特阿多利域的境界线 Γ 上任一点 ξ，不见得存在区域 B_ξ(Мергелян[1] 指出：存在 B_ξ 的点 ξ 处处稠密于 Γ 上). 因此定理 9.3 只是由 $f(z)$ 的逼近度得出 $f(z)$ 的局部性质，下面将证明由 $f(z)$ 的逼近度得出它在 $\bar{D}$ 中的连续性，为此，先证明下面引理：

引理 9.6 设 n 次多项式 $P_n(z)$ 满足

$$\max_{z \in \bar{D}} |P_n(z)| = M$$

那么

$$\max_{z \in \bar{D}} |P_n^{(r)}(z)| \leqslant \frac{M e\, r!}{\left[D\!\left(\Gamma; 1+\dfrac{1}{n}\right)\right]^r}$$

其中 $D\!\left(\Gamma; 1+\dfrac{1}{n}\right) = \inf\limits_{\xi \in \Gamma} D\!\left(\xi; 1+\dfrac{1}{n}\right)$，$\Gamma$ 是 D 的境界线.

证明 设 z 是 Γ 上的任意一点，则由 Cauchy 积分公式

$$P_n^{(r)}(z) = \frac{r!}{2\pi \mathrm{i}} \int_{|t-z| = D\left(\Gamma; 1+\frac{1}{n}\right)} \frac{P_n(t)}{(t-z)^{r+1}} \mathrm{d}t$$

Tschebyscheff 逼近定理

基于 Мергелян 论文[1] 第一章引理 3 知

$$\max_{t \in L_{1+\frac{1}{n}}} |P_n(t)| \leq M\left(1 + \frac{1}{n}\right)^n$$

于是

$$|P_n^{(r)}(z)| \leq \frac{r! \, M\left(1+\frac{1}{n}\right)^n}{\left[D\left(\Gamma;1+\frac{1}{n}\right)\right]^r} < \frac{r! \, Me}{\left[D\left(\Gamma;1+\frac{1}{n}\right)\right]^r}$$

由最大模原理推知

$$\max_{z \in \overline{D}} |P_n^{(r)}(z)| \leq \frac{Me\, r!}{\left[D\left(\Gamma;1+\frac{1}{n}\right)\right]^r}$$

引理证毕.

定理 9.7 设 $\Omega(x)$ 是实变数 x 单调不增的函数，且满足

$$\lim_{x \to \infty} \Omega(x) = 0 \qquad ㊽$$

$f(z)$ 是定义于 $\overline{D}$ 中的函数，若

$$\rho_n(f, D) < \left\{D\left(\Gamma;1+\frac{1}{n}\right)\right\}^r \Omega(n) \qquad ㊾$$

其中 $r \geq 0$ 是整数. 如果

$$\int^{\infty} \frac{\Omega(x)}{D\left(\Gamma;1+\frac{1}{x}\right)} \mathrm{d}D\left(\Gamma;1+\frac{1}{x}\right) < \infty \qquad ㊿$$

那么，$f(z)$ 在 $\overline{D}$ 中具有 r 级连续微商，且 $f^{(r)}(z)$ 的连续模 $\omega_r(\delta)$ 满足

$$\omega_r(\delta) \leq$$

$$C\left[-\delta \int_{n_1}^{D^{-1}\left(\frac{\delta}{4}\right)} \frac{\Omega(x)}{\left[D\left(\Gamma;1+\frac{1}{x}\right)\right]^2} \mathrm{d}D\left(\Gamma;1+\frac{1}{x}\right) - \right.$$

附录 IX　C. H. Мергелян 定理的推广

$$\int_{D^{-1}(\delta)}^{\infty} \frac{\Omega(x)}{D\left(\Gamma;1+\frac{1}{x}\right)} \mathrm{d}D\left(\Gamma;1+\frac{1}{x}\right) \Big] \quad \text{㊹}$$

其中 C, n_1 是常数，$D^{-1}(\delta)$ 是 $D\left(\Gamma;1+\frac{1}{x}\right)$ 的反函数.

如果对于任意的正整数 N 都有 $Q_N(x)$ 满足㊶㊸使

$$\rho_n(f,D) < \left\{D\left(\Gamma;1+\frac{1}{n}\right)\right\}^N \Omega_N(n)$$

成立，则 $f(z)$ 在 $\overline{D}$ 中有无限级微商存在.

定理 9.7 的证明，完全和定理 9.3 一样，只要在定理 1 的证明中把 $D\left(\xi;1+\frac{1}{x}\right)$ 改为 $D\left(\Gamma;1+\frac{1}{x}\right)$，$\overline{B}_\xi$ 改为 $\overline{D}$，而在应用引理 9.2 的地方，故用引理 9.6 便得.

推论 9.8　当区域 D 的境界线 Γ 是解析的约当曲线时，若

$$\rho_n(f,D) < \frac{M}{n^{r+\alpha}}$$

这里 $r \geq 0$ 是整数，$M > 0, 0 < \alpha \leq 1$ 是常数，当 $0 < \alpha < 1$ 时

$$f^{(r)}(z) \in \mathrm{Lip}(\overline{D},\alpha)$$

当 $\alpha = 1$ 时，$f^{(r)}(z)$ 的连续模 $\omega_r(\delta)$ 满足

$$\omega_r(\delta) \leq K\delta\ln\frac{1}{\delta}$$

这里 K 是正常数.

事实上，此时

Tschebyscheff 逼近定理

$$\frac{A}{n} < D\left(\Gamma;1+\frac{1}{n}\right) < \frac{B}{n}\text{\textcircled{1}}$$

这里 A 和 B 是与 n 无关的常数,于是

$$\rho_n(f,D) < M_1\left[D\left(\Gamma;1+\frac{1}{n}\right)\right]^{r+\alpha}$$

这里 M_1 是常数,由定理 9.7 即可推得结论.

推论 9.9 当区域 D 的境界线 Γ 是解析的约当曲线时,若

$$\rho_n(f,D) < \frac{M}{n^r[\ln n]^{1+\alpha}}$$

这里 $r \geq 0$ 是整数,$\alpha > 0$ 是常数,则 $\omega_r(\delta)$ 满足

$$\omega_r(\delta) < \frac{h}{\left(\ln\frac{1}{\delta}\right)^\alpha}$$

其中 $h > 0$ 是常数.

若区域 D 是约当区域,它的境界线由有限条解析的约当弧组成,如果这些弧在交点处所张外角的最大者为 $\lambda\pi$,定义 t 如下

$$t = \lambda, 当 \lambda > 1 时$$
$$t = 1, 当 \lambda \leq 1 时$$

我们称这样的约当曲线为 t 型曲线.

A 是已给正数,当 $\dfrac{A}{t}$ 不为正整数时,记 $\left[\dfrac{A}{t}\right] = r$ 为不超过 $\dfrac{A}{t}$ 的最大整数.

推论 9.10 设 D 是 t 型曲线 Γ 所界的约当区域,

---

① 例如见 [3] p.87.

附录 Ⅸ С.Н.Мергелян 定理的推广

M 是正常数,如果

$$\rho_n(f,D) < \frac{M}{n^A}$$

则当 $\frac{A}{t}$ 不为正整数时

$$f^{(r)}(z) \in \mathrm{Lip}\left(\overline{D};\frac{A}{t}-r\right)$$

若 $\frac{A}{t}$ 为正整数 r 时,则 $f^{(r-1)}(z)$ 连续,且于 $\overline{D}$ 中成立

$$\omega_{r-1}(\delta) \leqslant K\delta\ln\frac{1}{\delta}$$

事实上,当 Γ 为 t 型曲线时

$$D\left(\Gamma;1+\frac{1}{n}\right) \geqslant K(\Gamma)\frac{1}{n^t}①$$

这里 $K(\Gamma)$ 是仅与 Γ 有关的常数. 由此容易推得结论.

推论 9.11 设 D 是 t 型曲线 Γ 所界的约当区域,如果

$$\rho_n(f,D) < \frac{M}{n^A[\ln n]^{1+\alpha}}$$

其中 $\frac{A}{t}=r \geqslant 0$ 是整数,$\alpha > 0$,那么在 $\overline{D}$ 中成立

$$\omega_r(\delta) \leqslant \frac{h}{\left[\ln\frac{1}{\delta}\right]^\alpha}$$

这里 h 是正常数.

推论 9.12 设 D 是任意的有限区域,如果

$$\rho_n(f,D) < \frac{M}{n^A}$$

---

① 例如见[4] p.34-35.

当 $\dfrac{A}{2}$ 不为正整数时,记 $\left[\dfrac{A}{2}\right] = r$,则

$$f^{(r)}(z) \in \text{Lip}D\left(\overline{D}, \dfrac{A}{2} - r\right)$$

当 $\dfrac{A}{2} = r$ 为正整数时,则 $f^{(r-1)}(z)$ 在 $\overline{D}$ 中连续,且

$$\omega_{r-1}(\delta) < K\delta\ln\dfrac{1}{\delta}$$

在 $\overline{D}$ 中成立,$K > 0$ 是常数.

事实上,此时

$$D\left(\varGamma; 1 + \dfrac{1}{n}\right) \geq \dfrac{\lambda_1}{n^2}①$$

由此容易推得结论.

推论 9.13　设 D 是任意的有限区域,如果

$$\rho_n(f, D) < \dfrac{M}{n^A(\ln n)^{1+\alpha}}$$

这里 $\dfrac{A}{2} = r \geq 0$ 是整数,$\alpha > 0$,则

$$\omega_r(\delta) \leq \dfrac{h}{\left(\ln\dfrac{1}{\delta}\right)^\alpha}$$

这里 h 是正常数.

附记　推论 9.8,9.10 已见于 [4],推论 9.12 比 Мергелян 论文[1]第三章定理 3.2 的结果稍许好一些.

---

① 例如见 [1] p.27.

附录 IX　C. H. Мергелян 定理的推广

参考文献

[1] МЕРГЕЛЯН С Н. Некоторые вопросы конструктивной теории функций[C]. СССР:Труэы мамем. 1951(37). цсмцмума цм. В. А. Смеклова, АН СССР, 37(1951).

[2] POUSSIN D L V. Leçons sur làpproximation des fonctions d'une variable réelle[M]. [S. I.]:[s. n.], 1919.

[3] SEWEEL W E. Generalized derivatives and approximation by polynomials[J]. Trans. Amer. Math. Soc., 1937(41):84-123.

[4] SEWEEL W E. Degree of approximation by polynomials in the complex domain[M]. New Jersey:Princeton University press, 1942.

平方逼近

附录 X

1. 函数按最小二乘法的逼近

对于函数 $f(x)$ 在区间 $[a,b]$ 上以多项式按其在离散点 $a \leq x_0, x_1, \cdots, x_n \leq b$ 处给出的特殊值的逼近,我们曾依照内插节的个数选择了多项式的次数使得内插多项式可唯一确定. 当我们说到了函数 $f(x)$ 以多项式逼近的可能时,我们曾顾及到下一事实,即逼近多项式与 $f(x)$ 的偏差随着内插节个数的增加趋于零,并且此多项式的次数也因而随着节的个数的增加而增高. 然而节的个数的增加会使计算大加复杂,因而便有这样的情形,即考虑与 $f(x)$ 在区间 $[a,b]$ 的离散点处近似地重合的较低次(小于 n)的多项式更来得有利些. 对于大多数由经验所发现的函数关系毕竟用不着达到准确的重合,因为从表中或从图上所取的函数值,本身就有错误.

附录 X 平方逼近

因此，我们研究下一问题的解：设已知函数 $f(x)$ 的 $n+1$ 个准确（近似）值，且要在次数 $m<n$ 的诸多项式

$$\varphi_m(x)=a_0x^m+a_1x^{m-1}+\cdots+a_{m-1}x+a_m \qquad (455)$$

中去找那样一个多项式，使得它对于区间 (a,b) 内的自变量 x 的某些给定值，近似地取 $f(x)$ 的给定值. 当然，此时需要这样选择未知系数 $a_0,a_1,\cdots,a_m$，使得偏差的代数和

$$\sum_{v=0}^{n}[f(x_v)-\varphi_m(x_v)]$$

的绝对值是极小.

为了表示在此情形下，逼近方法完成到什么程度，只需指出，具有相反符号的随意大的误差可相互对消，因而虽然偏差的和可以是非显著的（按绝对值），但一个函数与另一个函烽在某些区间的偏差仍能是显著的.

再者，为简单计，作为逼近的量度，我们可取偏差在内插节处的绝对值之和

$$\sum_{v=0}^{n}|f(a_v)-\varphi_m(x_v)| \qquad (456)$$

且当任意选择系数 $a_0,a_1,\cdots,a_m$ 时，力求使此和为最小. 此处在和中出现的已是带相反符号的大的误差的绝对值，而此和也由此会显著地增大了. 因此，如果和 (456) 是小的，则最大差的绝对值 $|f(x_v)-\varphi_m(x_v)|$ 也将是小的.

在大多数情形，为了较好的逼近，对于作逼近函数我们利用偏差的平方，即利用量

Tschebyscheff 逼近定理

$$\sum_m = \sum_{v=0}^{n} \varepsilon_v^2 \qquad ⑮⑦$$

作为逼近的准确性的量度. 此处

$$\begin{cases} \varepsilon_0 = a_0 x_0^m + a_1 x_0^{m-1} + \cdots + a_{m-1} x_0 + a_m - f(x_0) \\ \varepsilon_1 = a_0 x_1^m + a_1 x_1^{m-1} + \cdots + a_{m-1} x_1 + a_m - f(x_1) \\ \vdots \\ \varepsilon_n = a_0 x_n^m + a_1 x_n^{m-1} + \cdots + a_{m-1} x_n + a_m - f(x_n) \end{cases} \qquad ⑮⑧$$

选择偏差的二次方由概率理论的观点可证明是正确的. 在方次的这样选择下,实验的偶然误差对于所作的逼近函数有最小的影响.

今考虑误差的平方的算术平均的平方根

$$\Delta_{m,n} = \sqrt{\frac{\sum_{v=0}^{n}[f(x_v) - \varphi_m(x_v)]^2}{n+1}}$$

量 $\Delta_{m,n}$ 叫作平均平方误差. 如果 $\Delta_{m,n}$ 不过 $\varepsilon > 0$,则我们说,函数 $\varphi_m(x)$ 在区间 $[a,b]$ 上平均逼近函数 $f(x)$ 准确到 ε. 完全明显,如果

$$|f(x_v) - \varphi_m(x_v)| < \varepsilon \quad (v = 0, 1, \cdots, n)$$

则 $\Delta_{m,n} < \varepsilon$. 因此,如果 $\varphi_m(x)$ 与 $f(x)$ 的偏离对于区间 $[a,b]$ 的所有离散点都是小的,则平均平方误差 $\Delta_{m,n}$ 也将是小的.

今我们着重的是下一问题:如果能够借助函数 $\varphi_m(x)$ 平均逼近任一函数准确到 ε,这是否意味着,不论 ε 是怎样,在区间 $[a,b]$ 的所有点处,偏离都满足不等式

$$|f(x_v) - \varphi_m(x_v)| < \varepsilon$$

很显然,$\Delta_{m,n}$ 不但取决于函数 $f(x)$ 和 $\varphi_m(x)$,而且

附录 X 平方逼近

也取决于为着构成偏差的平方和引用怎样一些离散点. 我们指出,如果对于固定区间$[a,x_k]$和$[x_{k+1},b]$的不同离散(图10.1)却考虑差$f(x)-\varphi_m(x)$,则可以指出小区间(x_k,x_{k+1})的这样一些点,对于它们,偏差$|f(x)-\varphi_m(x)|$将大于平均平方误差$\Delta_{m,n}$,但有可能,对于这个,其原因乃是在形成偏差的平方和时,没有计及区间$[x_k,x_{k+1}]$的点.

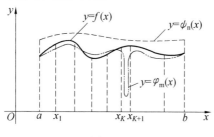

图 10.1

今将区间$[a,b]$分成n个相等部分,并计算当$n\to\infty$时,对在区间$[a,b]$内分段连续函数$f(x)$的$\Delta_{m,n}$的值. 于是

$$\lim_{n\to\infty}\Delta_{m,n}=\sqrt{\frac{\int_a^b[f(x)-\varphi_m(x)]^2\mathrm{d}x}{b-a}} \qquad ㊾$$

所得的极限叫作函数$\varphi_m(x)$与$f(x)$的平均平方偏差. 此时,可以任意改变函数$f(x)$或$\varphi_m(x)$在包含于$[a,b]$内的充分小的ε区间内的状态(完全与这些函数在ε区间外的状态无关),使得被积函数在$[a,b]$内的变化实际上并没有使在极限等式(5)右端的积分值改变,而在ε区间的点处$\varphi_m(x)$与$f(x)$的偏差的绝对值大于平均平方偏差(5).

Tschebyscheff 逼近定理

因此,如果函数 $\varphi_m(x)$ 与 $f(x)$ 的平均平方偏差是小的,则由此并不能推断在通常意义下,对 $[a,b]$ 的所有 x 偏差 $|f(x)-\varphi_m(x)|$ 也是小的. 这当然并不是说反对平均逼近方法的使用. 反之,应该把它看作一个论断,即仅仅一个连续性,一般说来,对于在通常意义下以任意多项式(具有任意准确程度)来逼近 $f(x)$ 是不充分的.

现在我们可以说明实线 $y = f(x), a \leqslant x \leqslant b$(图 10.1)以虚线 $\varphi_m(x)$ 和 $\psi_m(x)$ "平均逼近"的几何意义. 这就是,从一些逼近曲线中(现在是二条虚线),平均最优的逼近是由曲线 $y = \varphi_m(x)$ 给出的,虽然在某一小区间内,它与 $y = f(x)$ 偏离很大.

今要在系数 $a_0, a_1, \cdots, a_m$ 的任意选择下从偏差的平方和 $\sum_m$ 中去求最小的那一个;这样,和 $\sum_m$ 应是极小,因而现在剩下的是去解对于 $m+1$ 个变量的函数 $\sum_m$ 为极小的通常问题. 根据所谓最小二乘法的方法去求系数 a_v 便给出在区间 $[a,b]$ 上平均逼近函数 $f(x)$ 的所求的多项式 $\varphi_m(x)$.

顺便指出,以最小二乘法来作多项式 $\varphi_m(x)$,当 $f(x_v)$ 的值是由实验而得到时尤为有用. 今先来求使表达式 ㊺ 成为极小的系数 $a_0, a_1, \cdots, a_m$.

注意到逼近多项式 $\varphi_m(x)$ 的表达式,便可将 $\sum_m$ 的表达式写作下形

$$\sum_m = \sum_{v=0}^{n} [a_0 x_v^m + a_1 x_v^{m-1} + \cdots + a_{m-1} x_v + a_m - f(x_v)]^2$$

此等式的右端是非负的. 它当逼近多项式在离散点

$x_v(v=0,1,\cdots,n)$ 处的所有值与函数 $f(x)$ 在这些点处的给定值(确实)重合时,变为零.因此,函数 $\sum_m$ 的极小的存在应认定是已确立的.欲求此等系数 a_v 的值使得对于它们,$\sum_m$ 取得极值,只需解方程组

$$\frac{1}{2}\frac{\partial \sum_m}{\partial a_{m-k}} \equiv \sum_{v=0}^{n}[a_0 x_v^m + a_1 x_v^{m-1} + \cdots + a_{m-1}x_v + a_m - f(x_v)]x_v^k = 0$$

由此便得到具有 $m+1$ 个未知元的 $m+1$ 个线性方程

$$\begin{cases} (n+1)a_m + a_{m-1}\sum_{v=0}^{n}x_v + a_{m-2}\sum_{v=0}^{n}x_v^2 + \cdots + a_0\sum_{v=0}^{n}x_v^m = \sum_{v=0}^{n}f(x_v) \\ a_m\sum_{v=0}^{n}x_v + a_{m-1}\sum_{v=0}^{n}x_v^2 + a_{m-2}\sum_{v=0}^{n}x_v^3 + \cdots + a_0\sum_{v=0}^{n}x_v^{m+1} = \sum_{v=0}^{n}x_v f(x_v) \\ \vdots \\ a_m\sum_{v=0}^{n}x_v^m + a_{m-1}\sum_{v=0}^{n}x_v^{m+1} + a_{m-2}\sum_{v=0}^{n}x_v^{m+2} + \cdots + a_0\sum_{v=0}^{n}x_v^{2m} = \sum_{v=0}^{n}x_v^m f(x_v) \end{cases} \quad ⑩$$

解这些线性方程组,便求得逼近多项式 $\varphi_m(x)$ 的系数 $a_0,a_1,\cdots,a_m$.

可以证明(但不推拟述出此说明),方程组 ⑩ 有唯一解,即 $\sum_m$ 的极值只能是极小.关于方程组 ⑩ 的行列式是异于零的证明,可在冈查洛夫的著作中找到.

Tschebyscheff 逼近定理

因此,最小二乘法使我们得到唯一的一个 m 次多项式,从逼近的观点来看,它在具有这个次数的所有多项式中是最优的. 它给出对给定函数 $f(x)$ 的平均最优的逼近.

作为最小二乘法的基础的原则是勒让得在 1806 年发表的,虽然高斯从 1795 年就已利用了此方法,最小二乘法在高斯的研究发表后,得到了其最终与完善的形式.

⑩ 的各方程分别叫作关于 $a_m, a_{m-1}, \cdots, a_0$ 的正规方程. 对于作正规方程组我们得到下列规则:欲得到关于某一个未知元例如 a_k 的正规方程,应当将关系式 ⑱ 中的每个右端表达式乘以在每一表达式中 a_k 的系数,将这些乘积相加并使所得的和等于零.

作为一个例子,我们作对应于 a_0 的正规方程. 将 ⑱ 中的每一个右端表达式乘以在它里面的 a_0 的系数,即将这些表达式的第一个(上面一个)乘以 x_0^m,第二个乘以 x_1^m 等,将最后一个乘以 x_n^m 且使所得诸表达式的和等于零. 这便给出方程组 ⑩ 的最后一个正规方程.

在函数 $f(x)$ 以二次多项式

$$\varphi_2(x) = a_2 + a_1 x + a_0 x^2$$

逼近的情形下,正规方程具有形式

$$a_2(n+1) + a_1 \sum_{v=0}^{n} x_v + a_0 \sum_{v=0}^{n} x_v^2 = \sum_{v=0}^{n} f(x_v)$$

$$a_2 \sum_{v=0}^{n} x_v + a_1 \sum_{v=0}^{n} x_v^2 + a_0 \sum_{v=0}^{n} x_{3v} = \sum_{v=0}^{n} x_v f(x_v)$$

$$a_2 \sum_{v=0}^{n} x_v^2 + a_1 \sum_{v=0}^{n} x_v^3 + a_0 \sum_{v=0}^{n} x_v^4 = \sum_{v=0}^{n} x_v^2 f(x_v)$$

解它们,便得到所求的系数 a_0, a_1, a_2.

附录 X 平方逼近

在最小二乘法的很多应用中,最重要的就是依照在实验中所得的数据去作出内插公式.此时所得的值 $f(x_v)$ 易有误差.当利用通常内插方法时,我们必须从大量观测所得的值 $f(x_v)$ 中(假定 x_v 的值没有受到实验的误差)选择某一些使得内插多项式的次数比所选择的值 $f(x_v)$ 的个数少 1.简单的舍去 $f(x_v)$ 的某些值是不合适的,因为这样相当于默认,正是所舍去的一些数受到最高程度的实验的误差,然而所舍去的 $f(x_v)$ 值可能对于描述所研究的现象是最能表其特征的.在按最小二乘法作内插公式时,便可保留所有的 $f(x_v)$,因此最后的公式,一般说来,应是较好的表示依观测所得数据的所求的函数关系.

2. 周期函数借助于三角多项式的平方逼近

在研究数学科学的各种问题时,经常要采用三角多项式.如果我们预先知道关于某一现象的周期性,则对于描述此现象的函数自然以由正弦和余弦所组成的三角多项式来逼近较为合宜.这种多项式适用于有需要去处理振动过程的所有情形.设周期为 2π 的周期函数 $f(x)$ 在区间 $(0,2\pi)$ 的等距离点

$$x_v = \frac{2v\pi}{n} \quad (n > 2m; v = 0,1,\cdots,n-1)$$

处的 n 个值是已知的.

今将函数在分点的给定值分别记作

$$f(x_0), f(x_1), \cdots, f(x_n)$$

而且由于此函数的周期性, $f(x_0) = f(x_n)$.今考虑 m 次的三角多项式(m 固定)

Tschebyscheff 逼近定理

$$\varphi_m(x) = a_0 + \sum_{v=1}^{n}(a_v\cos vx + b_v\sin vx) \qquad \text{(461)}$$

并企图借助于具有适当选择的系数的这个多项式去逼近函数 $f(x)$.

今我们有 $n > 2m$ 个常数,因而在以多项式 $\varphi_m(x)$ 逼近 $f(x)$ 时,应当满足 n 个条件. 因此需要去确定 $2m+1$ 个常数 a_v 和 b_v 使得函数 $\varphi_m(x)$ 在 n 个给定点 x_v 处与 $f(x)$ 的给定值重合.

多项式 (461) 的系数应选择得使平均平方偏差为最小,为此必须确定 $2m+1$ 个系数 a 和 b 使之满足 n 个方程

$$\sum_{v=0}^{n-1}[f(x_v) - \varphi_m(x_v)] = 0 \qquad \text{(462)}$$

$$\sum_{v=0}^{n-1}[f(x_v) - \varphi_m(x_v)]\cos \lambda x_v = 0 \qquad \text{(463)}$$

$$\sum_{v=0}^{n-1}[f(x_v) - \varphi_m(x_v)]\sin \lambda x_v = 0 \qquad \text{(464)}$$

其中 λ 只取整数值 $1,2,\cdots,m$.

今去推导计算系数 a_v 和 b_v 的公式. 因为多项式在点 x_v 处取值 $\varphi_m(x_v)$,所以根据关系式 (461),将有

$$\varphi_m(x_0) = \sum_{v=0}^{m} a_v$$

$$\varphi_m(x_1) = a_0 + \sum_{v=1}^{m} a_v\cos\frac{2v\pi}{n} + \sum_{v=1}^{m} b_v\sin\frac{2v\pi}{n}$$

$$\varphi_m(x_2) = a_0 + \sum_{v=1}^{m} a_v\cos\frac{4v\pi}{n} + \sum_{v=1}^{m} b_v\sin\frac{4v\pi}{n}$$

$$\vdots$$

$$\varphi_m(x_{n-1}) = a_0 + \sum_{v=1}^{m} a_v\cos\frac{2v(n-1)\pi}{n} +$$

附录 X 平方逼近

$$\sum_{v=1}^{m} b_v \sin \frac{2v(n-1)\pi}{n}$$

利用方程 ⑫,便得

$$na_0 + a_1 \sum_{v=0}^{n-1} \cos v \frac{2\pi}{n} + a_2 \sum_{v=0}^{n-1} \cos v \frac{4\pi}{n} + \cdots +$$

$$a_m \sum_{v=0}^{n-1} \cos v \frac{2m\pi}{n} + b_1 \sum_{v=0}^{n-1} \sin v \frac{2\pi}{n} +$$

$$b_2 \sum_{v=0}^{n-1} \sin v \frac{4\pi}{n} + \cdots + b_m \sum_{v=0}^{n-1} \sin v \frac{2m\pi}{n} = \sum_{v=0}^{n-1} f(x_v)$$

⑯

由 ⑬ 得到

$$a_0 \sum_{v=0}^{n-1} \cos v \frac{2\lambda\pi}{n} + a_1 \sum_{v=0}^{n-1} \cos v \frac{2\pi}{n} \cos v \frac{2\lambda\pi}{n} +$$

$$a_2 \sum_{v=0}^{n-1} \cos v \frac{4\pi}{n} \cos v \frac{2\lambda\pi}{n} + \cdots +$$

$$a_m \sum_{v=0}^{n-1} \cos v \frac{2m\pi}{n} \cos v \frac{2\lambda\pi}{n} + b_1 \sum_{v=0}^{n-1} \sin v \frac{2\pi}{n} \cos v \frac{2\lambda\pi}{n} +$$

$$b_2 \sum_{v=0}^{n-1} \sin v \frac{4\pi}{n} \cos v \frac{2\lambda\pi}{n} + \cdots +$$

$$b_m \sum_{v=0}^{n-1} \sin v \frac{2m\pi}{n} \cos v \frac{2\lambda\pi}{n} = \sum_{v=0}^{n-1} f(x_v) \cos v \frac{2\lambda\pi}{n}$$

⑯

按 ⑭ 便有

$$a_0 \sum_{v=0}^{n-1} \sin v \frac{2\lambda\pi}{n} + a_1 \sum_{v=0}^{n-1} \cos v \frac{2\pi}{n} \sin v \frac{2\lambda\pi}{n} +$$

$$a_2 \sum_{v=0}^{n-1} \cos v \frac{4\pi}{n} \sin v \frac{2\lambda\pi}{n} + \cdots +$$

$$a_m \sum_{v=0}^{n-1} \cos v \frac{2m\pi}{n} \sin v \frac{2\lambda\pi}{n} +$$

Tschebyscheff 逼近定理

$$b_1 \sum_{v=0}^{n-1} \sin v \frac{2\pi}{n} \sin v \frac{2\lambda\pi}{n} + b_2 \sum_{v=0}^{n-1} \sin v \frac{4\pi}{n} \sin v \frac{2\lambda\pi}{n} + \cdots +$$

$$b_m \sum_{v=0}^{n-1} \sin v \frac{2m\pi}{n} \sin v \frac{2\lambda\pi}{n} = \sum_{v=0}^{n-1} f(x_v) \sin v \frac{2\lambda\pi}{n} \quad \text{⑯}$$

今要证,对于 $\mu = 1, 2, \cdots, m$,下列等式成立

$$\sum_{v=0}^{n-1} \cos \mu \frac{2v\pi}{n} = 0, \quad \sum_{v=0}^{n-1} \sin \mu \frac{2v\pi}{n} = 0 \quad \text{⑱}$$

$$\sum_{v=0}^{n-1} \cos \mu \frac{2v\pi}{n} \sin \lambda \frac{2v\pi}{n} = 0 \quad \text{⑲}$$

$$\left. \begin{array}{l} \sum\limits_{v=0}^{n-1} \cos \mu \dfrac{2v\pi}{n} \cos \lambda \dfrac{2v\pi}{n} \\ \sum\limits_{v=0}^{n-1} \sin \mu \dfrac{2v\pi}{n} \sin \lambda \dfrac{2v\pi}{n} \end{array} \right\} = \begin{cases} 0, & \text{当}\ \mu \neq \lambda \\ \dfrac{n}{2}, & \text{当}\ \mu = \lambda \end{cases} \quad \text{⑳}$$

利用方程 ⑮ − ⑯ 以及关系式 ⑱ − ⑳,便得

$$\begin{cases} a_0 = \dfrac{1}{n} \sum\limits_{v=0}^{n-1} f\left(\dfrac{2v\pi}{n}\right) \\ a_k = \dfrac{2}{n} \sum\limits_{v=0}^{n-1} f\left(\dfrac{2v\pi}{n}\right) \cos \dfrac{2vk\pi}{n} \\ b_k = \dfrac{2}{n} \sum\limits_{v=0}^{n-1} f\left(\dfrac{2v\pi}{n}\right) \sin \dfrac{2vk\pi}{n} \\ (k = 1, 2, \cdots, m) \end{cases} \quad \text{㉑}$$

对于奇函数

$$f(x) = -f(2\pi - x)$$

我们令 $n = 2r + 1$. 于是

$$f(x_1) = -f(x_{2r})$$
$$f(x_2) = -f(x_{2r-1})$$

等等.

如果函数 $f(x)$ 是能使

附录 X　平方逼近

$$f(x_0) = f(x_{2r+1}) = 0$$

则所有系数 $a_0, a_1, \cdots, a_m$ 变为零，而我们便得到仅由正弦所组成的逼近三角多项式

$$f(x) \sim \frac{2}{n}\sum_{v=0}^{n-1} f\left(\frac{2v\pi}{n}\right)\sin\frac{2v\pi}{n}\sin x +$$

$$\frac{2}{n}\sum_{v=0}^{n-1} f\left(\frac{2v\pi}{n}\right)\sin\frac{2\cdot 2v\pi}{n}\sin 2x + \cdots +$$

$$\frac{2}{n}\sum_{v=0}^{n-1} f\left(\frac{2v\pi}{n}\right)\sin\frac{m\cdot 2v\pi}{n}\sin mx$$

以三角多项式亦可逼近非周期函数．此时，逼近多项式只在长为 2π 的线段上而不是在全直线上近似地表示 $f(x)$．

所得的结果可以推广到连续区间的情形，这只要假定在区间 $(0, 2\pi)$ 内任一点处函数 $f(x)$ 的值是已知的．我们要 $f(x)$ 在此区间内为可积．由公式 ㊼，在极限时，得出公式

$$a_0 = \frac{1}{2\pi}\int_0^{2\pi} f(x)\,dx$$

$$a_k = \frac{1}{\pi}\int_0^{2\pi} f(x)\cos kx\,dx$$

$$b_k = \frac{1}{\pi}\int_0^{2\pi} f(x)\sin kx\,dx$$

因此，在当 $n \to \infty$ 时取极限的结果，便得到函数 $f(x)$ 的傅里叶系数．三角多项式就成为傅里叶无穷级数，但如以下将见，仅仅是函数 $f(x)$ 的可积性，对于此级数的收敛并不是充分的．为了建立保证函数 $f(x)$ 借助于带系数 ㊼ 的三角多项式 ㊻ 可能逼近到任意准确程度的这一条件，还要查明展开成收敛的傅里叶级数

的函数. 如果函数 $f(x)$ 的傅里叶级数收敛, 则由前 $2m+1$ 项所组成的它的部分和, 对于充分大的 n 值, 将与带系数 ㊀ 的 $\varphi_m(x)$ 恒等. 为证明这点, 只需指出, 在傅里叶系数公式中所列的积分可为和 ㊀ 代替.

直到现在, 在最小二乘方的观点上, 我们作出了不同的逼近多项式.

3. 借助于线性无关函数组的逼近表示

最小二乘法也可有效地用在当逼近函数不是多项式的情形. 我们仅仅要求逼近函数是给定线性无关函数组的线性组合

$$\varphi(x) = a_0\varphi_m(x) + a_1\varphi_{m-1}(x) + \cdots + a_{m-1}\varphi_1(x) + a_m\varphi_0(x) \qquad ㊁$$

函数组 $\varphi_0(x), \varphi_1(x), \cdots, \varphi_m(x)$ 叫作在区间 $[a,b]$ 上线性无关, 若恒等式

$$c_0\varphi_m(x) + c_1\varphi_{m-1}(x) + \cdots + c_{m-1}\varphi_1(x) + c_m\varphi_0(x) \equiv 0 \qquad ㊂$$

其中 $c_0, c_1, \cdots, c_m$ 为常数. 当

$$c_0 = c_1 = \cdots = c_m = 0$$

时, 而且仅当此时才成立.

之前, 我们选取了简单的线性无关函数组

$$1, \quad x, \quad x^2, \quad \cdots, \quad x^m, \quad \cdots$$
$$1, \quad \cos x, \quad \cos 2x, \quad \cdots, \quad \cos mx, \quad \cdots$$
$$\sin x, \quad \sin 2x, \quad \cdots, \quad \sin mx, \quad \cdots$$

且以这些函数的线性组合来逼近函数 $f(x)$, 即以通常的 m 次代数的或三角的多项式来逼近函数. 此处, 作出在函数 $f(x)$ 借助于 $m+1$ 个给定的任意线性无关函数

附录 X　平方逼近

$\varphi_0(x), \varphi_1(x), \cdots, \varphi_m(x)$ 来逼近的意义下的推广.

欲以函数 $\varphi(x)$ 逼近函数 $f(x)$，我们假定函数 $f(x)$ 在区间 $[a,b]$ 的离散点 x_v 处的 $n > m$ 个准确值或近似值是已知的并且函数

$$\varphi_0(x), \varphi_1(x), \cdots, \varphi_m(x)$$

在这些 x_v 点处的同样个数的值也是已知的. 今选取系数 (系数) $a_0, a_1, \cdots, a_m$ 使得 $\varphi(x)$ 与 $f(x)$ 的平均平方偏差为最小，即使得表达式

$$\sum_m \equiv \sum_{v=0}^n [f(x_v) - a_0 \varphi_m(x) - a_1 \varphi_{m-1}(x) - \cdots - a_m \varphi_0(x)]^2$$

为最小. 为此，我们还要解通常关于 $m+1$ 个变量的函数 $\sum_m$ 的极小问题，这就归结于去解下列方程组

$$\begin{cases} \sum_{v=0}^n [f(x_v) - a_0 \varphi_m(x) - a_1 \varphi_{m-1}(x) - \cdots - a_m \varphi_0(x)] \varphi_0(x_v) = 0 \\ \sum_{v=0}^n [f(x_v) - a_0 \varphi_m(x) - a_1 \varphi_{m-1}(x) - \cdots - a_m \varphi_0(x)] \varphi_1(x_v) = 0 \\ \quad\quad\quad\quad \vdots \\ \sum_{v=0}^n [f(x_v) - a_0 \varphi_m(x) - a_1 \varphi_{m-1}(x) - \cdots - a_m \varphi_0(x)] \varphi_m(x_v) = 0 \end{cases}$$

它是使 $\sum_m$ 对所有参数 $a_k (k=0,1,\cdots,m)$ 的偏导数等于零而得到.

特别是当

$$\varphi_0(x) \equiv 1, \varphi_1(x) \equiv x, \cdots, \varphi_m(x) \equiv x^m$$

Tschebyscheff 逼近定理

时，便又得到方程组 ⑩.

在解方程组 ⑭ 的结果中，我们一旦求得参数 a_0, $a_1, \cdots, a_m$ 最合用的值，我们便能在区间 (a,b) 的所有以上固定的点处，计算偏差

$$\varepsilon_v = f(x_v) - \varphi(x_v)$$

而由它们也能计算偏差的平方和

$$\sum_m = \sum_{v=0}^{n} \varepsilon_v^2$$

这使我们可以判断 $\varphi(x)$ 逼近函数 $f(x)$ 到怎样的准确程度. 偏差的平方和也可按下法计算：将第一个（上面一个）正规方程乘以 a_m，第二个乘以 a_{m-1} 等，将最后一个乘以 a_0 并将所得的乘积加起来；这就给出

$$\sum_{v=0}^{n} \varphi^2(x_v) = \sum_{v=0}^{n} f(x_v)\varphi(x_v)$$

在另一方面

$$\sum_{v=0}^{n} [f(x_v) - \varphi(x_v)]^2 = \sum_{v=0}^{n} f^2(x_v) - 2\sum_{v=0}^{n} f(x_v)\varphi(x_v) + \sum_{v=0}^{n} \varphi^2(x_v)$$

因之

$$\sum_{v=0}^{n} [f(x_v) - \varphi(x_v)]^2 = \sum_{v=0}^{n} f^2(x_v) - \sum_{v=0}^{n} \varphi^2(x_v)$$

故得

$$\sum_m = \sum_{v=0}^{n} f^2(x_v) - \sum_{v=0}^{n} \varphi^2(x_v)$$

此方程可用作检验，即根据它可以知道所进行的数值计算的正确性.

在运用最小二乘法时，通常对不同的观测数据，附

附录 X 平方逼近

加着不同的权(这与在概率论中对于研究不一样准确的量度时所作的相似). 因此(假定只是因变量受到量度的误差), 我们给实验的数据以权且确定系数 $a_0, a_1, \cdots, a_m$ 使得带权和

$$\sum_{m,n} = \sum_{v=0}^{n} p_v [a_0 \varphi_m(x_v) + a_1 \varphi_{m-1}(x_v) + \cdots + a_m \varphi_0(x_v) - f(x_v)]^2$$

尽可能的小. 此处 $p_v > 0$ 为附加在偏差上的不同的权. 通常将权正规化, 使得

$$\sum_{v=0}^{n} p_v = 1$$

今假定提出了关于给定函数 $f(x)$ 借助于线性相关函数 $\varphi_0(x), \varphi_1(x), \cdots, \varphi_m(x)$ 来逼近的问题. 由于这些函数线性相关, 我们便可选择常数 $c_0, c_1, \cdots, c_m$ 不全等于零而使恒等式 ㊸ 成立. 于是函数 $\varphi_m(x)$ 在点 x_v 的值便能以其余函数在这些点的值表出

$$\varphi_m(x_v) = -\frac{c_1}{c_0}\varphi_{m-1}(x) - \cdots - \frac{c_{m-1}}{c_0}\varphi_1(x) - \frac{c_m}{c_0}\varphi_0(x)$$

㊺

如果将方程组 (20) 的第一个 (上面一个) 方程乘以 $-\frac{c_m}{c_0}$, 第二个乘以 $-\frac{c_{m-1}}{c_0}$, 最后, 倒数第二个乘以 $-\frac{c_1}{c_0}$, 将所得的等式相加, 则根据 ㊺ 便知, 方程组的的最后一个方程是由首 $m-1$ 得出的; 因而实际上, 我们是有 m 个未知元的 $m-1$ 个 (或更少个) 方程. 由于方程的个数不等于未知系数 a_v 的个数, 所以方程组 ㊹ 便没有一个确定的解.

在以后, 我们恒将假定, 但不特别提及, 逼近函数

Tschebyscheff 逼近定理

㊼ 是由一组线性无关的函数组成的.

4. 平方逼近的 Tschebyscheff 公式

在前一节中,我们以函数 $\{\varphi_v(x)\}$ 的线性组合逼近函数 $f(x)$ 且按最小二乘法来计算逼近多项式的系数.这使我们要去解 $m+1$ 个未知元的线性方程组 ㊼,此方程组是可用通常的代数方法来解的.应用代数的方法去解方程组 ㊼,当方程的个数超过四个或五个时,便非常累赘(因为对于去求为保证所选取的准确度的内插多项式的次数,常要去作出形如 ㊼ 的方程组并去解它).因此,在实际计算上,便采用 Tschebyscheff 方法.

设 $f(x_0), f(x_1), \cdots, f(x_n)$ 为函数 $f(x)$ 对一组点
$$x_0 < x_1 < x_2 < \cdots < x_n$$
的观测值.在 Tschebyscheff 的著作中,在一般情形下,叙述并解决了函数 $f(x)$ 以 $m < n$ 次的多项式来逼近的问题.此问题首先是他在发表于 1855 年的研究报告"关于连分式"中叙述的.

很有意思的是由 Tschebyscheff 所引入的逼近多项式使下形的有权和
$$\sum_{v=0}^{n} p(x_v) [f(x_v) - Q_m(x_v)]^2$$
成为极小,即保证 $f(x)$ 在有任一非负的,在内插区间的离散点 $x_v (v = 0, 1, \cdots, n)$ 处给定的权函数 $p(x)$ 时的平方逼近.在包含点 $x_v (v = 0, 1, \cdots, n)$ 的区间上逼近函数 $f(x)$ 的多项式 $Q_m(x)$ 有下形
$$Q_m(x) = A_0 P_{0,n}(x) + A_1 P_{1,n}(x) + \cdots + A_m P_{m,n}(x)$$

其中,$P_{k,n}(x)$ 是递推地确定的多项式. 它们对权 $p(x)$ 在下一意义下正交,即当 $k \neq m$ 时

$$\sum_{v=0}^{n} p(x_v) P_{v,n}(x_v) P_{m,n}(x_v) = 0$$

由 Tschebyscheff 所得的对任一 m 次多项式是准确的内插公式,最终可写作下形

$$f(x) = \sum_{k=0}^{m} \left[\frac{\sum_{v=0}^{n} p(x_v) P_{k,n}(x_v) f(x_v)}{\sum_{v=0}^{n} p(x_v) P_{k,n}^2(x_v)} \right] P_{k,n}(x)$$

Tschebyscheff 的结果可很简单的转用到连续区间的情形. 此处代替 $\sum$ 的是使积分

$$\int_a^b p(x) [f(x) - Q_m(x)]^2 \mathrm{d}x$$

成为极小. 结果,函数 $f(x)$ 可用按次数渐增的正交多项式所成的级数的部分和来逼近.

  Tschebyscheff 得到了刚刚所述的内插公式后,在以"关于内插"为题的研究报告中写道:"这个级数,依着在多项式中所能保持的项的个数,以高次或低次的多项式的形式给出被内插量的表达式,而这些多项式,由解一组在关于所求表达式的次数的每一特殊假定下所特别作出的方程就可得到,且它们的系数与根据最小二乘法所求得者相同. 容易见到,不论求到这些表达式的哪一次,都可使用我们的级数得到简化,这些表达式可从这个级数直接得到,只要从零次开始以至相继的所有各次. 但这还没有讲完在内插时使用它的全部功用:这个级数也特别适用于为了去观察在对于内插的所求表达式中要用多少项,也就是要中止在哪一个

Tschebyscheff 逼近定理

方次上.

对于根据最小二乘法的通常内插法,为此需要根据在关于它的次数的每一特殊假定下所求得的公式来列出所有给定的量.然而想知道这样的计算要重复多少次才能得出以充分的准确度来表示数据的表达式是很困难的,因为这种表达式的次数难于预先猜定.在这一方面,我们的公式显示出重大的方便,即按照它在对于内插的所求表达式中一项一项的加添,我们便可直接见到,在依照它来确定所有给定的量时,为这些表达式所表示的平方误差的和是怎样的依次递减;由此也可容易得到这些误差的平均平方,根据它便可判定在所求的表达式中需要取多少个项才够".

这些话表明,对于实际的内插,当内插多项式的次数是依照为计算而选的准确程度来确定时,Tschebyscheff 公式有怎样的重大意义.

Tschebyscheff 公式在探讨观测的结果时,有广泛的应用.比尔松在其研究报告中曾将 Tschebyscheff 公式用于数理统计中回归方程的计算.现在,Tschebyscheff 内插公式为全世界的统计学界所使用.此公式在统计中对于相关、方差以及调和分析有广泛的应用.

对值 x_v 的级数和函数 $p(x)$ 的形式加以特殊假定,Tschebyscheff 得到对于实际有重要意义的有趣公式.因此,我们详细地来叙述一下等距离节的情形.为简单起见,我们假定在区间 $0 \leqslant x \leqslant 1$ 中来考虑函数 $f(x)$,而且是对于等距离节的情形在有同样的权时来解内插问题.

附录 X　平方逼近

今考虑等距离点组 $x_v = \dfrac{v}{n}(v=0,1,\cdots,n)$. 在去作内插公式以前，先将这个点组借助于满足条件 $t = \dfrac{x-x_0}{h}$ 的新变量 t（其中 h 是表的步度）转换成点组 $0,1,\cdots,n$.

今采用最二乘法. 作为逼近函数，我们取以上所考虑的多项式 $Q_m(x)$，设

$$P_{s,n}(x) = 1 + a_1 x^{(1)} + a_2 x^{(2)} + \cdots + a_s x^{(s)}$$

为关于 x 的 s 次多项式，而 a_k 和 A_k 为待定系数.

今选择系数 A_k 使得表达式

$$\sum\nolimits_m = \sum_{x=0}^{n} [f(x) - A_0 P_{0,n}(x) - A_1 P_{1,n}(x) - \cdots - A_m P_{m,n}(x)]^2$$

取尽可能小的值，其中 $m < n$.

使 $\sum_m$ 对参数 $A_k(k=0,1,\cdots,m)$ 的偏导数等于零而得的方程组具有形式

$$\frac{1}{2}\frac{\partial \sum_m}{\partial A_k} \equiv \sum_{x=0}^{n} [f(x) - Q_m(x)] P_{k,n}(x) = 0$$

今选择系数 a_k 使得多项式 $P_{k,n}(x)$ 为 Tschebyscheff 正交多项式：

$$\sum_{x=0}^{n} P_{m,n}(x) P_{k,n}(x) = 0, m \neq k$$

这使我们能以最简单的方式由方程 $\dfrac{\partial \sum_m}{\partial A_k} = 0$ 来确定系数 A_k. 我们有

Tschebyscheff 逼近定理

$$A_k = \frac{\sum_{x=0}^{n} P_{k,n}(x) f(x)}{\sum_{x=0}^{n} P_{k,n}^2(x)}$$

因此,要作所求的公式便成为作 Tschebyscheff 正交多面式 $P_{k,n}(x)$. 今作这些多项式,使得对于 $s = 0, 1, \cdots, m-1$

$$\sum_{x=0}^{n} (x+s)^{(s)} P_{m,n}(x) = 0$$

为了以最简单方法来确定系数 $a_1, a_2, \cdots, a_m$, 将 $P_{m,n}(x)$ 按阶乘多项式 $x^{(s)}$ 的展开式的两端乘以阶乘多项式 $(x+s)^{(s)}$ 并将所得的等式由 $x=0$ 到 $x=n$ 求和. 我们得到下一方程

$$\sum_{x=0}^{n} (x+s)^{(s)} P_{m,n}(x) = \frac{(n+s+1)^{(s+1)}}{s+1} +$$

$$a_1 \frac{(n+s+1)^{(s+2)}}{s+2} + \cdots +$$

$$a_m \frac{(n+s+1)^{(s+m+1)}}{s+m+1} = 0$$

利用容易导出的关系式

$$x^{(k)}(x+s)^{(s)} = (x+s)^{(s+k)} = \frac{\Delta(x+s)^{(s+k+1)}}{s+k+1}$$

便可证实上式的正确.

由此,逐项除以 $(n+s+1)^{(s+1)}$, 便得 $(s = 0, 1, \cdots, m-1)$

$$\frac{1}{s+1} + \frac{n^{(1)}}{s+2} a_1 + \frac{n^{(2)}}{s+3} a_2 + \cdots + \frac{n^{(m)}}{s+m+1} a_m = 0$$

此等式可以写成

$$\frac{1}{s+1} + \frac{n^{(1)}}{s+2} a_1 + \frac{n^{(2)}}{s+3} a_2 + \cdots + \frac{n^{(m)}}{s+m+1} a_m =$$

附录 X 平方逼近

$$\frac{Q(s)}{(s+m+1)^{(m+1)}}$$

其中 $Q(s)$ 为在 $s = 0, 1, \cdots, m-1$ 时变为零的多项式. 因此, $Q(s)$ 可准确地确定到只差一常数因子 c

$$Q(s) = c \cdot s^{(m)}$$

欲确定 c, 我们将倒数第二个等式乘以 $s+1$ 并在所得的关系式中令 $s = -1$. 于是得到

$$c = (-1)^m$$

因之

$$\sum_{k=0}^{m} \frac{(s+m+1)^{(m+1)}}{s+k+1} n^{(k)} a_k = (-1)^m s^{(m)}$$

当 $s = -k-1 (k \leqslant m)$ 时, 在此等式的左端, 除包含 a_k 的项外, 所有其他各项都消失, 因而

$$a_k = \frac{(-1)^k}{n^{(k)}} \binom{m}{k} \binom{m+k}{k}$$

因此, 对每一整数 m, 存在唯一的一个多项式 $P_{m,n}(x)$ 与阶乘多项式 $(x+s)^{(s)}$ 正交 $(s = 0, 1, \cdots, m-1)$. 它有形式

$$P_{m,n}(x) = \sum_{k=0}^{m} (-1)^k \binom{m}{k} \binom{m+k}{k} \frac{x^{(k)}}{n^{(k)}}$$

特别地, 由计算可给出

$$P_{0,n}(x) = 1$$

$$P_{1,n}(x) = 1 - 2\frac{x}{n}$$

$$P_{2,n}(x) = 1 - 6\frac{x}{n} + 6\frac{x(x-1)}{n(n-1)}$$

$$P_{3,n}(x) = 1 - 12\frac{x}{n} + 30\frac{x(x-1)}{n(n-1)} -$$

Tschebyscheff 逼近定理

$$20\frac{x(x-1)(x-2)}{n(n-1)(n-2)}$$

$$P_{4,n}(x) = 1 - 20\frac{x}{n} + 90\frac{x(x-1)}{n(n-1)} -$$

$$140\frac{x(x-1)(x-2)}{n(n-1)(n-2)} +$$

$$70\frac{x(x-1)(x-2)(x-3)}{n(n-1)(n-2)(n-3)}$$

$$P_{5,n}(x) = 1 - 30\frac{x}{n} + 210\frac{x(x-1)}{n(n-1)} -$$

$$560\frac{x(x-1)(x-2)}{n(n-1)(n-2)} +$$

$$630\frac{x(x-1)(x-2)(x-3)}{n(n-1)(n-2)(n-3)} -$$

$$252\frac{x(x-1)(x-2)(x-3)(x-4)}{n(n-1)(n-2)(n-3)(n-4)}$$

今往证 $P_{k,n}(x)$ 的正交性. 任一 k 次的多项式 $P_{k,n}(x)$ 可按阶乘多项式展开(6). 因此

$$P_{k,n}(x) = \sum_{s=0}^{k} B_s \sum_{x=0}^{n} (x+s)^{(s)}$$

将此等式的两端乘以 $P_{m,n}(x)$ 并由 $x=0$ 到 $x=n$ 对 x 求和,便得

$$\sum_{x=0}^{n} P_{m,n}(x) P_{k,n}(x) =$$

$$\sum_{s=0}^{k} B_s \sum_{x=0}^{n} (x+s)^{(s)} P_{m,n}(x) = 0, k < m$$

因此,多项式组 $\{P_{k,n}(x)\}$ 的正交性得证.

也可证明

$$\sum_{x=0}^{n} P_{m,n}^2(x) = \frac{(n+m+1)(m+n)^{(m)}}{(2m+1)n^{(m)}}$$

附录 X 平方逼近

对于 $n=5,6,\cdots,20, m=1,2,3,4,5$ 与 $x=0,1,\cdots,n$, 多项式 $P_{m,n}(x)$ 的值的表可在米尔纳一书的附录中找到. 为了得到应怎样来利用这些表的概念, 我们考虑对 $n=9$ 的表 10.1(10 个点). 所有另外一些表都是与此表同样作成的. 为简单起见, 在以后, 代替 $P_{k,9}(x)$ 就写作 $P_k(x)$.

表 10.1 多项式 $P_k(x)$ 的值的表 ($k=1,2,3,4,5$)

x	$P_1(x)$	$P_2(x)$	$P_3(x)$	$P_4(x)$	$P_5(x)$
0	9	6	42	18	6
1	7	2	-14	-22	-14
2	5	-1	-35	-17	1
3	3	-3	-31	3	11
4	1	-4	-12	18	6
5	-1	-4	12	18	-6
6	-3	-3	31	3	-11
7	-5	-1	35	-17	-1
8	-7	2	14	-22	14
9	-9	6	-42	18	-6
S_k	330	132	8 580	2 860	780

在所引入的表中没有函数值 $P_0(x)\equiv 1$. 在表示 $P_k(x)(k=1,2,3,4,5)$ 的行中, 仅列有表示多项式 $P_k(x)$ 值的分数的分子; 它们的分母是等于表的最上面一个数 (对应 $x=0$ 这一列的数). 例如

$$P_3(3)=-\frac{31}{42}, P_4(5)=\frac{18}{18}=1, P_5(6)=-\frac{11}{6}$$

在表的每一行下面有数 S_k, 它能用来计算量 $\sum_{x=0}^{n} P_k^2(x)$, 此量等于带下标 k 的一行中最下面一个数

Tschebyscheff 逼近定理

与此行最上面一个数的平方之比. 例如

$$\sum_{x=0}^{9} P_4^2(x) = \frac{2\,860}{18^2}$$

在表 33 中所列各数不同于多项式 $P_{k,n}(x)$ 的真确值的这一情况并不妨碍用它们来构成逼近多项式

$$Q_m(x) = \sum_{k=0}^{m} \left[\frac{\sum_{x=0}^{n} P_{k,n}(x) f(x)}{\sum_{x=0}^{n} P_{k,n}^2(x)} \right] P_{k,n}(x)$$

事实上, 由这个公式可见, 对于作 $Q_m(x)$ 只需有准确地确定到只差一常数因子的多项式 $P_{k,n}(x)$ 就行.

在转到去作逼近之前, 先简略的叙述一下检验数值计算的问题. 按照在前节所得的关于偏差的平方和的公式, 便可写出

$$\sum_m = \sum_{x=0}^{n} f^2(x) - \sum_{x=0}^{n} Q_m^2(x)$$

今如将 $Q_m(x)$ 按多项式 $P_{k,n}(x)$ 的展开式平方之并利用多项式 $P_{k,n}(x)$ 的正交性质, 便得

$$Q_m^2(x) = \sum_{k=0}^{n} A_k^2 P_{k,n}^2(x), \text{其中 } A_k = \frac{c_k}{S_k}$$

而且

$$c_k = \sum_{x=0}^{n} P_{k,n}(x) f(x), S_k = \sum_{x=0}^{n} P_{k,n}^2(x)$$

因此

$$\sum_{x=0}^{n} Q_m^2(x) = \sum_{k=0}^{n} \frac{c_k^2}{S_k}$$

故

$$\sum_m = \sum_{m=0}^{n} f^2(x) - \sum_{k=0}^{n} \frac{c_k^2}{S_k}$$

附录 X 平方逼近

此关系式可作计算的检验.

作为一个例子,我们来作五次多项式,它是在区间 $[0,2.7]$ 内逼近某一微分方程的积分. 满足此方程的 x(自变量)和 y(所求函数)的数值是由数值积分法(表 10.2)得到的.

表 10.2 所求积分的数值

x	y	x	y
0.0	1.300	1.5	0.037
0.3	1.245	1.8	-0.600
0.6	1.095	2.1	-1.295
0.9	0.855	2.4	-1.767
1.2	0.514	2.7	-1.914

将我们的计算列成表 10.3 的形式是方便的.

表 10.3 作多项式 $Q_5(x)$ 的表

x	t	$y(t)$	$P_0(t)$	$P_1(t)$	$P_2(t)$	$P_3(t)$	$P_4(t)$	$P_5(t)$	$\bar{y}(t)$	$\varepsilon(t)$
0.0	0	1.300	1	9	6	42	18	6	1.310	10
0.3	1	1.245	1	7	2	-14	-22	-14	1.236	-9
0.6	2	1.095	1	5	-1	-35	-17	1	1.098	3
0.9	3	0.855	1	3	-3	-31	3	11	0.868	13
1.2	4	0.514	1	1	-4	-12	18	6	0.514	0
1.5	5	0.037	1	-1	-4	12	18	-6	0.017	-20
1.8	6	-0.600	1	-3	-3	31	3	-11	-0.602	-2
2.1	7	-1.295	1	-5	-1	35	-17	-1	-1.263	32
2.4	8	-1.767	1	-7	2	14	22	-1	-1.793	-26
2.7	9	-1.914	1	-9	6	-42	18	-6	-1.908	6
		S_k	10	330	132	8580	2860	780		
		c_k	-0.530	66.802	-7.497	-14.659	14.515	-1.627		
		A_k	-0.0530	0.20243	-0.05680	-0.00486	0.00508	-0.00209		

Tschebyscheff 逼近定理

由表 10.3 可见,带步度 0.3 的等距离点组 x_v 转成为诸点 $0,1,\cdots,9$,且 t 被取作为自变量. 在表示 $y(t)$ 的一行中,列入所求积分的逼近值,因而 Tschebyscheff 内插公式成为

$$y(t) = \sum_{k=0}^{5} = A_k P_k(t) =$$
$$-0.053\ 0P_0(t) + 0.202\ 43P_1(t) -$$
$$0.056\ 80P_2(t) - 0.004\ 86P_3(t) +$$
$$0.005\ 08P_4(t) - 0.002\ 09P_5(t)$$

很显然,此处 $P_k(t)$ 是对应表 10.1 的数据的多项式.

在所有固定点 $0,1,\cdots,9$ 处计算逼近多项式的值后(它们列在表示 $\bar{y}(t)$ 的一行中),便可计算偏差 $\varepsilon(t) = y(t) - \bar{y}(t)$(偏差列在表示 $\varepsilon(t)$ 的一行中).

5. 非线性的依从于一个或几个参数的函数的逼近

3 的论证,能使我们作出一些方程,用以确定函数 $\varphi(x)$ 的参数 $a_0, a_1, \cdots, a_m$(函数对于 $a_0, a_1, \cdots, a_m$ 是线性的) 的一些值,对于这些值,此函数与给定函数 $f(x)$ 在固定点处的偏差的平方和是成为极小的,这个论证也可在一般情形下应用. 今提出下一问题:对于参数 $a_0, a_1, \cdots, a_m$ 的怎样一些值,关于这些参数是非线性的函数 $\varphi(x, a_0, a_1, \cdots, a_m)$ 在最小二乘法的意义下给出函数 $f(x)$ 的最优逼近.

此后,我们去解决下一问题. 对于单值的且关于 x 及变的参数 $a_0, a_1, \cdots, a_m$ 是非线性的给定函数 $\varphi(x, a_0, a_1, \cdots, a_m)$,要去如此选择参数 $a_0, a_1, \cdots, a_m$,使此函数与给定值

附录 X 平方逼近

$$f(x_0), f(x_1), \cdots, f(x_n)$$

在给定点 $x_0, x_1, \cdots, x_n$ 处的偏差的平方和为最小.

设 $\bar{a}_0, \bar{a}_1, \cdots, \bar{a}_m$ 是由解方程组

$$f(x_i) - \varphi(x_i, a_0, a_1, \cdots, a_m) = 0 \quad (i = 0, 1, \cdots, n)$$

的结果所得的 $a_0, a_1, \cdots, a_m$ 的逼近值. 我们假定在满足此方程组的值 $x_i, a_0, a_1, \cdots, a_m$ 的邻域 D 内(假定这些值的存在是预先知道的)函数 $\varphi(x, a_0, a_1, \cdots, a_m)$ 对所有参数的首二阶偏导数存在且连续.

由此此后仅限于在邻域 D 中, 所以, 如果 $\bar{a}_0, \bar{a}_1, \cdots, \bar{a}_m$ 不超出此邻域的范围, 便可写出 $\varphi(x_i, a_0, a_1, \cdots, a_m)$ 按修正值

$$\eta_k = a_k - \bar{a}_k \quad (k = 0, 1, \cdots, m)$$

的幂次的展开式, 其中 $\bar{a}_k$, 如上所说是表示 a_k 的逼近值. 于是

$$\varphi(x_i, a_0, a_1, \cdots, a_m) = \varphi(x_i, \bar{a}_0, \bar{a}_1, \cdots, \bar{a}_m) +$$

$$\eta\left(\frac{\partial \varphi_i}{\partial a_0}\right)_0 + \eta_1\left(\frac{\partial \varphi_i}{\partial a_1}\right)_0 + \cdots +$$

$$\eta_m\left(\frac{\partial \varphi_i}{\partial a_m}\right)_0 + 剩余 (i = 0, 1, \cdots, n)$$

其中 $\left(\frac{\partial \varphi_i}{\partial a_0}\right)_0$ 是 $\varphi(x_i, a_0, a_1, \cdots, a_m)$ 对 a_0 的导数在点 $x_i, \bar{a}_0, \bar{a}_1, \cdots, \bar{a}_m$ 处的值. 量

$$\left(\frac{\partial \varphi_i}{\partial a_1}\right)_0, \left(\frac{\partial \varphi_i}{\partial a_2}\right)_0, \cdots, \left(\frac{\partial \varphi_i}{\partial a_m}\right)_0$$

也都有完全同样的意义.

今还要在所得的展开式中舍去剩余项并把由其在点 $x_i(i=0,1,\cdots,n)$ 处的值所给定的函数 $f(x)$, 用函数

Tschebyscheff 逼近定理

$$\varphi(x_i, \bar{a}_0, \bar{a}_1, \cdots, \bar{a}_m) +$$
$$\eta_0 \left(\frac{\partial \varphi_i}{\partial a_0}\right)_0 + \eta_1 \left(\frac{\partial \varphi_i}{\partial a_1}\right)_0 + \cdots + \eta_m \left(\frac{\partial \varphi_i}{\partial a_m}\right)_0$$

借助最小二乘法来逼近. 换言之, 即还要使表表式

$$\delta = \sum_{i=0}^{n} \left[f(x_i) - \varphi(x_i, \bar{a}_0, \bar{a}_1, \cdots, \bar{a}_m) - \eta_0 \left(\frac{\partial \varphi_i}{\partial a_0}\right)_0 - \eta_1 \left(\frac{\partial \varphi_i}{\partial a_1}\right)_0 - \cdots - \eta_m \left(\frac{\partial \varphi_i}{\partial a_m}\right)_0 \right]^2$$

成为极小. 如使 δ 对所有参数 $\eta_0, \eta_1, \cdots, \eta_m$ 的偏导数等于零, 则便可构成正规方程组. 解此方程组, 便得出我们所要的修正值 $\eta_0, \eta_1, \cdots, \eta_m$. 如果这些修正值相当小, 则计算即可终止. 在相反的情形, 就须要以第二次逼近

$$\bar{a}_v + \eta_v \quad (v = 0, 1, \cdots, m)$$

作为初次逼近再作计算.

虽然我们的论证是有其局限性的, 但就在这样形式下, 它们常常是有用的. 在很多情形下, 修改实验数据仅限于以上所引入的小的邻域 D 内也就完全可以了.

6. 分段连续函数的逼近

今研究在某一区间 $[a, b]$ 上给定的任一分段连续函数 $f(x)$ 借助于函数序列 $\{\varphi_k(x)\}$ 的线性组合

$$F_m(x) = A_0 \varphi_0(x) + A_1 \varphi_1(x) + \cdots + A_m \varphi_m(x) \quad (a \leqslant x \leqslant b)$$

的平方逼近问题, 其中 $m \geqslant 0$ 为固定的整数, 并利用偏差 $f(x) - F_m(x)$ 的平方的积分

附录 X 平方逼近

$$\delta_m = \int_a^b [f(x) - F_m(x)]^2 dx \qquad ㊗$$

作为逼近的准确性的量度. 如在极限等式(5)中以 δ_m 表示 $(b-a)\lim_{n\to\infty}\Delta_{m,n}^2$,我们便得出表达式 ㊗. 因此,欲使 $F_m(x)$ 与 $f(x)$ 的平均平方偏差有最小值,我们应当选择系数 A_v 使得积分 ㊗ 成为极小.

因此,今研究积分 ㊗ 的极值问题. 由微分便得出下形的线性代数方程组(关于系数 A_v 的)

$$\frac{1}{2}\frac{\partial \delta_m}{\partial A_v} \equiv \int_a^b [f(x) - F_m(x)]\varphi_v(x) dx = 0$$

即

$$\int_a^b F_m(x)\varphi_v(x)] dx = \sum_{\mu=0}^m A_\mu \int_a^b \varphi_\mu(x)\varphi_v(x) dx = \int_a^b f(x)\varphi_v(x) dx$$

$$(v = 0, 1, \cdots, m) \qquad ㊆$$

上一等式给出求逼近函数 $F_m(x)$ 的所有系数的实际可能性. 还要证明,所得的方程组有解并且此解是唯一的.

作为逼近的准确性的量度也可利用偏差 $f(x) - F_m(x)$ 的 s 次方的积分(s 为固定数,大于0)

$$\delta_{ms} = \int_a^b |f(x) - F_m(x)|^s dx$$

此处也有 $\delta_{ms} > 0$,因而极小存在. 因此,量 δ_{ms} 对于系数 $A_0, A_1, \cdots, A_m$ 的某些值达到极小,所以我们的问题仍是去选择系数 A_v 使得上一积分成为极小.

带任意幂次 s 的幂次逼近并无实际价值,因为一般说来,在选择系数 $A_0, A_1, \cdots, A_m$ 要使 δ_{ms} 成为极小时,它们会使计算的性质有不可克服的困难. 因此,今

Tschebyscheff 逼近定理

后仅仅研究 $s = 2$ 的情形, 即只研究平方逼近.

方程组 ㊼ 的行列式

$$G(\varphi_0, \varphi_1, \cdots, \varphi_m) =$$

$$\begin{vmatrix} \int_a^b \varphi_0^2(x)\,\mathrm{d}x & \int_a^b \varphi_0(x)\varphi_1(x)\,\mathrm{d}x & \cdots & \int_a^b \varphi_0(x)\varphi_m(x)\,\mathrm{d}x \\ \int_a^b \varphi_1(x)\varphi_0(x)\,\mathrm{d}x & \int_a^b \varphi_1^2(x)\,\mathrm{d}x & \cdots & \int_a^b \varphi_1(x)\varphi_m(x)\,\mathrm{d}x \\ \vdots & \vdots & & \vdots \\ \int_a^b \varphi_m(x)\varphi_0(x)\,\mathrm{d}x & \int_a^b \varphi_m(x)\varphi_1(x)\,\mathrm{d}x & \cdots & \int_a^b \varphi_m^2(x)\,\mathrm{d}x \end{vmatrix} \quad ㊽$$

就是此方程组的格拉姆行列式. 此行列式异于零, 这是因为函数 $\varphi_0(x), \varphi_1(x), \cdots, \varphi_m(x)$ 线性无关. 这样我们已证明, 以上所提到的极小问题的解是唯一的.

在以后, 将假定函数 $f(x)$ 和 $F_m(x)$ 在黎曼意义下为可积; 因此, 我们将认定这些函数是有界的且有有限个间断点或甚至其间断点形成测度 (在勒贝格意义下) 为零的集合.

当给定函数 $f(x)$ 在区间 $[a, b]$ 的不同部分以不同的准确程度被逼近的情形, 引入表示在部分区间内的逼近的准确性的非负的权函数 $p(x)$ $(a \leqslant x \leqslant b)$ 是合适的.

与上面一样, 就在权函数 $p(x)$ 存在时, 我们也可提出求系数 $A_0, A_1, \cdots, A_m$ 使得量

$$\delta_m = \int_a^b [f(x) - F_m(x)]^2 p(x)\,\mathrm{d}x \quad ㊾$$

成为极小的问题. 用以确定系数 $A_0, A_1, \cdots, A_m$ 的线性代数方程的行列式也仍是格拉姆行列式; 为了将它写出, 只需在 ㊽ 中到处以 $\int_a^b p(x)\varphi_v(x)\varphi_\mu(x)\,\mathrm{d}x$ 代替

附录 X 平方逼近

$\int_a^b \varphi_v(x)\varphi_\mu(x)\,\mathrm{d}x$. 这使我们能肯定,带权积分 ㊙ 成为极小的问题也是有唯一一个解的.

为了完全起见,我们指出,还可能有平方逼近法的一个概括. 在积分 ㊙ 中,我们以 $\mathrm{d}\varphi(x)$ 代替 $p(x)\mathrm{d}x$ 而将积分

$$\delta_m = \int_a^b [f(x) - F_m(x)]^2 \mathrm{d}\varphi(x) \qquad ㊽$$

在斯提尔捷斯的意义下来了解. 函数 $f(x)$ 和 $F_m(x)$ 假定是在 $[a,b]$ 上连续. $[f(x) - F_m(x)]^2$ 对 $\varphi(x)$ 在区间 $[a,b]$ 上的斯提尔捷斯积分存在,这乃是由于对区间 $[a,b]$ 的任一 x,微分权 $p(x) \geqslant 0$,所以 $\varphi(x)$ 为一非递减的有界函数.

如果 $f(x) - F_m(x)$ 在区间 $[a,b]$ 上连续,而 $\varphi(x)$ 在此区间上具有有界的可积(按黎曼)的导数 $p(x)$,则

$$(S)\int_a^b [f(x) - F_m(x)]^2 \mathrm{d}\varphi(x) =$$
$$(R)\int_a^b [f(x) - F_m(x)]^2 p(x)\mathrm{d}x$$

其中 (S) 和 (R) 着重指出,对应它们的积分各是在斯提尔捷斯和黎曼意义下来考虑.

对于所作的假定,斯提尔捷斯积分转变成为通常黎曼积分. 欲使斯提尔捷斯积分实际上有较广泛的意义,我们可提出函数 $f(x)$ 在区间 $[a,b]$ 内对积分权 $\varphi(x)$ 的平方逼近问题,而并不认定权必是连续的.

设若我们应使积分 ㊽ 成为极小,假定函数 $\varphi(x)$ 在区间 (a,b) 内有有限个点处有第一类间断,而在这些点之间有有界的可积的导数 $\varphi'(x)$. 函数 $F_m(x)$ 和

Tschebyscheff 逼近定理

$f(x)$ 仍使之连续. 我们就限于考虑这样的函数 $f(x)$, $F_m(x)$ 和 $\varphi(x)$, 以 $a_v(v=0,1,\cdots,n)$ 表示 $\varphi(x)$ 间断的内点. 于是便有下一公式

$$(S)\int_a^b [f(x) - F_m(x)]^2 \mathrm{d}\varphi(x) =$$

$$(R)\int_a^b [f(x) - F_m(x)]^2 \varphi'(x) \mathrm{d}x +$$

$$\sum_{v=0}^m [f(a_v) - F_m(a_v)]^2 [\varphi(a_v + 0) - \varphi(a_v - 0)]$$

其中 $(h > 0)$.

$$\varphi(a_v + 0) = \lim_{h \to 0} \varphi(a_v + h)$$

$$\varphi(a_v - 0) = \lim_{h \to 0} \varphi(a_v - h)$$

因而这意味着,我们必须要使上一等式右端的表达式成为极小.

7. 用以确定平方逼近的系数的方程组

在以后,不失一般性,我们可取区间 $[-1,1]$ 作为自变量变化的区间. 我们可将任一有限区间 $a \leqslant x \leqslant b$ 变成 $[-1,1]$, 只要借方程

$$x = \frac{a+b}{2} + \frac{b-a}{2}t$$

作变量代换. 因此, 1 中的函数 $f(x)$ 和 $\varphi_m(x)$ 将是 t 的函数

$$f(x) = f\left(\frac{a+b}{2} + \frac{b-a}{2}t\right)$$

$$\varphi_m(x) = \varphi_m\left(\frac{a+b}{2} + \frac{b-a}{2}t\right) =$$

$$\sum_{v=0}^m a_v\left(\frac{a+b}{2} + \frac{b-a}{2}t\right)^v = \sum_{v=0}^m A_v t^v$$

附录 X　平方逼近

而平方逼近问题可归结于去选择系数 $A_0, A_1, \cdots, A_m$ 使得

$$\begin{cases} \sum_m = \int_{-1}^{+1} \left[f\left(\dfrac{a+b}{2} + \dfrac{b-a}{2}t\right) - \sum_{v=0}^{m} A_v t^v \right] \mathrm{d}t = 0 \\ \sum_m = \int_{-1}^{+1} \left[f\left(\dfrac{a+b}{2} + \dfrac{b-a}{2}t\right) - \sum_{v=0}^{m} A_v t^v \right] t \mathrm{d}t = 0 \\ \qquad\qquad\qquad \vdots \\ \sum_m = \int_{-1}^{+1} \left[f\left(\dfrac{a+b}{2} + \dfrac{b-a}{2}t\right) - \sum_{v=0}^{m} A_v t^v \right] t^m \mathrm{d}t = 0 \end{cases}$$

⑱

逼近多项式有形式

$$\varphi(x) = A_0 + A_1 \frac{2x-a-b}{b-a} + A_2 \left(\frac{2x-a-b}{b-a}\right)^2 + \cdots + A_m \left(\frac{2x-a-b}{b-a}\right)^m$$

但是也可不必回到旧的变量,而对于计算利用逼近等式

$$f\left(\frac{a+b}{2} + \frac{b-a}{2}t\right) = A_0 + A_1 t + A_2 t^2 + \cdots + A_m t^m$$

就可以了. 今如引入记号

$$2J_0 = \int_{-1}^{+1} f\left(\frac{a+b}{2} + \frac{b-a}{2}t\right) \mathrm{d}t$$

$$2J_1 = \int_{-1}^{+1} t f\left(\frac{a+b}{2} + \frac{b-a}{2}t\right) \mathrm{d}t$$

$$\vdots$$

$$2J_m = \int_{-1}^{+1} t^m f\left(\frac{a+b}{2} + \frac{b-a}{2}t\right) \mathrm{d}t$$

并算出在方程组 ⑱ 中所写出的积分

Tschebyscheff 逼近定理

$$\int_{-1}^{+1} t^\lambda \mathrm{d}t \quad (\lambda = 0, 1, \cdots, m)$$

则我们得到为了确定系数 $A_0, A_1, \cdots, A_m$ 的方程组

$$A_0 + \frac{1}{3}A_2 + \frac{1}{5}A_4 + \cdots - J_0 = 0$$

$$\frac{1}{3}A_1 + \frac{1}{5}A_3 + \frac{1}{7}A_5 + \cdots - J_1 = 0$$

$$\frac{1}{3}A_0 + \frac{1}{5}A_2 + \frac{1}{7}A_4 + \cdots - J_2 = 0$$

$$\frac{1}{5}A_1 + \frac{1}{7}A_3 + \frac{1}{9}A_5 + \cdots - J_3 = 0$$

$$\frac{1}{5}A_0 + \frac{1}{7}A_2 + \frac{1}{9}A_4 + \cdots - J_4 = 0$$

$$\vdots$$

今考虑当 $m = 0, 1, \cdots, 5$ 时的特殊情形. 当 $m = 0$ 时, 得到

$$A_0 = J_0$$

当 $m = 1$ 时, 便有

$$A_0 = J_0, A_1 = 3J_1$$

对于 $m = 2$, 则有

$$A_0 = \frac{3}{4}(3J_0 - 5J_2), A_1 = 3J_1, A_2 = \frac{15}{4}(3J_2 - J_0)$$

如今 $m = 3$, 则

$$A_0 = \frac{3}{4}(3J_0 - 5J_2), A_1 = \frac{15}{4}(5J_1 - 7J_3)$$

$$A_2 = \frac{15}{4}(3J_2 - J_0), A_3 = \frac{35}{4}(5J_3 - 3J_1)$$

令 $m = 4$, 便得

附录 X 平方逼近

$$A_0 = \frac{15}{64}(15J_0 - 70J_2 + 63J_4)$$

$$A_1 = \frac{15}{4}(5J_1 - 7J_3)$$

$$A_2 = \frac{105}{32}(-5J_0 + 42J_2 - 45J_4)$$

$$A_3 = \frac{35}{4}(5J_3 - 3J_1)$$

$$A_4 = \frac{315}{64}(3J_0 - 30J_2 + 35J_4)$$

最后,当 $m = 5$ 时,便有

$$A_0 = \frac{15}{64}(15J_0 - 70J_2 + 63J_4)$$

$$A_1 = \frac{105}{64}(35J_1 - 126J_3 + 99J_5)$$

$$A_2 = \frac{105}{32}(-5J_0 + 42J_2 - 45J_4)$$

$$A_3 = \frac{315}{32}(-21J_1 + 90J_3 - 77J_5)$$

$$A_4 = \frac{315}{64}(3J_0 - 30J_2 + 35J_4)$$

$$A_5 = \frac{693}{64}(15J_1 - 70J_3 + 63J_5)$$

8. 平方误差的计算

今简短地叙述一下积分 ⑱ 的计算,因为它的值给出平方误差,因而使给出关于平均逼近性质的表示. 今将方程组 ⑱ 的第一个(上面一个)乘以 A_0,第二个乘

Tschebyscheff 逼近定理

以 A_1 等最后一个乘以 A_m 并交所有这些积积加起来,我们得到

$$\int_{-1}^{+1} f\left(\frac{a+b}{2} + \frac{b-a}{2}t\right)(A_0 + A_1 t + \cdots + A_m t^m)\,\mathrm{d}t =$$

$$\int_{-1}^{+1}(A_0 + A_1 t + \cdots + A_m t^m)^2\,\mathrm{d}t$$

但

$$\int_{-1}^{+1}\left[f\left(\frac{a+b}{2} + \frac{b-a}{2}t\right) - A_0 - A_1 t - \cdots - A_m t^m\right]^2 \mathrm{d}t =$$

$$\int_{-1}^{+1} f^2\left(\frac{a+b}{2} + \frac{b-a}{2}t\right)\mathrm{d}t =$$

$$-2\int_{-1}^{+1} f\left(\frac{a+b}{2} + \frac{b-a}{2}t\right)(A_0 + A_1 t + \cdots + A_m t^m)\,\mathrm{d}t +$$

$$\int_{-1}^{+1} f(A_0 + A_1 t + \cdots + A_m t^m)^2\,\mathrm{d}t$$

因而得出我们所要的极小

$$\sum\nolimits_m = \int_{-1}^{+1} f^2\left(\frac{a+b}{2} + \frac{b-a}{2}t\right)\mathrm{d}t -$$

$$\int_{-1}^{+1}(A_0 + A_1 t + \cdots + A_m t^m)^2\,\mathrm{d}t$$

对于计算同一个极小,也可求得

$$\sum\nolimits_m = \int_{-1}^{+1} f^2\left(\frac{a+b}{2} + \frac{b-a}{2}t\right)\mathrm{d}t -$$

$$\int_{-1}^{+1} f\left(\frac{a+b}{2} + \frac{b-a}{2}t\right)(A_0 + A_1 t + \cdots + A_m t^m)\,\mathrm{d}t =$$

$$\int_{-1}^{+1} f^2\left(\frac{a+b}{2} + \frac{b-a}{2}t\right)\mathrm{d}t - 2Q_m$$

其中

$$Q_m = A_0 J_0 + A_1 J_1 + \cdots + A_m J_m$$

附录 X　平方逼近

特别地,由计算可得

$$Q_0 = J_0^2$$

$$Q_1 = J_0^2 + J_1^2$$

$$Q_2 = J_0^2 + 3J_1^2 + \frac{5}{4}(3J_2 - J_0)^2$$

$$Q_3 = J_0^2 + 3J_1^2 + \frac{5}{4}(3J_2 - J_0)^2 + \frac{7}{4}(5J_3 - 3J_1)^2$$

$$Q_4 = J_0^2 + 3J_1^2 + \frac{5}{4}(3J_2 - J_0)^2 + \frac{7}{4}(5J_3 - 3J_1)^2 +$$

$$\frac{9}{64}(35J_4 - 30J_2 + J_0)^2$$

$$Q_5 = J_0^2 + 3J_1^2 + \frac{5}{4}(3J_2 - J_0)^2 + \frac{7}{4}(5J_3 - 3J_1)^2 +$$

$$\frac{9}{64}(3J_0 - 30J_2 + 35J_4)^2 + \frac{11}{64}(63J_5 - 70J_3 + 15J_1)^2$$

9. 多个自变量函数的平方逼近

现在我们可以去平均逼近多个自变量的函数,这是与对单变量的函数所作的相类似. 设 D 为某一平面二维域. 今考虑在闭区域 D 内有定义的并在此域的每一点处连续的函数 $f(x,y)$. 我们可对此函数在域 D 内以多项式

$$\varphi(x,y) = a_{00} + a_{10}x + a_{01}y + a_{20}x^2 + a_{11}xy + a_{02}y^2 + \cdots$$

来逼近而使二重积分

$$\iint_D [f(x,y) - \varphi(x,y)]^2 \mathrm{d}x\mathrm{d}y \qquad ⑱$$

的值尽可能的小. 为了确定使此积分成为最小的系数

Tschebyscheff 逼近定理

a_{pq} 的一些值,我们使此积分对所有系数 a_{pq} 的偏导数都等于零. 我们得到线性方程组

$$\begin{cases} \iint_D [f(x,y) - a_{00} - a_{10}x - a_{01}y - \\ \quad a_{20}x^2 - a_{11}xy - a_{02}y^2 - \cdots]\mathrm{d}x\mathrm{d}y = 0 \\ \iint_D x[f(x,y) - a_{00} - a_{10}x - a_{01}y - \\ \quad a_{20}x^2 - a_{11}xy - a_{02}y^2 - \cdots]\mathrm{d}x\mathrm{d}y = 0 \\ \iint_D y[f(x,y) - a_{00} - a_{10}x - a_{01}y - \\ \quad a_{20}x^2 - a_{11}xy - a_{02}y^2 - \cdots]\mathrm{d}x\mathrm{d}y = 0 \\ \quad\quad\quad\quad \vdots \end{cases} \quad ㊽$$

如果方程组 ㊽ 有唯一一个解,则由此确定了系数 a_{pq},便得到所求的多项式 $\varphi(x,y)$.

计算积分 ㊼ 的极小,即计算平方误差也是很容易的. 事实上,如果将方程组 ㊽ 的第一个(上面一个)乘以 a_{00},第二个乘以 a_{10},第三个乘以 a_{01} 等并将所得的乘积加起来,则便有

$$\iint_D f(x,y)\varphi(x,y)\mathrm{d}x\mathrm{d}y = \iint_D \varphi^2(x,y)\mathrm{d}x\mathrm{d}y$$

利用这些等式,容易知道,积分 ㊼ 的最小值等于

$$\iint_D f^2(x,y)\mathrm{d}x\mathrm{d}y - \iint_D f(x,y)\varphi(x,y)\mathrm{d}x\mathrm{d}y =$$
$$\iint_D f^2(x,y)\mathrm{d}x\mathrm{d}y - \iint_D \varphi^2(x,y)\mathrm{d}x\mathrm{d}y$$

每一个所得的差给出关于平方误差的表示,即给出关于函数 $f(x,y)$ 以函数 $\varphi(x,y)$ 平均逼近的性质. 所得的结果对于在 D 内连续的函数 $f(x,y)$ 也是成立的. 当 $f(x,y)$ 有间断点存在时,则需要要求以上所有被考虑

的二重积分存在.

作为一个例子,我们借助于三次多项式来逼近函数 $f(x,y)$. 今取由直线 $x = \pm 1, y = \pm 1$ 所围成的正方形作为域 D. 我们得到

$$\varphi(x,y) = a_{00} + a_{10}x + a_{01}y + a_{20}x^2 + a_{11}xy + a_{02}y^2 + a_{30}x^3 + a_{21}x^2y + a_{12}xy^2 + a_{03}y^3$$

其中系数 $a_{00}, a_{10}, \cdots, a_{08}$ 由下列等式确定

$$a_{00} + \frac{1}{3}a_{20} + \frac{1}{3}a_{02} = J_{00}$$

$$\frac{1}{3}a_{10} + \frac{1}{5}a_{30} + \frac{1}{9}a_{12} = J_{10}$$

$$\frac{1}{3}a_{01} + \frac{1}{9}a_{21} + \frac{1}{5}a_{03} = J_{01}$$

$$\frac{1}{3}a_{00} + \frac{1}{5}a_{20} + \frac{1}{9}a_{02} = J_{20}$$

$$\frac{1}{9}a_{11} = J_{11}$$

$$\frac{1}{3}a_{00} + \frac{1}{9}a_{20} + \frac{1}{5}a_{02} = J_{02}$$

$$\frac{1}{5}a_{10} + \frac{1}{7}a_{30} + \frac{1}{15}a_{12} = J_{30}$$

$$\frac{1}{9}a_{01} + \frac{1}{15}a_{21} + \frac{1}{15}a_{03} = J_{21}$$

$$\frac{1}{9}a_{10} + \frac{1}{15}a_{30} + \frac{1}{15}a_{12} = J_{12}$$

$$\frac{1}{3}a_{01} + \frac{1}{15}a_{21} + \frac{1}{7}a_{03} = J_{13}$$

$$J_{00} = \frac{1}{4}\int_{-1}^{+1}\int_{-1}^{+1} f(x,y)\,\mathrm{d}x\mathrm{d}y$$

Tschebyscheff 逼近定理

$$J_{10} = \frac{1}{4}\int_{-1}^{+1}\int_{-1}^{+1} xf(x,y)\,\mathrm{d}x\mathrm{d}y$$

$$J_{01} = \frac{1}{4}\int_{-1}^{+1}\int_{-1}^{+1} yf(x,y)\,\mathrm{d}x\mathrm{d}y$$

$$J_{11} = \frac{1}{4}\int_{-1}^{+1}\int_{-1}^{+1} xyf(x,y)\,\mathrm{d}x\mathrm{d}y$$

$$\vdots$$

$$J_{03} = \frac{1}{4}\int_{-1}^{+1}\int_{-1}^{+1} y^3 f(x,y)\,\mathrm{d}x\mathrm{d}y$$

计算了逼近多项式的系数,便知

$$a_{00} = \frac{7}{2}J_{00} - \frac{15}{4}(J_{20} + J_{02})$$

$$a_{01} = \frac{45}{2}J_{01} - \frac{105}{4}J_{03} - \frac{45}{4}J_{21}$$

$$a_{10} = \frac{45}{2}J_{10} - \frac{105}{4}J_{30} - \frac{45}{4}J_{12}$$

$$a_{02} = \frac{45}{4}\left(J_{02} - \frac{1}{3}J_{00}\right)$$

$$a_{11} = 9J_{11}$$

$$a_{20} = \frac{45}{4}\left(J_{20} - \frac{1}{3}J_{00}\right)$$

$$a_{30} = \frac{175}{4}\left(J_{30} - \frac{3}{3}J_{10}\right)$$

$$a_{21} = \frac{135}{4}\left(J_{21} - \frac{1}{3}J_{01}\right)$$

$$a_{12} = \frac{135}{4}\left(J_{12} - \frac{1}{3}J_{10}\right)$$

$$a_{06} = \frac{175}{4}\left(J_{03} - \frac{1}{3}J_{01}\right)$$

参考文献

[1] ADAMS R A. Sobolev Spaces[M]. New York: Academic Press, 1975.

[2] AHLBERG J H, ITO T. A collocation method for two-point boundary value problems [J]. Math Comp, 1975(29):761-776.

[3] AHUÉS M. Raffinement des élérncnts propres d'un opérateur compact sur tin espace de Banach par des méthodes de tape Newton à jacobien approché[M]. Unpublished manuscript, Univ. de Grenoble, 1982.

[4] AHUÉS M, TELIAS M. Petrov-Galerkin schemes for the steady state convection-diffusion equation[J]. In Finite Elenuants in Water Resources (K. P. Holz, U. Meissner, W. Zulkc, C. A. Brebbia, G. Pinder and W. Gray, eds.). Springer-Verlag, Berlin and New York, 1982:2-3,2-12.

[5] AHUÉS M, TELIAS M. Quasi-Newton iterative refinement techniques for the eigenvalue problem of compact linear operators[J]. R. R. IMAG Univ. de Grenoble, 1982:325.

[8] AHUÉS M, CHATELIN F, D'ALMEIDA F, et al.

In Treatment of Integral Equations by Numerical Methods[M]. London: Academic Press, 1983.

[10] AHUÉS M, D'ALMEIDA F, TELIAS M. Iterative refinement for aproximate eigenelements of compact operators[J]. RAIRO Anal, 1983.

[11] ALBRECHT J, COLLATZ L. Numerical Treatment of Integral Equations [M]. Basel: Birkhaeuser, 1980.

[12] ANDERSSEN R S, PRENTER P M. A formal comparison of methods proposed for the numerical solution of first kind integral equations[J]. J. Austral. Math. Soc. Ser. B22, 1981:491-503.

[13] ANDERSSEN R S, DE HOOG F R, LUKAS M A, EDS. The Application and Numerical Solution of Integral Equations [M]. Netherlands: Sijthoff & Noordhoff, Alphen an den Rijn,1980.

[14] ANDREW A L. Eigenvectors of certain matrices [J]. Linear Algebra Appl, 1973(7):151-162.

[15] ANDREW A L. Iterative computation of derivatives of eigenvalues and eigenvectors [J]. J. Inst. Math. Appl, 1979(24):209-218.

[16] ANDREW A L, ELTON G C. Computation of eigenvectors corresponding to multiple eigenvalues [J]. Ball. Austral. Math. Soc. 1971(4):419-422.

[17] ANDRUSHKIN R I. On the approximate solution of K-positive eigenvalue problems $Tu-\lambda Su=0$ J[J].

Math. Anal. Appl. 1975(50):511-529.

[18] ANSELONE P M. Collectively Compact Operator Approximation Theory.[M]New Jersey: Prernice-Hall, 1971.

[19] ANSELONE P M. Nonlinear operator approximation[M]. J. Albrecht , L. Collatz. In Moderne Methoden der numerischen Mathematik, Basel: Birkhabser, 1976:17-24.

[20] ANSELONE P M, AMORGE R. Compactness principle in non linear operator approximation theory [J]. Numer. Funct. Anal. Optim. 1979 (1): 589-618.

[22] ANSELONE P M, GONZALEZ-FERNANDEZ M J. Uniformly convergent approximate solutions of Fredholm integral equations[J]. J. Math. Anal. Appl. 1965(10): 519-536.

[23] ANSELONE P M, KRABS W. Approximate solution of weakly singular integral equations[J]. J. Integral Equations, 1979(1):61-75.

[24] ANSELONE P M, LEE J W. Spectral properties of integral operators with non-negative kernels [J]. Linear Algebra Appl. 1974(9):67-87.

[25] ANSELONE P M, LEE J W. Double approximation methods for the solution of Fredholm integral equations[M]. L. Collatz, H. Werner, and G Meinardus. eds. In Numerische Methoden der Approximations Theorie. Basel:Birkhaeuser, 1976:9-34.

[26] ARNOLD D N, WENDLAND W L. On the asymptotic convergence of collocation methods [J]. Prepr: Hochschule Darmstadt, 1982.

[27] ARNOLDI W E. The principle of minimized iterations in the solution of the matrix eigenvalue problem[J]. Ouart. Appl. math, 1951(9):17-29.

[30] ATKINSON K E. The numerical solution of Fredholm integral equations of the second kind[J]. SIAM J. Numer. Anal. 1967a(4):337-348.

[31] ATKINSON K E. The numerical solution of the eigenvalue problem for compact integral operators [J]. Trans. Amer. Math. Soc. 1967b(129): 458-465.

[32] ATKINSON K E. The numerical solution of Fredholm intergral equations of the second kind with singular kernels[J]. Numer. Math. 1972(19): 248-259.

[33] ATKINSON K E. Iterative variants of the Nyström method for the numerical solution of integral equations[J]. Numer. Math. 1973(22):17-31.

[34] ATKINSON K E. Convergence rates for approximate eigenvalues of compact integral equations [J]. SIAM J. Numer. Anal. 1975(12):213-222.

[35] ATKINSON K E. A survey of Numerical Methods for the Solution of Fredholm Integral Equations of the Second Kind[M]. Pennsylvanis:SIAM, Phila-

参考文献

delphia, 1976.

[36] ATKINSON K E. An automatic program for linear Fredholm integral equations of the second kind [J]. ACM Trans. Math. Software, 1976(b)(2): 154-171.

[37] ATKINSON K E, GRAHAM I G, SLOAN I H. Piecewise continuous collocation for integral equations[M]. New South Wales: Kensington, 1982.

[38] AUBIN J P. Approximation of Elliptic Boundary Value Problems [M]. New York: Wiley (Interscience), 1972.

[39] AZIZ A K, ed. The Mathematical Foundations of the Finite Element Method with Applications to Partial Differential Equations[M]. New York: Academic Press, 1972.

[40] BABUSKA I. Error bounds for finite element method[J]. Nemer. Math, 1971(16):322-333.

[41] BAKUSKA I. The finite element method with Lagrangian multipliers [J]. Numer. Math, 1973 (20):179-192.

[42] BABUSKA L, AZIZ A K. Survey letures on the mathematical foundations of the finite element method [M]. A. K. Aziz. In the Mathematical Foundations of the Finite Element Method with Applications to Partial Differential Equations. New York: Academic Press, 1972.

[43] BABUSKA I, OSBORN J E. Numerical treatment

of eigenvalue problems for differential equations with discontinuous coefficients[J]. Math. Comp, 1978(32):991-1023.

[44] BABUSKA I, RHEINBOLDT W. Error estimates for adaptive finite element computations[J]. SIMA J. Numer. Anal, 1978(15):736-754.

[45] BAKER C T H. The deferred approach to the limit for eigenvalues of integral equations[J]. SIAM J. Numer. Anal, 1971(8):1-10.

[46] BAKER C T H. The Numerical Treatment of Integral Equations[M]. London and New York: Press (Clarendon), 1977.

[47] BAKER C T H, HODGSON G S. Asymptotic expansions for integration formulae in one and more dimensions[J]. SIAM J. Numer Anal, 1971(8): 473-480

[48] BANACH S, STEINHAUS H. Sur le principe de la condensation des singularités [J]. Fund. Math, 1927(9):51-57.

[49] BANK R E. Analysis of a mulitilevel inverse iteration procedure for eigenvalue problems[J]. Res. Rep. No. 199, Computer Science, Yale Univ., 1980.

[50] BANK R E, ROSE D J. Analysis of a multilevel iterative method for nonlinear finite element equatins [M]. Connecticut: Res. Rep. No. 202 COmputer Science. Yale Univ., Connecticut, 1981.

参考文献

[51] BARTELS R H, STEWART G W. Algorthm 432, solution of the matrix equation $AX + XB = C$ [J]. Comm ACM, 1972(15):820-826.

[52] BATHÉ K J, WILSON E L. Large eigenvalue problems in dynamic analysis[J]. ASCE J. Engrg. Mech. Div, 1972(98):1471-1485.

[53] BATHÉ K J, WILSON E L. Eigensolution of large structure systems with small band width[J]. ASCE J. Engrg. Mech. Div 1973(99), 467-480.

[54] BATHÉ K J, WILSON E L. Solution methods for eigenvalue problems in structural mechanics[J]. Internat. J. Numer. Methods Engrg. 1973b(6): 213-226.

[55] BATHÉ K J, WILSON E L. Numerical Methods in Finite Element Analysis[M]. New Jersey,: Prentice-Hall, Englewood Cliffs, 1976.

[56] BATHÉ K J, Ramaswamy S. An accelerated subspace iteration method[J]. Comput. Methods Appl. Mech, Engrg, 1980(23):313-331.

[57] BAUER F L. Das Verfahren der Treppeniteration und verwandte Verfahren zur Lösung algebraisher Eigenwertprobleme[J]. Z. Angew. Math. Phys. 1957(8):214-235.

[58] BAUER F L. On modern matrix iteration processes of Bernouilli and Graeffe type[J]. J. Assoc. Comput. Mach, 1958(5):246-257.

[59] BAUER F L, FIKE C T. Norms and exclusion the-

orems[J]. Numer. Math, 1960(2):137-141.

[60] BAVELY A C, STEWART G W. An algorithm for computing reducing subspaces by block diagonalization[J]. SIAM J. Numer. Anal, 1979(16): 359-367.

[61] BEGIS D, PERRONNET A. The Club MODULEF, a library of computer procedures for finite element analysis [M]. Rep. INRIA-MODULEF 73, INRIA, Le Chesnay, 1982.

[62] BERGER D, GRUBER R, TROYON F. A finite element approoach to the computation of the magnethhydrodynamic spectrum of straight noncircular plasma equilibria [J]. Comput. Phys. Commun, 1976(11):313-323.

[63] BERGER W A, MILLER H G, KREUZER K G, et al. An iterative method for calculating low lying eigenvalues of an Hermitian operator[J]. J. Phys. A, 1977(10):1089-1095.

[64] BERGER W A, KRUZER K G, MILLER H G. An algorithm for obtaining an optimalized projected Hamiltonian and its ground state[J]. Z. Physik. A, 1980(298):11-12.

[65] BIRKHOFF G, DE BOOR C, SWARTZ B, WENDROFF B. Rayleigh-Ritz approximation by piecewise polynomials [J]. SIAM J. Numer. Anal, 1966(3):188-203.

[66] BIRKHOFF G, DE BOOR C, SWARTZ B, WEN-

参考文献

DROFF B. Rayleigh-Ritz approximation by piecewise polynomials [J]. SIAM J. Numer. Anal, 1966(3):188-303.

[67] BJÖRCK A. Solving linear least squares problems by Gram-Schmidt orthogonalization [J]. BIT, 1967a(7):1-21.

[68] BJÖRCK A. Iterativ refinement of linear least squares solution: I [J]. BIT, 1967b(7):251-278.

[69] BJÖRCK A. Iterative refinement of linear least squares solution: H [J]. BIT, 1968(8):8-30

[70] BJÖRCK A, GOLUB G H. Numerical methods for computing angles between linear subspaces [J]. Math. Comp, 1973(27):579-594

[71] BJÖRCK A, PLEMMONS R J. Large Scale Matrix Problems [M]. New York: American Elsevier, 1980.

[72] BLAND S. The two-dimensional oscillating airfoil in a wind tunnel in subsonic flow [J]. SIAM J. Appl. Math, 1970(18):830-848.

[73] BLUM E K, GELTNER P B. Numerical solution of eigentuple-elgenvector problems in Hilbert space by a gradient method [J]. Numer. Math, 1978(31):231-246.

[74] BOWDLER H, MARTIN R S, REINSCH C, et al. The OR and OL algorithms for symmetric matrices [J]. Numer. Math, 1968(11):293-306.

[75] BRAKHAGE H. Áber die unmerische Behandlung von Integralglei-chungen nach der Quadraturformel-methode[J]. Numer. Math, 1960(2):183-196.

[76] BRAKHAGE H. Zur Fehlerabschätzung für die numerische Eigen-wertbestimmung bei Integralgleichungen[J]. Numer. Math, 1961(3):174-179.

[77] BRAMBLE J H, OSBORN J E. Rate of convergence estimates for nonselfadjoint eigenvalue approximations[J]. Math. Comp, 1973(27):525-549.

[78] BRAMBLE J H, SCHATZ A H. Rayleigh-Ritz-Galerkin methods for Dirichlet's proble using subspaces without boundary conditions[J]. Comm. Pure Appl. Math, 1970(23):653-675.

[79] BRANDT A. Multilevel adaptive solutions to boundary value problems [J]. Math. Comp, 1977(31):333-390.

[80] BREZINSKI C. Computation of the eigenelements of a matrix by the ε-algorithm[J]. Linear Algebra Appl, 1975(11):7-20.

[81] BREZZI F. On the existence, uniqueness and approximation of saddle-point problems arising from Lagrangian multipliers[J]. RAIOR Anal. Numér, 1974(2):129-151.

[82] BREZZI F. Sur la méthode des éléments finis hybrides pour le probléme biharmonique[J]. Numer. Math, 1975(24):103-131.

参考文献

[83] BROWDER F E. Approximation-solvability of nonlinear functional equations in normed linear spaces [J]. Arch. Rational Mech. Anal, 1967(26):33-42.

[84] BROWDER F E, PETRYSHYN W V. The topological degree and Galerkin approximations for noncompact operators in Banach spaces[J]. Bull. Amer. Math. Soc, 1968(74):641-646.

[85] BRUHN G, WENDLAND W L. Áber die näherungweise Lösung von linearen Funktionalgleichungen[M]. L. Collatz, G. Meinardus, and H. Unger, eds. In Funktionalanalysis Approxi-mationstheorie Numerrische Mathematik. Basel:Birkhaeuser, 1967.

[86] BRUNNER H. The application of the variation of constants formulas in the numerical analysis of integral and integro-differential equations[J]. Utilitas Mathematica, 1981(19):255-290.

[87] BUCKNER H. Die Praktische Behandlung[M]. Berlin and New York: BITshr cmf shrmmb Springer-Verlag, 1952.

[88] BULIRSCH R, STOER J. Asymptotic upper and lower bounds for results of extrapolation methods [J]. Numer. Math, 1966(8):93-101.

[89] BUNCH J R, NIELSEN C P. Rank-one modification of the symmetric eigenproblem[J]. Numer. Math, 1978(31):31-48.

[90] BUTSCHER W, KAMMER W E. Modification of Davidson's method for the calculation of eigenvalues and eigenvectors of large real symmetric matrices[J]: "Root-homing procedure." J. Comput. Phys, 1976(20):313-325.

[91] BUUREMA H J. A geometric proof of convergence for the QR method[M]. Ph. D. Thesis, Univ. of Groningen, 1970.

[92] CACHARD F. Etude numérique de réseaux de file d'attente[M]. Thèse Doct-Ing., Univ. de Grenoble, 1981.

[93] CANOSA J, GOMES DE OLIVEIRA R. A new method for the solution of the Schrödinger equation [J]. J. Comput. Phys, 1970(5):188-207.

[94] CANUTO C. Eigenvalue approximations by mixed-methods. RAIRO Anal[J]. Numéer, 1978(12): 27-50.

[95] CHAN S P, FELDMAN H, PARLETT B N. A program for computing the condition numbers of matrix eigenvalues without computing eigenvectors [J]. ACM Trans. Math. Software, 1977(3):186-203.

[96] CHAN T F, KELLER H B. Arc-length continuation and multigrid techniques for nonlinear elliptic eigenvalue problems [J]. SIAM J. Sci. Stat. Comp, 1982(3):173-194.

[97] CHANDLER G A. Superconvergence of numerical solutions of second kind integral equations [J].

参考文献

Ph. D. Thesis, Australia Natl. Univ., Canberra, 1979.

[98] CHANG P W, FINLAYSSON B A. Orthogonal collocation on finite elements for elliptic equations [J]. Math. Comput. Simulation, 1978(20):83-92.

[99] CHATELIN F. Méthodes d'approximation des valeurs propres d'opérateurs linéaires dans un espace de Banach [J]. I. Critère de stabilité. C. R. Hebd. Séances Acad. Sci. Ser. A, 1970a(271): 949-952.

[100] CHATELIN F. II Bornes d'eereur [J]. C. R. Hebd. Séances Acad. Sci. Ser. A, 1970b (271):1006-1009.

[101] CHATELIN F. Etude de la stabilité de méthodes d'approximation des éléments propres d'opérateurs linéaires [J]. C. R. Hebd. Séances Acad. Sci. Ser. A, 1971a(272):673-675.

[102] CHATELIN F. Etude de la continuité du spectre d'un opérateur linéaire [J]. C. R. Hebd. Séances Acad. Sci. Ser. A, 1972a(274):328-331.

[103] CHATELIN F. Etude de la continuitédu spectre d'un opérateur linéaire [J]. C. R. Hebd. Séances Acad. Sci. Ser. A, 1972a(274):328-331.

[104] CHATELIN F. Error bounds in QR and Jacobi al-

gorithms applied to hermitian or normal matrices [J]. Information Processing 71, Vol. 2, pp. 1254-1257. North-Holland Publ., Amsterdam, 1972b.

[105] CHATELIN F. Convergence of aapproximate methods to compute eigenelements of linear operators [J]. SIAM J. Numer. Anal, 1973 (10):939-948.

[106] CHATELIN F. La méthode de Galerkin. Ordre de convergence des éléments propres [J]. C. R. Hebd. Séances Acad. Sci. Ser. A, 1975(278): 1213-1215.

[107] CHATELIN F. Numerical computation of the eigenelements of linear intergral operators by iterations [J]. SIAM J. Numer. Anal. 1978 (15):112-1124.

[108] CHATELIN F. Sur les bornes d'erreur a posteriori pour les éléments propres d'opérateurs linéaires [J]. Numer. Math, 1979(32):233-246.

[109] CHATELIN F. The spectral approximation of linear operators with applications to the computation of eigenelements of differential and integral operators[J]. SIAM Rev, 1981(23):459-522.

[110] CHATELIN F. A posteriori bounds for the eigenvalues of matrices[M]. Computing (to appear), 1983.

[111] CHATELIN F, LEBBAR R. The iterated projec-

tion solution for the Fredholm integral equation of second kind[J]. J. Austral. Math. Soc. Ser. B, 1981(22):443-455(Special issue on integral equations).

[112] CHATELIN F, LEBBAR R. Superconvergence results for the iterated projection method applied to a second kind Fredholm integral equation and eigenvalue problem[J]. J. Integral Equations (to appear), 1983.

[113] CHATELIN F, LEMORDANT J. La méthode de Rayleigh-Ritz appliquée à des opérateurs différentiels elliptiques—ordres de convergence des éléments propres[J]. Numer. Math, 1975(23):215-222.

[114] CHATELIN F, LEMORDANT J. Error bounds in the approximation of eigenvalues of differential and integral operators[J]. J. Math. Anal. Appl. 1978(62):257-271.

[115] CHATELIN F, MIRANKER W L. Acceleration by appregaetion of successive approximation methods[J]. Linear Algebra Appl, 1982(43):17-47.

[116] CHATELIN F, MIANKER W L. Aggregation/disaggregation for eigenvalue problems[J]. SIAM J. Numer. Anal. (submitted), 1983.

[117] CHEN N F. The Rayleigh quotient iteration for non-normal matrices[M]. Ph. D. Thesis. Univ.

of California. Berkeley, 1975.
[118] CHENEY W. Intrduction to Approximation Theory[M]. New York: McGraw-Hill, 1966
[119] CHEUNG L M, BISHOP D M. The group-coordinate relaxation method for solving the generalized eigenvalue problem for large real symmetric matrices[M]. Comput. Phys. Commun, 1977(12): 247-250.
[120] CHRISTIANSEN S, HANSEN E B. Numerical solution of boundary value problems through integral equations [J]. Z. Angew. Math. Mech, 1978(58):T14-T15.
[121] CHRISTIANSEN J, RUSSEL R D. Error analysis for spline collocation methods with application to knot selection [J]. Math. Comp. 1978 (32): 415-419.
[122] CHU K W, SPENCE A. Defered correction for the integral equation eigenvalue problem[J]. J. Austral. Math. Soc. Ser. B, 1981 (22):478-490.
[123] CIARLET P B. The Finite Element Method for Elliptic Problems. [M] North-Holland Publ., Amsterdam, 1978.
[124] CIARLET P B. introduction à l'Analyse Numérique Matricielle et à l'Optimisation[M]. Masson, Paris, 1982.
[125] CIARLET P G, RAVIART P A. General La-

grange and Hermite interpolation in R^n with applications to finite element methods[J]. Arch. Rational. Mech. Anal. 1972(46):177-199.

[126] CIARLET P G, SCHULZT M H, VARGA R S. Numerical methods of high order accuracy for nonlinear boundary value problems[J]. III. Eigenvalue problems. Numer. Math, 1968(12):120-133.

[127] CLINE A K GOLUB G H, PLATZMAN G W. Calculation of normal modes of oceans using a Lanczos method[M]. In Sparse Matrix Computations (J. R. Bunch and D> J. Rose, eds.), New York: Academic Press, 1976:409-426.

[128] CLINE A K, MOLER C B, STEWART G W, et al. An estimate for the condition number of a matrix[J]. SIAM J. Numer. Anal, 1979(16):368-375.

[129] CLINT M, JENNINGS A. The evaluation of eigenvalues and eigenvectors of real symmetric matrices by simultaneous iterations[J]. Comput. J, 1970(13):76-80.

[130] CLINT M, JENNINGS A. A Simultaneous iteration method for the unsymmetric eigenvalue problem[J]. J. Inst. Math. Appl, 1971,8:111-121.

[131] CLINT M, JENNINGS A. A simultaneous iteration method for the unsymmetric eigenvalue problem[J]. J. Inst. Math. Appl. 1971,8:111-

Tschebyscheff 逼近定理

121.

[132] CODDINGTON E A, LEINSON N. Theory of Ordinary Differential Equations [M]. New York: McGraw-Hill, 1955.

[133] COLLAT L. Konvergenzbeweis und Fehlerabschätzung für das Differenzenverfahren bei Eigenwertproblemen gewöhnlicher Differen tiglgleichungen zweiter und vierte Ordnung [J]. Deutsche Math, 1937(2):189-215.

[134] COLLATZ L. The Numerical Treatment of Differential Equations, 3rd ed [M]. Berlin and New York: Springer-Verlag, 1966a.

[135] COLLATZ L. Functional Analysis and Numerical Mathematics [M]. New York: Academic Press, 1966b.

[136] COOPE J A R, SABO D W. A new approach to the determination of several eigenvectors of a large Hermitian matrix [J]. J. Comput. Phys. 1977 (23):404-424.

[137] CORR R B, JENNINGS A. Implementation of simultaneous iteration for vibration analysis [J]. Comput. & Structures, 1973(3):497-507.

[138] CORR R B, JENNINGS A. A simultaneous iteration algorithm for symmetric eigenvalue problems [J]. Internat. J. Numer. Methods Engrg, 1976 (10):647-663.

[139] COURANT R, HIBERT D. Methods of Mathemat-

ical Physics[M]. Vols. 1 and 2. New York: Wiley (Interscience), 1953.

[140] CRANDALL S H. Iterative procedures related to relaxation methods for eigenvalue problems[J]. Proc. Roy. Soc. London Ser. A, 1951(207): 416-423.

[141] CRUICKSHANK D M, WRIGHT K. Computable error bounds for polynomial collocation methods [J]. SIAM J. Numer. Anal, 1978(15): 134-151.

[142] CUBILLOS P O. On the numerical solution of Fredholm integral equations of the second kind [M]. Ph. D. Thesis, Univ. of Iowa, 1980.

[143] CULLUM J. The simultaneous computation of a few of the algebraically largest and smallest eigenvalues f a large, symmetric, sparse matrix[J]. BIT, 1978(18): 265-275.

[144] CULLUM J, DONATH W E. A block Lanczos algorithm for computing the q algebraically larges eigenvalues and a corresponding eigenspace for large, sparse symmetric matrices [J]. Proc. IEEE Conf. Decision Contr., Phoenix. Ariz, 1974: 505-509.

[145] CULLUM J, WILLOUGHBY R. The equivalence of the Lanczos and the conjugate gradient algorithms [M]. Tech. Rep. Rc 6903, IBM Research Center. Yorktown Heights, 1977

Tschebyscheff 逼近定理

[146] CULLUM J, WILLOUGHBY R. The Lanczos tridiagonalization and the conjugate gradient with local ε-orthogonality of the Lanczos vectors [M]. Tech. Rep. RC 7152, IBM Research Center, Yorktown Heights, 1978.

[147] CULLUM J, WILLOUGHBY R A. Fast modal analysis of large, sparse but unstructured symmetric matrices [J]. Proc. IEEE Conf. Decision Contr., San Diego, Calif., 1979a:45-53.

[148] CULLUM J, WILLOUGHBY R A. Lanczos and the computation in specifiend intervals of the spectrum of large, sparse real symmetric matrices [M]. In Sparse Matrix Proceedings 1978 (I. S. Duff and G. W. Stewart, eds.), pp. 220-225. SIAM, Philadelphia, Pennsylvania, 1979b.

[149] CULLM J, WILLOUGHBY R A. The Lanczos phenomenon—an interpretation based upon conjugate gradient optimization [J]. Linear Algebra Appl, 1980a(29):63-90.

[150] CULLUM J, WILLOUGHBY R A. Computing eigenvectors (and eigenvalues) of large, symmetric matrices using Lanczos tridiagonali-zation [J]. Proc. Numerical Analysis conf. (G. A. Watson, ed.), Lecture Notes in Mathematics. Vol. 773, pp. 46-63. Berlin and New York: Springer-Verlag, 1980b.

[151] DAHLQUIST G, BJÖRCK A. Numerical Methods

[M]. Prentice-Hall, Englewood Cliffs, New Jersey, 1974.

[152] Dahmen W. On multivariate B-splines[J]. SIAM J. Numer. Anal. 1980(17):179-191.

[153] D'ALMEIDA F. Etude numérique de la stabilité dynamique des modèles macroéconomiques—Logiciel pour MODULECO[M]. Thèse 3ème Cycle, Univ. de Grenoble, 1980.

[154] DANIEL J W, GRAGG W B, KAUFMAN L, et al. Reorthogonalization and stable algorithms for updating the Gram-Schmidt QR factorization[J]. Math. Comp, 1976(30):772-795.

[155] DAVIDOSN E R. The iterative calculation of a few of the lowest eigenvalues and corresponding eigenvectors of large real symmetric matrices[J]. J. Comput. Phys, 1975(17):87-94.

[156] DAVIS C. The rotation of eigenvectors by a perturbation[J]. I. J. Math. Anal. Appl, 1963(6):159-173

[157] DAVIS C. The rotation of eigenvectors by a perturbation[J] II. J. Math. Anal. Appl, 1965(11):10-27.

[158] DAVIS C, KAHAN W. The rotation of eigenvectors by a perturbation[J]. III. SIAM J. Numer. Anal, 1968(7):1-46.

[159] DAVIS C, KAHAN W, WEINBERGER H. Norm preserving dilations and their applications to Opti-

mal error bounds. SIAM J[J]. Numer. Anal., 1982(19):445-469.

[160] DAVIS G J, MOLER C B. Sensitivity of matrix eigenvalues[J]. Internat. J. Numer. Methods Engrg., 1978(12):1367-1373.

[161] DAVIS P J, RABINOWITZ P. Methods of Numerical Integration [M]. New York: Academic Press, 1974.

[162] DAY W B. More bounds for eigenvalues[J]. J. Math. Anal. Appl, 1974(46):523-532.

[163] DEAN P. The spectral distribution of a Jacobian matrix[J]. Proc. Cambridge Phil. Soc., 1956(52):752-755.

[164] DEAN P. Vibratioanl spectra of diatomic chains [J]. Proc. Roy. Soc. Ser. A, 1960(254):507-521.

[165] DEAN P. Vibrations of glass-like disordered chains[J]. Proc. Phys. Soc., 1964(84):727-744.

[166] DEAN P. The constrained quantum mechanical harmonic oscillator[J]. Proc. Phys. Soc., 1966(62):277-286.

[167] DEAN P. Atomic vibrations in solids [J]. J. Inst. Math. Appl., 1967(3):98-165.

[168] DEAN P. The vibrational probperties of disordered systems: numerical studies[J]. Rev. Modern Phys., 1972(44):127-168.

[169] DE BOOR C. On uniform approximation by splines[J]. J. Approx. Theory, 1968(1):219-235.

[170] DE BOOR C. On calculation with B-splines[J]. J. Approx. Theory 1972(6):50-62.

[171] DE BOOR C. A bound on the L_∞-norm of $L2$-approximation by splines in terms of a global mesh ratio[J]. Math. Comp., 1976(30):765-771.

[172] DO BOOR C, RICE J R. An adaptive algorithm for multivariate approximation giving optimal convergence rates[J]. J. Approx. Theory, 1979(25):337-39.

[173] DE BOOR C, SWARTZ B. Collocation at Gaussian poins[J]. SIAM J. Numer. Anal., 1973(10):582-606.

[174] DE BOOR C, SWARTZ B. Comments on the comparixon of global methods for linear two-point boundary value problems[J]. Math. Comp. 1977(31):916-921.

[175] DE BOOR C, SWARTZ B. Collocation approximation to eigen-values of an ordinary differential equation: The principle of the thing[J]. Math. Comp. 1980(35):679-694.

[176] DE BOOR C, SWARTZ B. Collocation approximation to eigenvalues of an ordinary differential equation: numerical illustrations[J]. Math. Comp., 1981a(36):1-19.

[177] DE BOOR C, SWARTZ B. Local piecewise polynomial projection methods for an ode which give high order convergence at knots [J]. Math. Comp., 1981b(36):21-33.

[178] DEHESA J S. The asymptotic eigenvalue density of rational Jacobi matrices[J]. I. J. Phys. A, 1978(9):223-226.

[179] DEHESA J S. The eigenvalue density of rational Jacobi matrices [J]. II. Linear Algebra Appl, 1980(33):41-55.

[180] DE HOOG F R, WEISS R. Asymptotic expansions for product integration[J]. Math. Comp., 1973(27):295-306.

[181] DELVES L M, ABD-ELAL L F. The fast Galerkin algorithm for the solution of linear Fredholm equations, algorithm 97[J]. Comput. J., 1977 (20):374-376.

[182] DELVES L M, WALSH J, EDS. Numerical Solutin of Integral Equations[M]. Oxford Univ. Press (Clarendon), London and New York, 1974.

[183] DELVES L M, ABD-ELAL L F, HENDRY J A. A fast Galerkin algorithm for singular kernel equations[J]. J. Inst. Math. Appl., 1979(23): 139-166.

[184] DE PREE J D, HIGGINS J A. Collectively compact sets of linear operators [J]. Math. Zeitschrift, 1970(115):366-370.

[185] DE PREE J D, KLEIN H S. Characterization of collectively compact sets of linear operators[J]. Pacif. J. Math., 1974(55):45-54.

[186] DESCLOUX J. Error bounds for an isolated eigenvalue obtained by the Galerkin method[J]. J. Appl. Math. Phys., 1979(30):167-176.

[187] DESCLOUX J. Essential numerical range of an operator with respect to a coercive form and the approximation of its spectrum by the Galerkin method[J]. SIAM J. Numer. Anal. 1981(18): 1128-1133.

[188] DESCLOUX J, GEYMONAT G. On the essential spectrum of an operator relative to the stability of a plasma in toroidal geometry[M]. Rep. Math. Dept., Ecole Polytechn. Féd. de Lausanne, 1979.

[189] DESCLOUX J, NASSIF N R. Stability analysis with error estimates for the approximation of the spectrum of self-adjoint operators on unbounded domains by finite element and finite difference methods [M]. Application to Schrödinger's equatkon. Rep. Math. Dept., Ecole Polytechn. Fed. de Lausanne, 1982.

[190] DESCLOUX J, TOLLEY M D. Approximation of the Poisson problem and of the eigenvalue problem for the Laplace operator by the method of the large singular finite elements[M]. Res. Rep.

No. 81-01, Angew. Math., Eidg. Techn. Hochschule Zürich, 1981.

[191] DESCLOUX J, NASSIF N, RAPPAZ J. Various results on spectral approximation [M]. Rep. Math. Dept., Ecole Polytechn. Féd. de Lausanne, 1977.

[192] DESCLOUX J, NASSIF N, RAPPAZ J. On spectral approxi-mation. Part 1 [J]. The problem of convergence. RAIRO Anal. Numér. 1978a(12): 97-112.

[193] DESCLOUX J, NASSIF N, RAPPAZ J. Part 2, Error estimates for the Galerkin method, RAIRO Anal[J]. Numér, 1978b(12):113-119.

[194] DESCLOUX J, LUSKIN M, RAPPAZ J. Approximation of the spectrum of closed operators—The determination of normal modes of a rotating basin [J]. Math. Comp., 1981(36):137-154.

[195] DIAZ J B, METCALF F T. A functional equation for the Ray leigh quotient for eigenvalues, and some applications [J]. J. Math. Mech., 1968 (17):623-630.

[196] DIETRICH G. On the efficient and accurate solution of the skew-symmetric elgenvalue problem. An arrangement of new and alrady known algorithmic formulations [J]. Comput. Methods Appl. Mech. Engrg., 1978(14):209-235.

[197] DIEUDONNÉ J. Foundations of Modern Analysis

[M]. New York: Academic Press, 1960.

[198] DOMB C, MARADUDIN A A, MONTROLL E W, et al. Vibration frequency of spectra of disordered lattices[J]. I. Moments of the spectra for disordered linear chains. Phys. Rev. 1959a (115):18-24; II. Spectra of disordered one-dimensional lattices. Phys. Rev. , 1959a(115): 24-34.

[199] DOMB C, MARADUDIN A A, MONTROLL E W, et al. The vibration spectra of disordered lattices[J]. J. Phys. Chem. Solids, 1959b(3): 419-422.

[200] DONGARRA J J, MOLER C B, BUNCH J R, et al. LINPACK User's Guide[M]. SIAM, Philadelphia, Pennsylvania, 1979.

[201] DONGARRA J J, MOLER C B, WILKINSON J H. Improving the accuracy of computed eigenvalues and eigenvectors[M]. Tech. Rep. ANL 81-43, Argonne Nat. Lab. , Illionois, 1981.

[202] DOUGLAS J, DUPONT T. Superconvergence for Galerkin methods for the two point boundary problem via local projections [J]. Numer. Math. , 1973(21):270-278.

[203] DOUGLAS J, DUPONT T. Galerkin approximations for the two point boundary problem using continuous piecewise polynomial spaces[J]. Numer. Math. , 1974(22):99-109.

[204] DOUGLAS J, DUPONT T, WHEELER M F. An L^∞-estimate and a superconvergence result for a Galerkin method for elliptic equations based on tensor products of piecewise polynomials [J]. RAIRO Anal. Numér., 1974(2):61-66.

[205] DOWSON H R. Spectral Theory of Linear Operators[M]. New York: Academic Press, 1975.

[206] DOWSON H R. Spectral Theory of Linear Operators[M]. New York: Academic Press, 1978.

[207] DUFF J S. A survey of sparse matrix research [J]. Proc. IEEE, 1977(65):500-535.

[208] DUFF I S, ED. Conjugate Gradient Methods and Similar Techniques [M]. Tech. Rep. R-9636, AERE Harwell, 1979.

[209] DUFF I S. Recent developments in the solution of large sparse linear equations [J]. In Computing Methods in Appled Sciences and Engineering (R. Glowinski and J. L. Lions, eds.), North Hoiland Publ., Amsterdam, 1980:407-426.

[210] DUFF I S. A sparse future. In Sparse Matrices and Their Uses [M]. (I. S. Duff, ed.). New York: Academic Press, 1981.

[211] DUFF I S. A survey of sparse matrix software [M]. Report CSS 21, AERE Harwell. To appear in Sources and Development of Mathematical Software (W. R. Cowell, ed.). Prentice-Hall, Englewood Cliffs, New Jersey, 1982.

参考文献

[212] DUFF I S, REID J K. On the reduction of sparse matrices to condensed forms by similarity transformations[J]. J. Inst. Math. Appl., 1975(15): 217-214.

[213] DUFF I S, REID J K. Performance evaluation of codes for sparse matrix problems [M]. In Performance Evaluation of Numerical Software (L. D. Fosdick, ed.), pp. 121-135. North-Holland Publ., Amsterdam, 1979.

[214] DUMONT-LEPAGE M C, GANI N, GAZEAU J P, et al. Spectrum of potentials $gr^{-(S+2)}$ via SL (2,R) acting on quaternions[J]. J. Phys. A, 1980(13):1243-1257.

[215] DUNFORD N, SCHWARTZ J T. Linear Operators. Part I: General Theory [M]. New York: Wiley (Interscience), 1958.

[216] DUNFORD N, SCHWARTZ J T. Linear Operators. Part II: Spectral Theory, Selfadjoint Operators in Hilbert Spaces [M]. New York: Wiley (Interscience), 1963.

[217] DUPONT T. A unified theory of superconvergence for Galerkin methods for two-point boundary problems[J]. SIAM J. Numer. Anal., 1976(13): 362-368.

[218] EDWARDS J T, LICCIARDELLO D C, THOULESS D J. Use of the Lanczos method for finding complete sets of eigenvalues of large

sparse matrices[J]. J. Inst. Math. Appl. 1979 (23):277-283.

[219] EGGERMONT P P. Collocation as a projection method and superconvergence for Volterra integral equations of the first kind [M]. Rep. Math. Dept., Univ. of Delaware, 1982a.

[220] EGGERMONT P P. Collocation for Volterra integral equations of the first kind with iterated kernel [J]. Rep. Math. Dept., Univ. of Delaware, 1982b.

[221] EINARSSON B. Bibliography on the evaluation of numerical software [J]. J. Commput. Appl. math. 1979(5):145-159.

[222] ELMAN H. Iterative methods for large, sparse nonsymmetric systems of linear equations [M]. Res. Rep. No. 229. COmputer Science Dept., Yale Univ., Connecticut, 1982.

[223] Erdelyi I. An iteratie least square algorithm suitable for computing partial eigensystems[J]. SIAM J. Numer. Anal. 1965(2):421-436.

[224] ERDÖS P, FELDHEIM E. Sur le mode de convergence de l'interpolation de Lagrance[J]. C. R. Hebd. Séances Acad. Sci., 1936 (203): 913-915.

[225] EVEQUOZ, H. Approximation spectrale liée à l' étude de la stabilité magnétohydrodynamique d' un plasma par une méthode d'élémetnts finis non

参考文献

conformes[M]. Thèse Math. Dept., Ecole Polytechn. Péd. de Lausanne, 1980.

[226] EVEQUOZ H, JACCARD Y. A nonconforming finite element method to compute the spectrum of an operator relative to the stability of a plasma in toroidal geometry [J]. Numer. Math., 1981 (36):455-465.

[227] FADDEEV D K, FADDEEVA V N. Computational Methods of Linear Algebra. Freeman[M]. San Francisco, California, 1963.

[228] FAIRWEATHER G. Finite Element Galerkin Methods for Differential Equations [M]. New York: Dekker, 1978.

[229] FAN K. On a theorem of Weyl concerning eigenvalues of linear transformations[J]. Proc. Nat. Acad. Sci. USA, 1949(35):652-655.

[230] FELER M G. Calculation of eigenvectors of large matrices[J]. J. Comput. Phys., 1974 (14): 341-349.

[231] FENNÊR T I, LOIZOU G. Some new bounds on the condition numbers of optimally scaled matrices [J]. J. Assoc. Comput. Mach., 1974 (21): 514-524.

[232] FICHERA G. Numerical and Ouantitative Analysis[M]. Pitman, London, 1978.

[233] FIEDLER M, PTÁK V. Estimates and iteration procedures for proper values of almost decompos-

able matrices[J]. Czechoslovak Math. J. , 1964 (39):593-608.

[234] FINLAYSSON B A. The Method of Weighted Residuals[M]. New York: Academic Press, 1972.

[235] FIX G J. Effects of quadrature errors in finite element approximation of steady state, eigenvalue and parabolic problems[M]. In The Mathematical Foundations of the Finite Element Method with Applications to Partial Differential Equations (A. K. Aziz, ed.), New York: Academic Press: 1972:525-556.

[236] FIX G J. Eigenvalue approximation by the finite element method[J]. Adv. in Math., 1973(10): 300-316.

[237] FIX G J. Hybrid finite element methods[J]. SIAM Rev, 1976(18):460-484.

[238] FIX G J, HEIBERGER R, An algorithm for the ill-conditioned generalized eigenvalue problem [J]. SIAM J. Nummer. Anal., 1972(9):78-88.

[239] FORSYTHE G E, HENRICI P. The cyclic Jacobi method for computing the principal values of a complex matrix[J]. Trans. Amer. Math. Soc., 1960(94):1-23.

[240] FORSYTHE G E, WASOW W. Finite Difference Methods for Partial Differential Equations [M]. New York: Wiley (Interscience), 1960.

[241] FOX L, GOODWIN E T. The numerical solution of nonsingular linear integral equations[J]. Philos. Trans. Roy. Soc. London, 1953(245):501-534.

[242] FRANCIS J G F. The QR transformation: a unitary analogue to the LR transformation. Parts I and II[J]. Comput. J. 1961-1962 (4):265-271, 332-345.

[243] FRANK R, VEBERHUBER C W. Iterated defect correction for differential equations[J]. Part I: Theoretical results. Computing, 1978(20):207-228.

[244] FREDHOLM I. Sur une nouvelle méthode pour la résolution du problème de Dirichlet[J]. Kung. Vet. Akad. Förh. Stockholm, 1900:39-46.

[245] FREDHOLM I. Sur une classe d'équations fonctionnelles[J]. Acta Math, 1903(27):365-390.

[246] FROMME J, GOLBERG M. Unsteady two dimensional airloads acting on oscillating thin airfoils in subsonic ventilated wind tunnels [M]. Rep. NASA Contract, Univ. of Nevada, Las Vegas, 1978.

[247] GANTMACHER F R. The Theory of Matrices [M]. New York: Chelsea, 1959.

[248] GARABEDIAN P R. Partial Differential Equations[M]. New York: Wiley, 1967.

[249] GEIER E. Eigenvalue and eigenvector calculation

Tschebyscheff 逼近定理

by simultaneous vector iteration (in German) [J]. Z. Angew. Math. Mech. 1977(57):T279- T281.

[250] GEKELER E. On the eigenvectors of a finite-difference approximation to the Sturm-Liouville eigenvalue problem [J]. Math. Comp., 1974 (28):973-979.

[251] GEORG K. On the convergence of an inverse iteration method for nonlinear elliptic eigenvalue problems[J]. Numer. Math., 1979(32):69-74.

[252] GERADIN M. Error bounds for eigenvalue analysis by elimination of variables[J]. J. Sound Vibr., 1971(19):111-132.

[253] GERADIN M. On the Lanczos method for solving large structural eigenvalue problems[J]. Z. Angew. Math. Mech., 1979(59):T127-T129.

[254] GERSCHGORIN S. On bounding the eigenvalues of a matrix (in German)[J]. Izv. Akad. Nauk SSSR Ser. Mat., 1931(1):749-754.

[255] GHEMIRES T. Utilisation du quotient de Rayleigh dans la méthode aux différences finies[M]. Thèse 3ème Cycle, Univ. de Grenoble, 1979.

[256] GLAZMAN I, LIUBITCH Y. Analyse Linéeire dans les Espaces de Dimensions Finies[M]. Mir, Moscow, 1972.

[257] GODUNOV S K, PROPKOPOV G P. A method of minimal iterations for evaluating the eigenvalues

of an elliptic operator (in Russian)[J]. Z. Vycisl. Mat. i Mat. Fiz., 1970(10):1180-1190.

[258] GODUNOV S K, RYABENKI V S. Theory of Difference Schemes[M]. North-Holland Publ., Amsterdam, 1964.

[259] GOHBERG I C, KREIN M G. The basic propositions on defect numbers root numbers and indices of linear operators [J]. Amer. Math. Soc. Transl., 1960(13):185-264.

[260] GOLDBERG S. Unbounded Linear Operators: Theory and Applications [M]. New York: McGraw-Hill, 1966.

[261] GOLDBERG S. Perturbations of semi-Fredholm operatiors by operators converging to zero compactly[J]. Proc. Amer. Math. Soc. 1974(45): 93-98.

[262] GOLUB G H, Plemmons R J. Large scale geodetic least squares adjustment by dissection and orthogonal decomposition[J]. Linear Algebra Appl., 1980(34):3-37.

[263] GOLUB G H, WILKINSON J H. Ill-conditioned eigensystems and the computation of the Jordan canonical form[J]. SIAM Rev. 1976(18):578-619.

[264] GOLUB G H, NASH S, VAN LOAN C. A Hessenberg-Schur method for the problem $AX+XB=C$ [J]. IEEE Trans. Automat. Control AC, 1979

(24):909-913.

[265] GOOS G, HARTMANIS J, EDS. EISPACK-Matrix Eigensystem Routines-Guide Extension [J]. Lecture Notes in Computer Science, Vol. 51. Springer-Verlag, Berlin and New York, 1978.

[266] GORDON R G. New method for constructing wave-functions for bound states and scattering [J]. J. Chem. Phys., 1969(51):14-25.

[267] GORDON R G. Quantum scattering using piecewise analytic solutions [M]. In Methods in COmputational Physics (B. Adler, S. Fernbach, and M. Rottenberg, eds.), Vol 10, pp. 81-109. New York: Academic Press, 1971.

[268] GOSE G. The Jacobi method for $Ax = \lambda Bx$ [J]. Z. Angew. Math. Mech., 1979(59):93-101.

[269] GOWER J C. A modified Leverrier-Faddeev algorithm for matrices with multiple eigenvalues [J]. Linear Algebra Appl., 1980(31):61-70.

[270] GRAHAM I G. The numerical solution of Fredholm integral equations of the second kind [M]. Ph. D. Thesis, Univ. of New South Wales, Kensington, 1980.

[271] GRAHAM I G. Collocation methods for two dimensional weakly singular integral equations [J]. J. Austral. Math. Soc. Ser. B, 1981(22):460-477.

[272] GRAHAM I G. Galerkin methods for second-kind

integral equations with singularities [J]. Math. Comp., 1982(39):519-533.

[273] GRAHAM I G, SLOAN I H. Onthe compactness of certain integral operators[J]. J. Math. Anal. Appl. 1979(68):580-594.

[274] GRÉGOIRE J P, NEDELEC J C, PLANCHARD J. A method of finding the eigenvalues and eigenfunctions of selfadjoint elliptic operators[J]. J. Comput. Methods appl. Mech. Engrg., 1976 (8):201-214.

[275] GREGORY R T, TARNCY D L. A Collection of Matrices for Testing Computational Algorithms [M]. New York: Wiley (Intercience), 1969.

[276] GRIFFITHS D F, LORENZ J. An analysis of the Petrov-Galerkin finite element method [J]. J. Comput. Methods Appl. Mech. Engrg., 1978 (14):65-92.

[277] GRIGORIEFF R D. Die Konvergenz des Rand- und Eigenwertproblems linearer gewöhnlicher Differenzengleichungen[J]. Numer. Math. 1970a (15):15-48.

[278] GRIGORIEFF R D. Áber die Koerzitivität gewöhnlicher Differenze-noperatoren und die Konvergenz des Mehrschrittverfahren [J]. Numer. Math., 1970b(15):196-218.

[279] GRIGORIEFF R D. Áber die Fredholm-Alternative bei linearen approximationsregulären Opera-

toren [J]. Applicable Anal. , 1972 (2) : 217-227.

[280] GRIGORIEFF R D. Zur Theorie linearer approximationsregulärer Operatoren I, II [J]. Math. Nachr. , 1973(55) :233-249,250-363.

[281] GRIGORIEFF R D. Áber diskete Approximation nichtlinearer Gleichungen I [J]. Art. Math. Nachr. 1975a(69) :253-272.

[282] GRIGORIEFF R D. Diskrete Approximation von Eigenwertproblemen. I. Qualitative Konvergenz [J]. Numer. Math. , 1975b(24) :355-374.

[283] GRIGORIEFF R D. Diskrete Approximation von Eigenwertroblemen. II. Konvergenzordnung [J]. Numer. Math. , 1975c(24) :415-433.

[284] GRIGORIEFF R D. Diskrete Approximation von Eigenwertproblemen. III. Asymptotische Entwicklung [J]. Numer. Math. , 1975d(25) :79-97.

[285] GRIGORIEFF R D. Jeggle H. Approximation von Eigenwert-problemen bei nichtlinearer Parameterabhängigkeit [J]. Manuscripta Math. , 1973(10) :245-271.

[286] GRUBER R. HYMMIA—band matrix package for solving eigenvalue problems [J]. Comput. Phys. Comm. , 1975(10) : 30-41.

[287] GRUBER R. Finite hybrid elements to compute the ideal megneto-hydrodynamic spectrum of an

axiasymmetric plasma [J]. J. Comput Phys., 1978(26):378-388.

[288] GUPTA K K. Eigenproblem solution by a combined Sturm sequence and inverse iteration technique. Internat [J]. J. Numer. Methods Engrg., 1973(7):17-42.

[289] GUPTA K K. Eigenproblem soution of damped structural systems. Internat [J]. J. Numer. Methods Engrg., 1974(8):877-911.

[290] GUPTA K K. On a finite dynamic element method for free vibration analysis of structures[J]. Comput. Method Appl. Mech. Engrg., 1976a(9):105-120.

[291] GUPTA K K. On a numerical solutin of the supersonic panel flutter eigenproblem Internat[J]. J. Numer. Methods Engrg., 1976b(10):637-645.

[292] GUPTA K K. On a numerical solution of the pastic buckling problem of structures[J]. Internat. J. Numer. Methods Engrg., 1978a(12):941-947.

[293] GUPTA K K. Development of a finite dynamic element for free vibration analysis of two-dimensional structures[J]. Internat. J. Numer. Methods Engrg., 1978b(12):1311-1327.

[294] GÁSSMAN, B. L_∞-bounds of L_2-projections on splines [M]. In Quantitativve Approoximation (R. A. de Vore and K. Scherer, eds.), New

York: Academic Press, 1980:153-162.

[295] HACKBUSCH W. On the computation of approximate eigenvalue and eigenfunctions of elliptic operators by means of a multigrid method[J]. SIAM J. Numer. Anal., 1979(16):201-215.

[296] HACKBUSCH W. Multigrid solutions to linear and nonlinear eigenvalue problems for integral and differential equations [M]. Rep. 80-3, Math. Inst., Univ. zu Köln, 1980.

[297] HACKBUSCH W. On the convergence of multigrid iterations[J]. Beitr. Numer. Math., 1981a (9):213-239.

[298] HACKBUSCH W. Error analysis of the nonlinear multigird method of the second kind[J]. Appl. Matem., 1981b(26):18-29.

[299] HACKSHUSCH W, HOFMANN G. Results of the eigenvalue problem for the plate equation[J]. Z. Angew. Math. Phys., 1980(31):730-739.

[300] HADAMARD J. Mémoire sur le problème d'analyse relatif à l'équilibre des plaques élastiques encastrées [J]. Memoires savants étrangers. Acad. Sci. Paris, 1908(33):1-28.

[301] HMMERLIN G. Zur numerischen Behandlung von homogenen Fredholmschen Integralgelichungen 2 [M]. Art mit Splines. In Spline Functions Karlsruhe 1975 (K. Böhmer, G. Meinardus, and W. Schempp, eds.) Lecture Notes in Mathemat-

ics, Vol. 501, pp. 92-98. Springer-Verlag, Berlin and New York, 1976.

[302] HMMERLIN G, SCHUMAKER L L. Error bounds for the approximation of Green's kernels by splines[J]. Numer. Math., 1979(33):17-22.

[303] H? MMERLIN G, Schumaker L L. Procedures for kernel approximation and solutin of Fredholm integral equations of the second kind. Numer. Math., 1980(34):125-141.

[304] HAHN H. Reelle Funktionen[M]. Punktfunktionen. New York: Chelsea, 1948.

[305] HAIRER E. On the order of iterated defect correction: An algebraic proof[J]. Nemer. Math., 1978(29):409-443.

[306] HALL G G. On the eigenvalues of molecular graphs[J]. Molecular Phys. 1977(33):551-557.

[307] HALMOS P H. Normal dilations and extensions of operators[J]. Summa Brasil. Math., 1950(2):125-134.

[308] HANSON R J. Integral equations of immunology[J]. Comm. ACM 15, 883-890. Harwell Subroutine Library Manual. Harwell. Oxfordshire., 1972.

[309] HASHIMOTO M. A method of solving large matrix equations reduced from Fredholm equations of

the second kind[J]. J. Assoc. Comput. Mach. , 1970(17):629-636.

[310] HEMKER P W. Galerkin's method and Lobatto points[M]. Rep. NW 24/75, Stichting Math. Cent. , Amsterdam, 1975.

[311] HEMKER P W. The defect correction principle [M]. BAIL II short course lecture notes, 1982a.

[312] HEMKER P W. Extensions of the defect correction principle[M]. BAIL II short couorse lecture notes. , 1982b.

[313] HEMKER P W, SCHIPPERS H. Multigrid methods for the solution of Fredholm equations of the second kind[J]. Math. Comp. , 1981(36):215-232.

[314] HENDERSHOTT M C. Ocean tides[J]. Trans. Amer. Geophys. Union, 1973(54):76-86.

[315] HENRICI P. Bounds for iterates, inverses, spectral variation and fields of values of non-normal matrices[J]. Numer. Math. , 1962(4):24-40.

[316] HENRICI P. Bounds for eigenvalues of certain tridiagonal matrices[J]. SIAM J. Appl. Math. , 1963(11):289-290;(12):497.

[317] HERBOLD R J, SCHULTZ M H, VARGA R S. The effect of quadrature errors in the numerical solution of boundary value problems by variational techniques[J]. Aequationes Math. , 1969(3):247-270.

[318] HERMAN H. Extension of Lanczos' method of fundamental eigenvalue approximation [J]. Trans. ASME Ser. E. J. Appl. Mech., 1975 (42):484-489.

[319] HESTENES M R, KARUSH W. A method of gradients for the calculation of the characteristic roots and vectors of a real symmetric matrix [J]. J. Res. Nat. Bur. Standards Sect. B., 1951(47): 471-478.

[320] HESTENES M R, STEIN M L. The solution of linear equations by minimization J[J]. Optim. Theory Appl., 1973(11):335-359.

[321] HESTENES M R, STIEFEL E. Method of conjugate gradients for solving linear systems[J]. J. Res. Nat. Bur. Standards Sect. B., 1952(49): 409-436.

[322] HIRAI I, YOSHIMURA T, TAKAMURA K. On a direct eigenvalue analysis for locally modified structures[J]. Internat. J. Numer. Methods Engrg., 1973(6):441-456.

[323] HODGE D H. A theoretical technique for analyzing aeroelastic stability of bearingless rotors[J]. AIAA J., 1979a(17):400-407.

[324] HODGES D H. Vibration and responses of nonuniform rotating beams with discontinuities [J]. AIAA-ASME Dyn. Mater. Conf. St. Louis. Missouri, 1979b:29-38.

[325] HODGE D H. Aeromechanical stability analysis for bearingless rotor helicopters [J]. I. Amer. Helicopter Soc. , 1979c(24) :2-9.

[326] HODGES D H. Torsion of pretwisted beams due to axial loading [J]. J. Appl. Mech. , 1980 (47) :393-397.

[327] HOUSEHOLDER A S. The Theory of Matrices in Numerical Analysis[M]. Ginn (Blaisdell), Boston, Massachusetts, 1964.

[328] HOUSEHOLDER A S, BAUER F L. On certain iterative methods for solving linear systems [J]. Numer. Math. , 1960(2) :55-59.

[329] HOUSTIS E N, PAPATHEODOROU T S. A collocation method for Fredholm integral equation of the 2nd kind [J]. Math. Comp. , 1978 (32) : 159-173.

[330] HOUSTIS E N, LYNCH R E, RICE J R, et al. Evaluation of numerical methods for elliptic partial differential equations [J]. J. Comput. Phys. , 1978(27) :323-350.

[331] HOWSON W P, WILLIAMS F W. Natural frequencies of frames with axially loaced Timoshenko members[J]. J. Sound Vibration, 1973 (26): 503-515.

[332] HSIAO G C, WENDLAND W L. A finite element method for some integral equations of the first kind[J]. J. Math. Anal. Appl. , 1977 (58) :

449-481.

[333] HSIAO G C, WENDLAND W L. The Aubin-Nistche lemma for integral equations[J]. J. Integral Equation, 1981(3):299-315.

[334] HUANG L. Some perturbation problems for generalized eigenvalues (in Chinese) [J]. Beijing Daxue Xuebao, 1978(4):20-25.

[335] HUBBARD B E. Bounds for eigenvalues of the Sturm-Liouville problem by finite difference methods[J]. Arch. Rational Mech. Anal., 1962(28):171-179.

[336] HUGHES T J R. Reduction sheme for some structural eigenvalue problems by a variational theorem [J]. Internat. J. Numer. Methods Engrg., 1976(10):8453852.

[337] HUSCYIN K. Standard forms of eigenvalue problems asscoiated with gyroscopic systems[J]. J. Sound Vibration, 1976(45):29-37.

[338] HUSEYIN K, ROORDA J. The loading-frequency relationship in multiple eigenvalue problems[J]. Trans. ASME Ser. E. J. Appl. Mech., 1971(38):1007-1011.

[339] IKEBE Y. The Galerkin method for the numerical solution of Fredholm integral equations of the second kind[J]. SIAM Rev. 1972(14):465-491.

[340] IMSL Library 3 Reference Manual. Internal[M]. Math. Statist. Libraries, Houston, Texas, 1975.

Tschebyscheff 逼近定理

[341] IRONS B M. Eigenvalue economizers in vibration problems[J]. J. Roy. Aero. Soc., 1963(67): 526-528.

[342] ISAACSON E, KELLER H. Analysis of Numerical Methods[M]. New York: Wiley, 1966.

[343] ISHIHARA K. Convergence of the finite element method applied to the eigenvalue problem $\Delta u + \lambda u = 0$[J]. Publ. Res. Inst. Math. Sci., 1977(13):47-60.

[344] ISHIHARA K. A mixed finite element method for the biharmonic eigenvalue problems of plate bending[J]. Publ. Res. Inst. Math. Sci., 1978(14):399-414.

[345] ISHIHARA K. On the mixed finite element approximation for the buckling of plates[J]. Numer. Math., 1979(33):195-210.

[346] IVANOV V V. The Theory of Approximate Methods and Their Application to the Numerical Solution of Singular Integral Equations[M]. Noordhoff, Groningen, The Netherlands, 1976.

[347] IWAI Z, KUBO Y. Determination of eigenvaluzs in Marshall's model reduction[J]. Internat. J. Control, 1979(30):823-836.

[348] IXARU L G. The error analysis of the algebraic method for solving the Schrödinger equation[J]. J. Comput. Phys., 1972(9):159-163.

[349] JACCARD Y. Approximation spectrale par la

méthode des éléments finis conformes d'une classe d'opérateurs non compacts et partiellement réguliers[M]. Thèse Math. Dept., Ecole Polytehn. Féd. de Lausanne, 1980.

[350] JACOBI C G J. Áber ein leichtes Verfahren die in der Theorie der Säcularstörungen vorkommenden Gleichungen numerisch aufzulösen[J]. Crelle J. Reine Angew. Math., 1846(30):51-94.

[351] JAMES R L. Uniform convergence of positive operators[J]. Math. Z. 1971(120):124-142.

[352] JEGGLE H. Áber die Approximation non linearen Gleichungen zweiter Art und Eigenwertprobleme in Banach-Räumen[J]. Math. Z., 1972(124):319-342.

[353] JEGGLE H, WENDLAND W L. On the discrete approximation of eigenvalue problems with holomorphic parameter dependence[J]. Proc. Roy. Sco. Edinburgh Sect. A., 1977(78):1-29.

[354] JENNINGS A. Mass condensation and simultaneous iteration for vibration problems[J]. Internal. J. Num. Methods. Engrg., 1973(6):543-552.

[355] JENNINGS A. Matrix Computations for Engineers and Scientists-Wiley[M]. New York, 1977a.

[356] JENNINGS A. Matrices[J]. Ancient and modern. Bull. Inst. Math. Appl., 1977b(13):117-123.

[357] JENNINGS A. Influence of the eigenvalue spec-

trum on the convergence rate of the conjugaete gradient method [J]. J. Inst. Math. Appl, 1977c(20):61-72.

[358] JENNINGS A. Eigenvalue methods for vibration analysis [J]. Shock Vibration Digest, 1980 (12):3-19.

[359] JENNINGS A. Eigenvalue methods and the analysis of structural vibrations [M]. In Sparse Matrices and Their Uses (I. S. Duff, ed.), New York: Academic Press, 1981:109-138.

[360] JENNINGS A, AGAR T J S. Progressive simultaneous inverse iteration for symmetric eigenvalue problems[M]. CE Report, Queen's Univ., Belfast, 1979.

[361] JENNINGS A, ORR D R L. Application of the simultaneous iteratin method to undamped vibration problems[J]. Internat. J. Numer. Methods Engrg., 1971(3):13-24.

[362] JENNINGS A. STEWART W J. Simultancous iteration for partiab eigensolution of real matrices [J]. J. Inst. Math. Appl., 1975 (15):351-361.

[363] JENNINGS A. HALLIDAY J, COIE M J. Solution of linear generalized eigenvalue problems containing singular matrices[J]. J. Inst. Math. Appl., 1978(22):401-410.

[364] JENSEN P S. The solution of large eigenproblems

by sectioning[J]. SIAM J. Numer. Anal. , 1972 (9):534-545.

[365] JENSSEN O. Eigenlunctions and spectrum of the hard-sphere collision operator [J]. Phys. Norveg. , 1972(6):180-191.

[366] JESPERSEN D. Ritz-Galerkin methods for singular boundary valueproblems[J]. SIAM J. Numer. Anal. , 1978(15):813-834.

[367] JOHNSEN T L. On the computation of natural modes of an unsupported vibrating structure by simultaneous iterations [J]. Comput. Methods Appl. Mech. Engrg. , 1973(2):305-322.

[368] JOHNSEN T L. A numerical method for eigenreduction of nonsymmetric real matrices[J]. Comput. & Struct. , 1978(3):399-402.

[369] KAGIWADA H H, KALABA R E, VERECKE B J. The invariant imbedding numerical method for Fredholm integral equations with degenerate kernels[J]. J. Approx. Theory, 1968 (1):355-364.

[370] KAGSTRÖM B. Methods for the numerical computation of matrix functions and the treatment of ill conditioned eigenvalue problems[M]. Report UMINE-59. 77, Univ. of Umea. , 1977a.

[371] KAGSTRÖM B. Bounds and perturbation bounds for the matrix exponential[J]. BIT, 1977b(17):39-57.

[372] KAGSTRÖM B. How to compute the Jordan normal form—the choie between similarity transformations and methods using chain relations[M]. Report UMINF-91. 81, Univ. of Umea, 1981.

[373] KAGSTRÖM B, RUHE A. An algorithm for numerical computation of the Jordan normal form of a complex matrix[J]. ACM Trans. Math. Sofware, 1980a(6):398-421.

[374] KAGSTRÖM B, Rube A. Algorthm 560 JNF, an algorithm for numerical computation of the Jordan normal form of a complex matrix [J]. ACM Trans. Math. Software, 1980b(6):437-443.

[375] KAHAN W, PARLETT B N. How far should you go with the Lanczos process[M]? In Sparse Matrix Computation (J. R. Bunch and D. J. Rose, eds.), New York: Academic Press, 1976:131-144.

[376] KAHAN W, PARLETT B N, JIANG E. Residual bounds on approximate eigensystems of nonnormal matrices[J]. SIAM J. Numer. Anal., 1981(19):470-484.

[377] KANIEL S. Estimates for some computational techniques in linear algebra[J]. Math. Comp., 1966(20):369-378.

[378] KANTOROVITCH L V. On a new method of approximate solution of partial differential equations (in Russian)[J]. Dokl. Aked. Nauk SSSR,

1934(4):532-436.

[379] KANTOROVITCH L V. Functional analysis and applied mathematics (in Russian). Usp. Mat. Nauk, 1948(3):89-185.

[380] KANTOROVITCH L V, AKILOV G P. Functoinal Analysis in Normed Spaces[M]. Pergamon, Oxford, 1964.

[381] KANTOROVITCH L V, KRYLOV V I. Approximate Methods of Higher Analysis [M]. New York: Wiley (Interscience), 1955.

[382] KARMA O O. Asymptotic error estimations for characteristicvalues of holomorphic Fredholm operator functions (in) Russian [J]. Ẑ. Vyĉisl. Mat. i Mat. Fiz., 1971(11):559-568.

[383] KARPEL M, NEWMAN M. Accelerated convergence for vibration modes using the substructure coupling method and fictitious coupling masses [J]. Israel J. Tech., 1975(13):55-62.

[384] KARPILOVSKAIA E B. On the convergence of an interpolation method for ordinary differential equations (in Russian)[J]. Usp. Mat. Nauk, 1953(8):111-118.

[385] KARPILOVSKZIA E B. On the convergence of the collocation method (in Russian)[J]. Dokl. Akad. Nauk SSSR, 1963(151):766-769.

[386] KARPILOVSKAIA E B. On the convergence of the collocation method for integro-differential e-

quations (in Russian) [J]. Ž. Vyĉisl. Mat. i Mat. Fiz., 1965(5):124-132.

[387] KATO T. On the upper and lower bounds of eigenvalues[J]. J. Phys. Soc. Japan, 1949(4): 334-339.

[388] KATO T. Perturbation theory for nullity deficiency and other quantities of linear operators[J]. J. Analyse Mathj., 1958(6):261-322.

[389] KATO T. Perturbation Theory for Linear Operators, 2nd ed[M]. Springer-Verlag, Berlin and New York, 1976.

[390] KAUCHER E, RUMP S M. E-methods for fixed point equations $f(x) = x$ [J]. Computing, 1982 (28):31-42.

[391] KELDYŜ M V. On the characteristic values and characteristic functions of certain classes of non-self-adjoint equations (in Russian) [J]. Dokl. Akad. Nauk SSSR, 1951(77):11-14.

[392] KELLER H B. On the accuracy of finite diference approximations to the eigenvalues of differential and integral operators[J]. Numer. Math., 1965 (7):412-419.

[393] KELLER H B. Approximation methods for nonlinear problems with application to two-point boundary value problems [J]. Math. Comp., 1975 (29):464-474.

[394] KIKUCHI F. On a mixed finite element scheme

for linear buckling analysis of plates[J]. In Computational Methods in Nonlinear Mechanics (J. T. Oden, ed.), Nerth-Holland Publ Amsterdam, 1980a:289-302.

[395] KIKUCHI F. Numerical analysis of a mixed finite element method for plate buckling problems ISAS Rep[J]. No 584, Univ of Tokyo, 1980(45): 165-190.

[396] KLEINERT P. Cluster approximation for the spectral density of mixed diatomic systems[J]. Phys. Status Solidi(B), 1979(91):455-465.

[397] KOLATA W G. Approximation in variationally posed eigenvalue problems[J]. Numer. Match., 1978(29):159-171.

[398] KOLATA W G. Eigenvalue approximation by the finite element method: the method of Lagrange multipliers[J]. Math. Comp. 1979(33):63-76.

[399] KOLATA W G, OSBORN J E. Nonselfadjoint spectral aproximation and the finite element method[M]. In Functional Analysis Methods in Numerical Analysis (M. Z. Nashed, ed.), Lecture Notes in Mathematics, Vol. 101, pp. 115-133. Springer-Verlag, Berlin and New York.

[400] KONDRASHEV V I. On some properties of functions of spaces Lp(in Russian)[J]. Dokl. Akad. Nauk SSSR, 1945(48):563-566.

[401] KRANSNOSELSKII M A, VAINIKKO G M, ZA-

BREIKO P P, et al. Approximate Solutions of Operator Equations [M]. Wolters-Noordhoff, Groningen, The Netherlands, 1972.

[402] KRASNOSELSKIT M A, ZABREIKO P P, PUSTILNIK E I, et al. Integral Operators in Spaces of Summable Functions [M]. Noordhoff Int, Leyden, The Netherlands, 1976.

[403] KREISS H O. Difference approximation for boundary and eigenvalue problems for ordinary differential equations [J]. Math. Comp, 1972 (26):605-624.

[404] KREUZER K G, MILLER H G, DREIZLER R M, et al. Extension of an iterative method to obtain low-lying eigenstates of unbounded Hermitian operators[J]. J. Phys. A., 1980(13):2645-2652.

[405] KREUZER K G, MILLER H G, BERGER W A. The Lanczos algorithm for self-adjoint operators [J]. Phys. Lett. A., 1981:429-432.

[406] KRYLOV A N. On the numerical solution of equations which in technical questions are determined by the frequency of small vibrations of material systems (in Russian) [J]. Izv. Akad. Nauk SSSR Otd. Mat. Estest., 1931(1):491-539.

[407] KRYLO V I. Approximate Calculation of Integrals [M]. New York: Macmillan, 1962.

[408] KUBLANOVSKAYA V N. On some algorithms for the solution of the complete eigenvalue problem (in Russian)[J]. Ẑ. Vyĉisl. Mat. i Mat. Fiz. , 1961(1):555-570.

[409] KULKARNI R P. Convergence and computation of approximate eigenelements[M]. Ph. D. Thesis, Math. Dept. , Indian Inst. Technology, Bombay, 1982.

[410] KULKARNI R P, LIMAYE B V. On error bounds in strong approximations for eigenvalue problems [J]. J. Austral. Math. Soc. Ser. B, 1981 (22):270-283.

[411] KULKARNI R P, LIMAYE B V. On Chatelin's algorithm for the computation of the eigenelements by iterations [M]. Rep. Math. Dept. Indian Inst. Technol, Bombay, 1982.

[412] KULKARNI R P, LIMAYE B V. On the steps of convergence of approximate eigenvectors in the Rayleigh-Schrödinger series[M]. Numer. Math. (to appear), 1983a.

[413] KULKARNI R P, LIMAYE B V. Geometric and semi-geometric approximation of spectral projections[M]. J. Math. Anal. Appl. (to appear), 1983b.

[414] KULISCH U, MIRANKER W L. Computer Arithmetic in Theory and Practice [M]. New York: Academic Press, 1981.

Tschebyscheff 逼近定理

[415] KURATOWSKI C. Topologie, 3rd ed[M]. Polska Akad. Nauk. Warsaw, 1961.

[416] KUTTLER J R. Remarks on a Stekloff eigenvalue problem[J]. SIAM J. Numer. Anal., 1972(9): 1-5.

[417] KUTTLER J R. Dirichlet eigenvalues. SIAM J. Numer. Anal[J]., 1979(16):332-338.

[418] KUTTLER J R, SIGILLITO V G. Inequalities for membrane and Stekloff eigenvalues[J]. J. Math. Anal. Appl., 1968(23):148-160.

[419] LAASONEN P. A Ritz method for simultaneous determination of several eigenvalues and eigenvectors of a big matrix[J]. Ann. Acad. Sci. Fenn. Ser. A I Math., 1959(265):3-16.

[420] LANCZOS C. An iterative method for the solution of the eigenvalue problem of linear differential and integral operators[J]. J. Res. Nat. Bur. Standards Sect. B, 1950(45):255-282.

[421] LANCZOS C. Solution of systems of linear equations by minimized iterations[J]. J. Res. Nat. Bur. Standards Sect. B, 1952(49):33-53.

[422] LANCZOS C. Linear Differential Operators[M]. New York: Van Nostrand, 1961.

[423] LAURENT P J. Approximation et Optimisation [M]. Hermann, Paris, 1972.

[424] LEBBAR R. Sur ies propriétés de superconvergence des solutions approchées de certaines

équations intégrales et différentielles[M]. Thèse 3ème Cycle, Univ. de Grenoble, 1981a.

[425] LEBBAR R. Superconvergence at the knots for the generalized eigenvectors of differential and integral operators[J]. R. R. IMAG No. 272, Univ. de Grenoble, 1981b.

[426] LEBBAR R. Superconvergence with adaptive mesh for weakly singular equations[M]. Unpublished manuscript, Univ. de Grenoble, 1982.

[427] LEBEDEV V I. An iteratie method with Chebyshev parameters for finding the maximum eigenvalue and corresponding eigenfunction[J]. U. S. S. R. Computational Math. and Math. Phys., 1977(17):92-101.

[428] LEE I W, PRENTER P M. An analysis of the numerical solution of Fredholm integral equations of the first kind[J]. Numer. Math., 1978(30):1-23.

[429] LEHMANN N J. On optimal eigenvalue localization in the solution of symmetric matrix problems [J]. Numer. Math., 1966(8):42-55.

[430] LEMORDANT J. Localisation d'un groupe de valeurs propres[M]. Note d'étude LA No. 46. Centre de Calcul Scientifique de l'Armement, CELAR, Bruz, 1977.

[431] LEMORDANT, J. Etide de l'ensemble des schémas d'approximation spectrals stable d'un

opérateur compact [M]. Sém. Anal. Numér. No. 319, Univ. de Grenoble, 1979.

[432] LEMORDANT J. Localisation de valeurs propress et calcul de sous-espaces invariants [M]. Thèse d'Etat, Univ. de Grenoble, 1980.

[433] LEROU R J L, DEKKER H. Exact computation of high-order perturbational eigensolutions and its application to the analysis of a spectral degeneracy in a bistable diffusion process [J]. Phys. Lett. A, 1981(83):371-375.

[434] LEWIS J G. Algorithms for sparse matrix eigenvalue problems [M]. Ph. D. Thesis, Report CS-77-595, Computer Sci. Dept., Staford Univ., Stanford, California, 1977.

[435] LEWIS J G, GRIMES R G. Practical Lanczos algorithms for solving structural engineering problems [M]. In Sparse Matrices and Their Uses (I. S. Duff, ed.). pp. 349-355. New York: Academic Press, 1981.

[436] LIN QUN. Some problems about the approximate solution for operator equations (in Chinese) [J]. Acta Math. Sinica, 1979(22):219-230.

[437] LIN QUN. How to increase the accuracy of lower order elements in nonloinear finite element methods [J]. In Computing Methods in Applied Sciences and Engineering (R. Glowinski and J. L. Lions, eds.), pp. 41-47. North-Holland Publ.,

Amsterdam, 1980.

[438] LIN QUN. Interative corrections for nonlinear eigenvalue problem of operator equations [M]. Res. Rep. IMS-1, Institute of Mathematical Sciences, Chengdu Branch of Academia Sinica, 1981a.

[439] LIN Qun. Iterative corrections for nonlinear operator equations with applications to difference method[J]. J. Sys. Sci. Math. Sci. 1981(1):139-146.

[440] LIN QUN. Deferred corrections for equations of the second kind J[J]. Austral. Math. Soc. Ser. B, 1981(22):456-459.

[441] LIN QUN. Itreative refinement of finite element approximations for eliptic problems [J]. RAIRO Anal. Numér., 1982(16):39-47.

[442] LIN QUN. Personal communication[M], 1982.

[443] LIN QUN, JANG LI-SHANG. Investigation of the system $\Delta U_1 = \sum_{j=1}^{n} U_i(\partial u_i/\partial x_j) + j_i, i = 1, \cdots, n$ [M]. Tech. Rep., Inst. Math. Acad. Sinica, Beijing, 1979.

[444] LIN QUN, LIU JIAQUAN. Extrapolation method for Fredholm integral equations with non-smooth kernels [J]. Numer. Math., 1980 (35):459-464.

[445] LINZ P. On the numerical computation of eigenvalues and eigenvectors of symmetric integral e-

quation[J]. Math. Comp., 1970(24):905-909.

[446] LINZ P. Error estimates for the computation of eigenvalues or selfadjoint operators[J]. BIT, 1972(12):528-533.

[447] LINZ P. A general theory for the approximate solution of operator equations of the 2nd kind[J]. SIAM J. Numer. Anal., 1977(14):543-554.

[448] LINZ P. Theoretical Numerical Analysts [M]. New York: Wiley, 1979.

[449] LIONS J L, MAGENES E. Problèmes aux Limites Non Homogènes et Applications [M]. Vol. 1. Dunod. Paris, 1968.

[450] LO W S. Spectral approximation theorems for bounded linear operators [J]. Bull. Austral. Math. Soc., 1973(8):279-287.

[451] LOCKER J, PRENTER P M. Optimal L^2 and L^∞ error estimates for continuous and discrete least squares methods for boundary value problems[J]. SIAM J. Numer. Anal., 1978(15):1151-1160.

[452] LOCKER J, PRENTER P M. On least squares methods for linear two-point boundary value problems [M]. In Functional Analysis Methods in Numerical Analysis (M. Z. Nashed, ed.), Lecture Notes in Mathematics, Vol. 701, pp. 149-168. Springer-Verlang, Berlin and New York, 1979.

[453] LOCKER J, PRENTER P M. Regularizaion with

differential operators[J]. I. General theory. J. Math. Anal. Appl. 1980(74):504-529.

[454] LOCKER J, PRENTER P M. Regularization with differential operators[J]. II. Weak least squares finite element solutions to first kind integral equations. SIAM J. Numer. Anal., 1981(17):247-267.

[455] LOKER J, PRENTER P M. Representors and superconvergence of least squares finite element approximates[J]. Numer. Funct. Anal. Optim. (to appear), 1983.

[456] LONGSINE D E, MCCORMICK S F. Simultaneous Rayleigh quotient minimization methods for $Ax=\lambda Bx$[J]. Linear Algehra Appl., 1980(34): 195-234.

[457] LORENTZ G G. Approximation of Functions [M]. New York: Holt, 1966.

[458] LUSKIN M. Convergence of a finite element method for the approximation of normal modes of the oceans[J]. Math. Comp., 1979(33):493-519.

[459] LUTHEY Z A. Piecewise analytical solutions method for the radial Schrödinger equation[J]. Ph. D. Thesis, Harvard Univ., Cambridge, Massachusette, 1974.

[460] MCCORMICK S F. Mesh refinement for integral equations[M]. In Numerical Treatment of Inte-

gral Equations (J. Albrecht and L. Collatz, eds.), pp. 183-190. Birkhaeuser, Basel, 1980.

[461] MCCORMICK S F. A mesh refinement method for $Ax = \lambda Bx$ [J]. Math. Comp. , 1981 (36) : 485-498.

[462] MCCORMICK S F, NOE T. Simultaneous iteration for the matrix eigenvalue problem[J]. Linear Algebra Appl. , 1977(16) :43-56.

[463] MELAURIN J W. General coupled equation spproach for solving the biharmonic boundary value problem [J]. SIAM J. Numer. Anal. , 1974 (11) :14-33.

[464] MAREK I. Approximations of the principal eigenelements in K-positive nonselfadjoint eigenvalue probiems, Math [J]. Systems Theory, 1971(5) :204-215.

[465] MARCKUK G I, AGOŜKOV V I. The selection of coordinate functions in the generalized Bubnov-Galerkin method (in Russian)[J]. Dokl. Akad. Nauk SSSR, 1977(232) :1253-1256.

[466] MASUR E F. Bounds and error estimates in structural eigenvalue problems[J]. J. Struct. Mech. , 1973(1) :417-438.

[467] MEAD D J, PARTHAN S. Free wave propagation in two-dimensional plates[J]. J. Sound Vibr. , 1979(64) :325-348.

[468] MEIROVITCH L. A new method of solution of the

eigenvalue problem for gyroscopic systems [J]. AIAA J., 1974(12):1337-1342.

[469] MERCIER B, RAPPAZ J. Eigenvalue approximation via nonconforming and hybrid finite element methods [M]. Rep. No. 33, Math. Appl., Ecole Polytechnique, Palaiseau, 1978.

[470] MERCIER B, OSBORN J RAPPAZ J, et al. Eigenvalue approximation by mixed and hybrid methods [J]. Math. Comp., 1981(36):427-453.

[471] MÉRIGOT M. Régularité des fonctions propres du laplacien dans un cône [J]. C. R. Hebd. Séances Acad. Sci. Ser. A, 1974(279):503-505.

[472] MIESCH A. Scaling variables and interpretation of eigenvalues in principal component analysis of geologic data[J]. J. Internal. Assoc. Math. Geol., 1980(12):523-538.

[473] MIKA J. Fundamental eigenvalues of the linear transport equation[J]. J. Quant. Spectrosc. Radiat. Transfer, 1971(11):879-891.

[474] MIKHLIN S G. Integral Equations and Their Applications to Some Problems of Mechanics, Mathematical Physics and Engineering, 2nd ed [J]. Pergamon, Oxfod, 1964.

[475] MIKHLIN S G. The Numerical Performance of Variational Methods [M]. Wolters-Noordhoff,

Groningen, The Netherlands, 1971.

[476] MIKHLIN S G, SMOLITSKII K L. Approximate Methods for Solutions of Differential and Integral Equations [J]. Amer. New York: Elsevier, 1967.

[477] MILLER H G, BERGER W A. An investigation of pseudocon-vergence in an iterative method for calculating the low-lying eigen vectors of a Hermitian matrix[J]. J. Phys. A., 1979(12):1693-1698.

[478] MILLS H W. The resolvent stability for spectral convergence with application to the finite element approximation of noncompact operators[J]. SIAM J. Numer. Anal., 1979a(16):695-703.

[479] MILLS W H. Optimal error estimates for the finite element spectral approximation of noncompact operators [J]. SIAM J. Numer. Anal., 1979b(16):704-718.

[480] MILLS W H. Finite element error estimates for singular veriational eigenvalue problems [M]. Res. Rep., Math. Dept., Pennsylvania State Univ., University Park., 1979c.

[481] MILLS W H. Convergence and errors for projective finite element approximation of variational eigenvalue problems [J]. Res. Rep. No. 8015, Math. Dept, Pennsylvania State Univ, University Park, 1980.

参考文献

[482] MINDLIN R D. Simple modes of vibrations of crystals[J]. J. Appl. Phys., 1956(27):1462-1466.

[483] MIRANKER W L. Galerkin approximations and the optimization of difference schemes for boundary value problems[J]. SIAM J. Numer. Anal., 1971(8):486-496.

[484] MIYOSHI T. A mixed finite element method for the solution of the von Kármán equations[J]. Numer. Math., 1976(26):255-269.

[485] MOCK M S. Projection methods with different trial and test spaces[J]. Math. Comp., 1976(30):400-416.

[486] MOISEIWITSCH B L. Integral Equations[M]. Longman, London, 1977.

[487] MOLER C B, STEWART G W. An algorithm for generalized matrix eigenvalue problems[J]. SIAM J. Numer. Anal., 1973(10):241-256.

[488] MOORE E H. On the reciprocal of the generalized algebraic matrix (abstract)[J]. Bull. Amer. Math. Soc., 1919-1920(26):394.

[489] MORO G, FREED J H. Calculation of ESR spectra and related Fokker-Planck forms by the use of the Lanczos algorithm[J]. J. Chem. Phys., 1981(74):3757-3773.

[490] MOSZYNSKI K. On the approximation of the spectral density function of a self-adjoint operator

[J]. Rep. No. 206, Inst. Math., Polish Academy of Seienrices, 1980.

[491] MUDA Y. A new relaxation method for obtaming the lowest eigenvalue and eigenvector of a matrix equation[J]. Internat. J. Numer. Methods Engrg., 1973(6):511-519.

[492] MUNTEANU M J, SCHUMAKER L L. Direct and inverse theorems for multidimensional spline approximation[J]. Indians Univ. Math. J., 1973(23):461-470.

[493] MUROYA Y. On a posteriori error estimates for Galerkin approxi-mations to the solutions of two-point boundary value problems[J]. Mem. School Sci. Engrg. Waseda Univ. 1979(43):163-169.

[494] MYSOVSKIH I P. Computation of the eigenvalues of integral equations by means of iterated kernels (in Russian) [J]. Dokl. Akad. Nauk SSSR, 1957(115):45-48.

[495] MYSOVSKIH I P. On error bounds for approximate methods of estimation of eigenvalues of hermitian kernels[J]. Amer. Math. Soc. Transl., 1964a(35):237-250.

[496] MYSOVSKIH I P. On error bounds for eigenvalues calculated by replacing the kernel by an approximating kernel [J]. Amer. Math. Soc. Transl., 1964b(35):251-262.

[497] NAG Library Reference Manual[J]. Numerical

Algorithms Group, Oxford.

[498] NAKAMURA S. Analysis of the coarse-mesh rebalancing effect on Chebyshev polynomial iterations[J]. Nuclear Sci. Engrg., 1976(61):98-106.

[499] NASH J C. The Hermitian matrix eigenproblem $Hx = eSx$ [J]. Comput. Phys. Comm., 1974(8):85-94.

[500] NASH J C. Compact Numerical Methods for Computers[M]. Linear Algebra and Function Minimization. Ney York: Wiley, 1979.

[501] NASHED M Z, WAHBA G. Convergence rates of approximate least-squeres solutions of linear integral and operator equations of the first kind[J]. Math. Comp., 1974(28):69-80.

[502] NATANSON I P. Theory of Functions of a Real Variable[M], Vols. 1 and 2. New York: Ungar, 1977.

[503] NATTERER F. Uniform convergence of Galerkin's method for splines on highly non uniform meshes [J]. Math. COmp., 1977(31):457-468.

[504] NAU R W. Computation of upper and lower bounds to the frequencies of clamped cylindrical shells [J]. Internat. J. Earthquake Engrg. Struct. Dynam., 1976(4):553-559.

[505] NEDELEC J C. Approximation des équations intégrales en mécanique et en physiqual [M].

Lecture Notes, Math. Appl., Ecole Polytechnique, Palaiseau, 1977.

[506] NELSO P, ELDER I T. Calculation of eigenfunctions in the context of intergration-to-blowup[J]. SIAM J. Numer. Anal., 1977(14):124-136.

[507] NEMAT-MASSER S, LANG K W. Eigenvalue problems for heat conduction in composite material[J]. Iranian J. Sci. Tech., 1979(7):243-269.

[508] NESBET R K. Large matrix techniques in quanfum chemistry and atomic physics[M]. In Sparse Matrices and Their Uses (I. S. Duff, ed.). pp. 161-174. New York: Academic Press, 1981.

[509] NIKOLAI P J. Algorithm 538 eigenvectors and eigenvalues of real generalized symmetric matrices by simultaneous iteration[J]. ACM Trans. Math. Software, 1979(5):118-125.

[510] NIKOLSKII S M. Approximatin of Functions of Several Variables and Imbedding Theorems[M]. Springer-Verlag, Berlin and New York, 1975.

[511] NISBET R M. Acceleration of the convergence in Nesbet's algorithm for eigenvalues and eigenvectors of large matrices[J]. J. Comput. Phys., 1972(10):614-619.

[512] NITSCHE J. Áber ein Variationprinzip zur Lösung von Dirichlet Problemen bei Verwendung von Teilräumen, die keinen Randbe-dingungen unter-

worfen sind[J]. Abh. Math. Sem. Univ. Hamburg, 1970(36):9-15.

[513] NITSCHE J. L_∞ convergence of finite element approximation [M]. Second Conf. Finite Elem., Rennes, 1975.

[514] NITSCHE J. On projection methods for the plate equation. In Numerical Analysis J[M]. Descloux and J. Marti, eds. pp. 49-61. Birkhaeuser, Bascl, 1977.

[515] NOBLE B. A bibliography on "Methods for solving integral equations" Tech[M]. Rep. 1176, 1177, MRC, Univ. of Wisconsin, Madison, Wisconsin, 1971.

[516] NOBLE B. Error analysis of collocation methods for solving Fredholm integral equations [M]. In Topics in Numerical Analysis (J. J. H. Miller, ed.). pp. 211-232. Academic Press, London, 1975.

[517] NYSTRÖM E J. Áber die praktische Auflösung von Integralgleichungen mit Anwendungen auf Randwertaufgaben[J]. Acta Math. 1930(54): 185-204.

[518] ODEN J T, REDDY J N. An Introduction to the Mathematical Theory of Finite Elements [M]. New York: Wiley (Interscience), 1976.

[519] OJAIVA I U, NEWMAN M. Vibration modes of large structure by an automatic-matrix-reduction

method[J]. AIAA J. 1970(8):1334-1239.

[520] O'LEARY D P, STEWART G W, VANDERG-RAFT J S. Estimating the largest eigenvalue of a positive definite matrix[J]. Math. Comp., 1979(33):1289-1292.

[521] OLIVEIRA ALEIXO F. Collocation and residual correction[J]. Numer. Math., 1980(36):27-31.

[522] ON EGA R J, KARCHER K E. Nonlinear dynamics of a pressurized water reactor core[J]. Nucl. Sci. Engrg., 1976(61):276-282.

[523] ORBACH O, CROWE C M. Convergence promotion in the simulation of chemical processes with recycle: The dominant eigenvalue method[J]. Canad. J. Chem. Engrg., 1971(49):509-513.

[524] ORTEGA J M. Numerical Analysis: A Second Course[M]. New York: Academic Press, 1972.

[525] ORTEGA J M, RHEINBOLDT W C. Iterative Solution of Nonlinear Equations in Several Variables [M]. New York: Academic Press, 1971.

[526] OSBORN J E. Approximation of the eigenvalues of a class of unbounded, nonselfadjoint operators [J]. SIAM J. Numer. Anal., 1967(4):45-54.

[527] OSBORN J E. Spectral approximation for compact operators[J]. Math. Comp., 1975(29):712-725.

[528] OSBORN J E. Approximation of the eigenvalues

of a non selfadjoint operator arishing in the study of the stability of stationary solutions of the Navier-Stokes equations [J]. SIAM J. Numer. Anal., 1976(13):185-197.

[529] OSTROWSKI A M. On the convergence of the Rayleigh quotient iteration for the computation of the characteristic roots and vectors [J]. Arch. Rational Mech. Anal., 1958-1959 (1): 233-241; (2): 423-428; (3): 325-347, 472-481; (4):153-165.

[530] PAIGE C C. Practical use of the symmetric Lanczos process with reorthogonalization [J]. BIT, 1970(10):183-195.

[531] PAIGE C C. The computatio of eigenvalues and eigenvectors of very large sparse matrices [M]. Ph. D. Thesis. London Vniv, 1971.

[532] PAIGE C C. Computational variants of the Lanczos method for the eigenproblem [J]. J. Inst. Math. Appl., 1972(10):373-381.

[533] PAIGE C C. Eigenvalues of perturbed Hermitian matrices[J]. Linear Algebra Appl, 1974(8):1-10.

[534] PAIGE C C. Accuracy and effectiveness of the Lanczos algorithm for the symmetric eigenproblem [J]. Linear Algebra Appl., 1980 (34):235-258.

[535] PAIGE C C. Properties of numerical algorithms

related to computing controllability [J]. IEEE Trans. Automat. Control AC-26, 1981:130-138.

[536] PAINE J W, ANDERSSEN R S. Uuiformly valid approximation of eigenvalues of Sturm-Liouville problems in geophysics [J]. Geophys. J. R. Austral. Soc., 1980(63):441-465.

[537] PALMER T W. Totally bounded sets of precompact linear operators [J]. Proc. Amer. Math. Soc., 1969(20):101-106.

[538] PAPATHOMAS T, WING O. Sparse Hessenberg reduction and the eigenvalue problem for large sparse matrices [J]. IEEE Trans. Circuits and Systems ACS-23, 1976:739-744.

[539] PARLETT B N. The origin and development of methods of LR type[J]. SIAM Rev., 1964(6): 273-295.

[540] PARLETT B N. Convergence of the QR algorithm [J]. Numer. Math., 1965(7):187-193.

[541] PARLETT B N. Global convergence of the basic QR algorithm on Hessenberg matrices[J]. Math. Comp., 1968(22):803-817.

[542] PARLETT B N. Présentation géométrique des méthodes de calcul des valeurs propres[J]. Numer. Math., 1973(21):223-233.

[543] PARLETT B N. The Rayleigh quotient iteration and some generalizations for nonnormal matrices [J]. Math. Comp., 1974(28):679-693.

[544] PARLETT B N. The Symmetric Eigenvalue Problem[M]. Prentice-Hall, New Jersey: Englewood Cliffs, 1980a.

[545] PARLETT B N. A new look at the Lanczos algorithm for solving symmetric systems of linear equations[J]. Linear Algebra Appl., 1980b(29): 323-346.

[546] PARLETT B N. Comment résoudre $(K-\lambda M)z = 0$? [M]. In Méthodes Numériques pour les Sciences de l'Ingénieur (E. Absi, R. Glowinski, P. Lascaux, and H. Veysseyre, eds.), Tome 1, pp. 97-106. Dunod, Paris, 1981.

[547] PARLETT B N, KAHAN W. On the convergence of a practical QR algorithm[J]. In Information Processing 68, Vol. 1, pp. 114-118. North-Holland Publ. Amsterdam, 1969.

[548] PARLETT B N, POOLE W G. A geometric theory for the QR, LU and power iterations[J]. SIAM J. Numer. Anal., 1973(10):389-412.

[549] PARLETT B N, REID J K. Tracking the progress of the Lanczos algorithm for large symmetric eigenproblems[J]. IMA J. Numer. Anal., 1981(2):135-156.

[550] PARLETT B N, REMSCH C. Balancing a matrix for calculation of eigenvalues and eigenvectors[J]. Numer. Math., 1969(13):293-304.

[551] PARLETT B N, SOCC D S. The Lanczos algo-

rithm with selective orthogonalization[J]. Math. Comp. , 1979(33):217,238.

[552] PARLETT B N, Taylor D. A look ahead Lanczos algorithm for unsymmetric matrices[M]. Techn. Rep. PAM-43, Center for Pure and Appl. Math. , Univ. of California, Berkeley, 1981.

[553] PENROSE R. A generalized inverse for matrices [J]. Proc. Cambridge Philos. Soc. , 1955(51): 406-413.

[554] PERECLL P, WHEELER M F. A C^1 finite element collocation method for elliptic equations[J]. SIAM J. Numer. Anal. , 1980(17):605-622.

[555] PEREYRA V, SCHERER G. Eigenvaluse of symmetric tridiagonal matrices: A fast, accurate and reliable algorithm [J]. J. Inst. Math. Appl. , 1973(12):209-222.

[556] PETERS G, WIKINSON J H. Eigenvectors of real and complex matrices by LR and QR triangularizations[J]. Numer. Math. , 1970a(16):181-204.

[557] PETERS G, WILKINSON J H. The least squares problem and pseudoinverses [J]. Comput. J. 1970b(13):309-316.

[558] PETERS G, WILKINSON J H. $Ax = \lambda Bx$ and the generalized eigenproblem[J]. SIAM J. Numer. Anal. , 1970c(7):479-492.

[559] PETRYSHYN W V. On a general iterative method

for the approximate solution of linear operator equations[J]. Math. COmp., 1963(17):1-10.

[560] PETRYSHYN W V. On the eigenvalue problem $Tu-\lambda Su = 0$ with unbounded and non symmetric operators T and S[J]. Philos. Trans. Roy. Soc. London Ser. A, 1967a(262):413-458.

[561] PETRYSHYN W V. Projection methods in nonlinear numerical functional analysis[J]. J. Math. Mech., 1967b(17):353-372.

[562] PETRYSHYN W V. On projectional-solvability and the Fredholm alternative for equations involving linear A-proper operators[J]. Arch. Rational Mech. Anal., 1968(30):270-284.

[563] PFEIFER E. Discrete convergence of multi-step methods in eigenvalue problems for ordinary second order differential equations[J]. U.S.S.R. Computational Math. and Math. Phys., 1979(19):64-73.

[564] PHILLIPS J L. The use of collocation as a projection method for solving linear operator equations[J]. SIAM J. Numer. Anal., 1972(9):14-28.

[565] PHILLIPS D R. The existence of determining equations and their application for finding fixed points of nonlinear operators and error bounds for eigenvalues estimates of compact linear operators and finite matrices[M]. Ph. D. Thesis, Univ. of Maryland, College Park, 1978.

[566] PITK? RANTA J. On the differential properties of solutions to Fred-holm equations with weakly singular kernels [J]. J. Inst. Math. Appl, 1979 (24):109-119.

[567] PLATZMAN G W. North Atlantic ocean: Preliminary description of normal modes [J]. Science, 1972a(178):156-157.

[568] PLATZMAN G W. Two dimensional free oscillations in natural basins [J]. J. Phys. Oceanogr., 1972b(2):117-138.

[569] PLATZMAN G W. Normal modes of the Atlantic and Indian oceans [J]. J. Phys. Oceanogr., 1975(5):201-221.

[570] PLATZMAN G W. Normal modes of the world ocean. Part I. Design of finite-element barotropic model [M]. J. Phys. Oceanogr., 1978(8):323-343.

[571] PLATZMAN G W. A Kelin wave in the eastern north Pacific ocean [J]. J. Geophys. Res., 1979 (84):2525-2528.

[572] POKRZYWA A. On the asymptotical behaviour of spectra in the method of orthogonal projections [M]. Rep. No. 161, Inst. Math., Polish Academy of Sciences, 1978.

[573] POKRZYWA A. Method of orthogonal projections and approximation of the spectrum of a bounded operator [J]. Studia Math., 1979(65):21-29.

参考文献

[574] POKRZYWA A. Spectra of compressions of an operator with compact imaginary part J[J]. Operator Theory, 1980(3):151-158.

[575] POLANSKY O E, GUTMAN I. On the calculation of the largest eigenvalue of molecular graph[J]. MATCH, 1979(5):149-159.

[576] POLSKII N I. Projection methods in applied mathematics[J]. Dokl. Akad. Nauk SSR, 1962 (143):787-790 [Soviet Math. Dolk. (3):488-491].

[577] PRENTER P M. A collocation method for the numerical solution of integral equations[J]. SIAM J. Numer. Anal., 1973(10):570-581.

[578] PRENTER P M. Splines and Variational Methods [M]. New York: Wiley (Interscience), 1975.

[579] PRIKAZCHIKOV V G. Strict estimate of convergece rate in an iterative method for calculation of eigenvalues[J]. U. S. S. R. Comput. Math. and Math Phys., 1975(15):1330-1333.

[580] PROSKUROWSKI W. On the numerical solution of the eigenvalue problem of the Laplace operator by a capacity matrix method [J]. Computing, 1978(20):139-151.

[581] PRUESS S. Estimating the eigenvalues of Sturm-Liouville problems by approximating the differential equation[J]. SIAM J. Numer. Anal., 1973a (10):55-68.

[582] PRUESS S. Solving linear boundary value problems by approximating the coefficients[J]. Math. Comp., 1973b(27):551-561.

[583] PRUESS S. High order approximation to Sturm-Liouville eigenvalues [J]. Numer Math, 1975(24):241-247.

[584] PTÁK V. Non discrete mathematical induction and iterative existence proofs[J]. Linear Algebra Appl., 1976(13):223-238.

[585] RAFFENETTI R C. A simultaneous coordinate relaxation for large, sparse matrix eigenvalue problems[J]. J. Comput. Phys., 1979(32):403-419.

[586] RAJU I S, RAO G V, MURTHY T V G K. Eigenvalues and eigenvectors of large order banded matrices[J]. Comput. & Structures, 1974(4):549-558.

[587] RAKOTCH E. Numerical solution for eigenvalues and eigenfunctions of a hermitian kernel and error estimates[J]. Math. Comp., 1975(29):794-805.

[588] RAKOTCH E. Numerical solution with large matrices of Fredholm's integral equation[J]. SIAM J. Numer. Anal., 1976(13):1-7.

[589] RAKOICH E. Improved error estimates for numerical solutions of symmetric integral equations[J]. Math. Comp., 1978(32):399-404.

[590] RALL L B. Computational Solution of Nonlinear Operator Equations [M]. New York: Wiley, 1969.

[591] RAMAMURTI V. Application of simultaeous iteration method to torsional vibration problems[J]. J. Sound Vibr., 1973(29):331-340.

[592] RAMASWAMY S. On the effectiveess of the Lanczos method for the solution of large eigenvalue problems[J]. J. Sound Vibr., 1980(73): 405-418.

[593] RAMSDEN J N, STOKER J R. Asemi-automatic method for reducing the size of vibration problems [J]. Internat. J. Numer. Methods Engrg., 1969(1):339-349.

[594] RANACHER R. Nonconforming finite element methods for eigenvalue problems in linear plate theory[J]. Numer. Math., 1979(33):23-42.

[595] RAO C R, MITRA S K. Generalized Inverse of Matrices and Its Applications [M]. New York: Wiley, 1971.

[596] RAPPAZ J. Approximation of the spectrum of a noncompact operator given by the megnetohydrodynamic stability of a plasma [J]. Numer. Math., 1977(28):15-24.

[597] RAPPAZ J. Some properties on the stability relatd to the approximation of eigenvalue problems[M]. In Computing Methods in Applied Sciences and

Engineering (R. Glowinski and J. L. Lions, eds.), pp. 167-174. North-Holland Publ., Amsterdam, 1982.

[598] RAYLEIGH LORD (STRUTT J W). The Theory of Sound[M]. Vol. 1 and 2. Macmillan, London and New York, 1894-1896.

[599] RAYLEIGH, LORD. On the calculation of the frequency of vibration of a system in its gravest mode, with an example from hydrodynamics[J]. Philos. Mag., 1899(47):566-572.

[600] REDONT P. Application de la théorie de la perturbatin des opérateurs linéaires à l'obtention de bornes d'erreur sur les éléments propres et à leur calcul[M]. Thèse Doct-Int, Univ. de Grenoble, 1979a.

[601] REDONT P. Sur la convergence régulière d'opérateurs[M]. Unpublished manuscript, Univ. de Grenoble, 1979b.

[602] REED M, SIMON B. Analysis of Operators[J]. New York: Academic Press, 1978.

[603] REGIN'SKA T. Approximate methods for solving differential equations on infinite intervals [J]. Apl. Mat., 1977(22):92-109.

[604] REGIN'SKA T. External approximation of eigenvalue problems [M]. Rep. No. 229, Inst. Math., Polish Academy of Sciences, 1980.

[605] REGIN'SKA T. Eigenvalue aproximation. In

Computational Mathematics[M]. Banach Center Publ., Vol. 10, Warsaw, 1981.

[606] REID J K. A survey of sparse matrix computation [M]. In Sparse Matrix Techniques (V. A. Barker, ed.), Lecture Notes in Mathematics, Vol. 572, pp. 41-48. Springer-Verlag, Berlin and New York, 1976.

[607] REID J K. A survey of sparse matrix computation [M]. In Electric Power Problems: The Mathematical Challenge (A. M. Erisman, K. W. Neves, and M. H. Dwarakanath, eds.), pp. 41-69. SIAM, Philadelphia, Pennsylvania, 1980.

[608] RELLICH F. Störungstheorie der Spektralzerlegung[J]. Math. Ann. 1936(113):600-619; 1936(113):667-685;1939(116):555-570;1940(117):356-382;1942(118):462-484.

[609] RHEINBOLDT W C. On measures of ill-conditioning for nonlinear equations[J]. Math. Comp. 1976(30):104-111.

[610] RHEINBOLDT W C. On a theory of mesh-refinement processes[J]. SIAM J. Numer. Anal., 1980(17):766-778.

[611] RICE J R. On the degree of convergence of non linear spline approximation[M]. In Approximation with Special Emphasis on Spline Functions (I. J. Schoenberg, ed.), pp. 349-365. New

York: Academic Press, 1969.

[612] RICHTER G R. Superconvergence of piecewise polynomial Galerkin approximations, for Fredholm integral equations of the second kind[J]. Numer. Math. , 1978(31):63-70.

[613] RICHTMYER R D, MORTON K W. Difference Methods for Initial Value Problems, 2nd ed[M]. New York: Wiley (Interscience), 1967.

[614] RIDDELL I J, DELVES L M. The comparison of routines for solving Fredholm integral equations of the second kind[J]. Comput. J. , 1680(23): 274-285.

[615] RIEHL J P, DIESTLER D J, WAGNER A F. Comparison of perturbatie and direct numerical integration techniques for the calculation of phase shifts for elastic scattering [J]. J. Comput. Phys. , 1974(15):212-225.

[616] RIESZ F, SZ-NAGY B. Lecons d'Analyse Fonctionnelle, 3rd ed[M]. Akadémiai Kiadó, Budapest, 1955.

[617] RITZ W. Áber eine neue Methode zur Lösung Gewisser Variationsprobleme der Mathematischen Physik [J]. J. Reine Angew. Math. , 1909 (135):1-61.

[618] RIVLIN T J. The Chebychev Prlynomials[M]. New York: Wiley, 1974.

[619] ROARK A L. On the eigenproblem for convolu-

tion integral equation[J]. Numer. Math., 1971(17):54-61.

[620] ROARK A L, SHAMPINE L F. On a paper of Roark and Wing[J]. Numer. Math., 1965(7):394-395.

[621] ROARK A L, WING G M. A method for computing the eigenvalues of certain integral equations[J]. Numer. Math., 1965(7):159-170.

[622] RODRIGUE G. A gradient method for the matrix eigenvalue problem $Ax = \lambda Bx$ [J]. Numer. Math., 1973(22):1-16.

[623] ROTHBLUM U G. Algebraic eigenspace of nonnegative matrices[J]. Linear Algebra Appl, 1975(12):281-292.

[624] RUGE J. Multigrid methods for differential eigenvalue and variational problems and multigrid simulation[M]. Ph. D. Thesis, Math. Dept, colorado State Univ., Fort Collins, Colorado, 1981.

[625] RUHE A. An algorithm for numerical determination of the structure of a general matrix[J]. BIT, 1970(10):196-216.

[626] RUHE A. Perturbation bounds for means of eigenvalues and invariant subspaces[J]. BIT, 1970b(10):343-354.

[627] RUHE A. Properties of a matrix with a very ill-conditioned eigenproblem[J]. Numer. Math., 1970c(15):57-60.

[628] RUHE A. SOR-methods for the eigenvalue problem with large sparse matrices [J]. Math. comp., 1974a(28):695-710.

[629] RUHE A. Iterative eigenvalue algorithms for large symmetric matrices[M]. In Eigenwerte Probleme (L. Collatz, ed.), pp. 97-115. Birkhaeuser, Basel, 1974b.

[630] RUHE A. Iterative eigenvalue algorithms based on convergent splittings[J]. J. Comput. Phys., 1975(19):110-120.

[631] RUHE A. Numerical methods for the solutin of large sparse eigenvalue problems[M]. In Sparse Matrix Techniques (V. A. Barker, ed.), Lecture Notes in Mathematics, Vol. 572, pp. 130-184. Springer Verlag, Berlin and New York, 1977.

[632] RUBE A. Implementation aspects of band Lanczos algorthms for computation of eigenvalues of large sparse symmetric matrices [J]. Math. Comp., 1979a(33):680-687.

[633] RUHE A. Eigenvalues in a APL environment, an algorithm based on Rayleigh quotient iteration [J]. APL. Quote Quad., 1979b(10):29-30.

[634] RUHE A. The relation between the Jacobi aigorithm and inverse iteration and a Jacobi algorithm based on elementary reflections[J]. BIT, 1980 (20):88-96.

参考文献

[635] RUHE A, ERICSSON T. The spectral transformation Lanczos method in the numerical solution of large, sparse, generalized symmetric eigenvalue problems[J]. Math. Comp., 1980(35):1251-1268.

[636] RUHE A, WIBERG T. The method of conjugate gradient used in inverse iteration[J]. BIT, 1972(12):543-554.

[637] RUMP S M, BÖHM H. Least significant bit evaluation of arithmetic expressions in single-precision[M]. Computing (to appear), 1982.

[638] RUSSEL R D. Collocation for systems of boundary value problems[J]. Numer. Math., 1974(23):119-133.

[639] RUSSEL R D. A comparison of collocation and finite differences for two-point boundary value problems[J]. SIAM J. Numer. Anal., 1977(14):19-39.

[640] RUSSEL R D, SHAMPINE L F. A collocation method for boundary value problems[J]. Numer. Math., 1972(19):1-28.

[641] RUSSEL R D, VARAH J M. A comparison of global methods for linear two-point boundary value problems[J]. Math. Comp., 1975(29):1007-1019.

[642] RUTISHAUSER H. Solution of eigenvalue problem with the LR-transformation[J]. Appl. Math.

Tschebyscheff 逼近定理

Ser. Nat. Bur. Standards, 1958(49):47-81.

[643] RUTISHAUSER H. Computational aspects of F. L[J]. Bauer's simultaneous iteration method. Numer. Math., 1969(13):4-13.

[644] RUTISHAUSER H. Simultaneous iteration method for symmetric matrices[J]. Numer. Math., 1970 (16):205-223.

[645] SAAD Y. Shifts of origin for the QR algorithm [M]. In Information Processing 74, pp. 527-531. North-Holland Publ., Amsterdam, 1975.

[646] SAAD Y. Calcul de vecteurs propres d'une grande matrice creuse par la méthode de Lancos[M]. In Méthodes Numériques pour les Sciences de l' Ingénieur (E. Absi and R. Glowinski, eds.), Dunod, Paris, 1979a.

[647] SAAD Y. Etude de la convergence du procédé d' Arnoldi pour le calcul des éléments propres de grandes matrices creuses non symétriques [M]. Sém. Anal. Numér. No. 321, Univ. de Grenoble, 1979b.

[648] SAAD Y. On the rates of convergence of the Lanczos and the block-Lanczos methods[J]. SIAM J. Numer. Anal., 1980a(17):687-706.

[649] SAAD Y. Variations on Arnoldi's method for computing eigenelements of large unsymmetric matrices [J]. Linear Algebra Appl., 1980b (34):269-295.

[650] SAAD Y. Krylov subspace methods for solving large unsymmetric linear systems [M]. Math. Comp., 1981(37):105-126.

[651] SAAD Y. The Lanczos biorthogonalizatin algorithm and other oblique projection methods for solving large unsymmetric systems[J]. SIAM J. Numer. Anal., 1982a(19):485-509.

[652] SAAD Y. Practical use of Krylov subspace methods for solving indefinite and unsymmetric linear systems[M]. To appear in SIAM J. Sci. Stat. Comp, 1982b.

[653] SAAD Y. Projection methods for solving large sparse eigenvalue problems [M]. Techn. Rep. No. 224, COmputer Science, Yale Univ., Connecticut, 1982c.

[654] SACKS-DAVIS R. Real norm-reducing Jacobi-type eigenvalue algorithm[J]. Austral. Comput. J., 1975(7):65-69.

[655] SAKAGUCHI R L, TABARROK B. Calculation of plate frequencies from complementary energy functions Internal[J]. J. Numer. Methods Engrg., 1970(2):283-293.

[656] SALA I. On the numerical solution of certain boundary value problems and eigenvalue problems of the 2nd and 4th order with the aid of integral equations[J]. Acta Polytech. Scand. Ser. D, 1963(9):1-24.

[657] SAMET A, LERMIT J, NOH K. On the intermediate eigenvalues of symmetric sparse matrices [J]. BIT, 1975(15):185-191.

[658] SCH? FER E. Spectral approximation for compact integral operators by degenerate kernel methods[M]. Numer. Funct. Anal. Optim., 1980(2):43-63.

[659] SCHIPPERS H. The automatic solution of Fredholm equation of the second kind[M]. Techn. Rep., Stichting Math. Cent., Amsterdam, 1981.

[660] SCHLESSINGER S. Approximating eigenvalues and eigenfunctions of symmetric kernels[J]. SIAM J. Appl. Math., 1957(5):1-14.

[661] SCHNEIDER C. Regularity of the solution of a class of weakly singular Fredholm integral equations of the second kind[J]. Integral Equations Operator Theory, 1979(2):62-68.

[662] SCHNEIDER C. Produktintegration mit nichtäquidistanten Stützstellen. [J] Numer. Math., 1980(35):35-43.

[663] SCHNEIDER C. Product integration for weakly singular integral equations[J]. Math. Comp., 1981(36):207-213.

[664] SCHRÖDINGER E. Ouantisierung als eigenwertproblem[J]. IV. Störungtheorie mit Anwendung auf den Starkeffekt der Balmerlinien. Ann. Phys-

ik, 1926(80):437-490.

[665] SCHWARZ H R. The eigenvalue problem $(A - \lambda B)x = 0$ for summetric matrices of high order [J]. Comput. Methods Appl. Mech. Engrg., 1974a(3):11-28.

[666] SCHWARZ H R. A method of coordinate overrelaxation for $(A - \lambda B)x = 0$ [J]. Numer. Math., 1974b(23):135-151.

[667] SCHWARZ H R. Two algorithms for treating $Ax = \lambda Bx$ [J]. Comput. Methods Appl. Mech. Engrg., 1977(12):181-199.

[668] SCOTT D S. How to make the Lanczos algorithm converge slowly Math [J]. Comp., 1979(33):239-247.

[669] SOTT D S. Solving sparse symmetric generalized eigenvalue problems without factorization [J]. SIAM J. Numer. Anal., 1981a(18):102-110.

[670] SCOTT D S. The Lanczos algorithm [M]. In Sparse Matrices and Their Uses (I. S. Duff, ed.), pp. 139-159. New York: Academic Press, 1981b.

[671] SCOTT R. Optimal L^∞ estimates for the finite elecment method on irregular meshes [J]. Math. Comp., 1976(51):100-123.

[672] SEBE T, NACHAMKIN J. Variational buildup of nuclear shelf model bases [J]. Ann. Physics, 1969(51):100-123.

Tschebyscheff 逼近定理

[673] SENETA E. Computing the stationary distribution for infinite Markov chains [J]. Linear Algebra Appl., 1980(34):259-269.

[674] SHAVITT I. Modification of Nesbet's algorithm for the iterative evaluation of eigenvalues and eigenvectors of large matrices J [J]. Comput. Phys., 1970(6):124-130.

[675] SHAVITT I, BENDER C F, PIPANO A, et al. The iterative calculation of several of the lowest or highest eigenvalues and corresponding eigenvectors of very large symmetric matrices [J]. J. Comput. Phys., 1973(11):90-108.

[676] SIMPSON A. A generalization of Kron's eigenvalue procedure [J]. J. Sound Vibr., 1973a (26):129-139.

[677] SIMPSON A. Kron's method: a consequence of the minimization of the primitive Lagrangian in the presence of displacement constraints [J]. J. Sound Vibr., 1973b(27):377-386.

[678] SIMPSON A. Eigenvalue and vector sensitivities in Kron's method [J]. J. Sound Vibr., 1973c (31):73-87.

[679] ŜINDLER A A. Certain theorems in the general theory of approximate methods of analysis and their application to the methods of collocation, moments and Galerkin [J]. Siberian Math. J., 1967(8):302-314.

[680] ŜINDLER A A. The rate of convergence of an enriched collocation method for ordinary differential equations[J]. Siberian Math. J., 1969(10): 160-163.

[681] SINGH S R. Some convergence properties of the Bubnov-Galerkin method[J]. Pacific J. Math., 1976(65):217-221.

[682] SLOAN I H. Convergence of degenerate-kernel methods[J]. J. Austral. Math. Soc. Ser. B, 1976a(19):422-431.

[683] SLOAN I H. Error analysis for a class of degenerate kernel methods[J]. Numer. Math., 1976b(25):231-238.

[684] SLOAN I H. Iterated Galerkin method for eigenvalue problems[J]. SIAM J. Numer. Anal., 1976c(13):753-760.

[685] SLOAN I H. Improvement by iteration for compact operator equations[J]. Math. Comp., 1976d(30):758-764.

[686] SLOAN I H. On the numerical evaluation of singular integrals[J]. BIT, 1978(19):91-102.

[687] SLOAN I H. On choosing the points in product integration[J]. J. MAth. Phys., 1980a(21): 1032-1039.

[688] SLOAN I H. A review of numerical methods for Fredholm equations of the second kind[J]. In The Applicatin and Numerical Solution of Integral

Equations (R. S. Anderssen, F. R. de Hoog, and M. A. Lukas, eds.), pp. 51-74. SIjthoff & Noordhoff, Alphen aan den Rijn. The Netherlands, 1980b.

[689] SLOAN I H. The numerical solution of Fredholm equations of the second kind by polynomial interpolation[J]. J. Integral Equations, 1980c(2): 265-279.

[690] SLOAN I H. Analysis of general quadrature methods for integral equations of the second kind[J]. Numer. Math., 1981(38):263-278.

[691] SLOAN I H. Superconvergence and the Galerkin method for integral equations of the second kind [J]. In Treatment of Integral Equations by Numerical Methods (C. T. H. Baker and G. F. Miller, eds.), pp. 197-208, Academic Press, London, 1983.

[692] SLOAN I H, BURN B J. Collocation with polynomials for integral equations of the second kind: a new approach to the theory[J]. J. Integal Euations, 1979(1):77-94.

[693] SLOAN I H, NOUSSAIR E, BURN B J. Projection method for equations of the second kind[J]. J. Math. Anal. Appl., 1979(69):84-103.

[694] SMITH B T, BOYLE J M, GARBOW B S, et al. Matrix Eigensystem Routines—EISPACK Guide [M]. Lecture Notes in Computer Science, Vol.

6, 2nd ed. Springer-Verlang, Berlin and New York, 1976.

[695] SMOOKE M D. Piecewise analytical perturbation series solutions of the radial Schrödinger equation [M]. Ph. D. Thesis, Harvard Univ., Cambridge, Massachusetts, 1978.

[696] SMOOKE M D. Error estimates for piecewise analytical perturbation series solutions of the radial Schrödinger equation [M]. I. Onedimensional case. Rep. SAND 80-8611, Sandia Livermore Lab., Livermore, California, 1980a.

[697] SMOOKE M D. Error estimates for piecewise analytical perturbation series solutions of the radial Schrödinger equation [M]. II. Multidimensional case. Rep. SAND 80-8610, Sandia Livermore Lab., Livermore California, 1980b.

[698] SOBOLVE S L. On a boundary problem for semi harmonic equations (in Russian) [J]. Mat. Sb., 1937(2):467-500.

[699] SOBOLEV S L. Some remarks on numerical solution of integral equations (in Russian) [J]. Izv. Akad. Nauk SSSR Ser. Mat., 1956(20):413-436.

[700] SPENCE A. On the convergence of the Nyström method for the integral equation eigenvalue problem[J]. Numer. Math., 1975(25):57-66.

[701] SPENCE A. (1978-1979) Error bounds and esti-

mates for eigenvalues of integral equations[J]. Numer. Math., 1978(29):133-147; 1979(32): 139-146.

[702] SPENCE A. Product integration for singular integrals and singular integral equations[M]. In Numerische Integration (G. Hämmerlin, ed.), pp. 288-300. Birkhaeuser, Basel, 1979.

[703] SPENCE A, MOORE G. A convergence analysis for turning points of nonlinear compact operator equations[M]. In Numerical Treatment of Integral Equations (J. Albrecht and L. Collatz, eds.), pp. 203-212. Birkhaeuser, Basel, 1980.

[704] SPENCE D A. An eigenvalue problem for elastic contact with finite friction[J]. Proc. Cambridge Phil. Soc., 1972(73):249-268.

[705] SRINIVASAN R S, SANKARAN S. Vibration of cantilever cylindrical shells[J]. J. Sound Vibr., 1975(40):425-430.

[706] SRIVASTAVA B P. Calculatin of bounds on the eigenvalue spectrum of anharmonic phonon collision operator[J]. Phys. Lett. A, 1975(54): 222-224.

[707] STETTER H J. Analysis of Discretization Methods for Ordinary Differential Equations[J]. Springer-Verlag, Berlin and New York, 1973.

[708] STETTER H J. The detect correction principle and discretization methods[J]. Numer. Math.,

1978(29):425-443.

[709] STEWART G W. Error bounds for approximate invariant subspaces of closed linear operators [J]. SIAM J. Numer. Anal. 1971(8):796-808.

[710] STEWART G W. On the sensitivity of the eigenvalue problem $Ax=\lambda Bx$ [J]. SIAM J. Numer. Anal., 1972(9):669-686.

[711] STEWART G W. Introduction to Matrix Computations[M]. New York: Academic Press 1973a.

[712] STEWART G W. Error and perturbation bounds for subspaces associated with certain eigenvalue problems[J]. SIAM Rev., 1973b(15):727-764.

[713] STEWART G W. The numerical treatment of large eigenvalue problems[M]. In Information Processing 74, pp. 666-672. North-Holland Publ., Amsterdam, 1975a.

[714] STEWART G W. The convergence of the method of conjugate gradients at isolated extreme points of the spectrum[J]. Numer. Math., 1975b(24):85-93.

[715] STEWART G W. Gerschgorin theory for the generalized eigenvalue problem $Ax=\lambda Bx$ [J]. Math. Comp., 1975c(29):600-606.

[716] STEWART G W. Methods of simultaneous iteration for calculating eigenvectors of matrices[M].

In Topics in Numerical Analysis II (J. J. H. Miller, ed.), pp. 185-196. New York: Academic Press, 1975d.

[717] STEWART G W. Simultaneous iteration for computing invariant subspaces of non-Hermaian matrices[J]. Numer. Math. , 1976a(25):123-136.

[718] STEWART G W. Abibliographical tour of the large, sparse generalized gienvalue problem [M]. In Sparse Matrix Computations (J. R. Bunch and D. J. Rose, eds.), pp. 113-130. New York: Academic Press, 1976b.

[719] STEWART G W. Perturbation bounds for the definite generalized eigenvalue problem[J]. Linear Algebra Appl. , 1979(23):69-85.

[720] STEWART W J, JENNINGS A. A simultaneous iteration algorithm for real matrices [J]. ACM Trans. Math. Software, 1981(7):184-198.

[721] STOER J, BULIRSCH R. Introduction to Numerical Analysis [M]. Springer-Verlag, Berlin and New York, 1980.

[722] STRAKHOVSKAYS L G. An iterative method for evaluating the first eigenvalue of an elliptic operator. U. S. S. R. Computational Math. and Math [J]. Phys. , 1977(17):38-101.

[723] STRANG G. Approximation in the finite element method. Numer[J]. Math. , 1972(19):81-98.

[724] STRANG G. Linear Algebra and Its Applications

[M]. New York: Academic Press, 1976.

[725] STRANG G, FIX G J. An Analysis of the Finite Element Method[M]. Prentice-Hall, Englewood Cliffs, New Jersey, 1973.

[726] STROUD A H. Approximate Calculation of Multiple Integrals [M]. Prentice-Hall, New Jersey: Englewood Cliffs, 1971.

[727] STRYGIN V V. Application of Bubnov-Galerkin method to find autooscillations (in Russian)[J]. Prikl. Mat. Meh. , 1973(37):1015-1019.

[728] STRYGIN V V, CYGANKOV A I. Applicatin of collocation and difference method to find autooscillations of differential-dif-ference equations (in Russian)[J]. Ẑ. Vyĉisl. Mat. i Mat. Fiz. , 1974(14):691-698.

[729] STUMMEL F. Diskrete Konvergenz linearer Operatoren[J]. Part I. Math. Ann. , 1970(190):45-92.

[730] STUMMEL F. Diskrete Konvergenz linearer Operatoren Part II[J]. Math Z. , 1971(120):213-264.

[731] STUMMEL F. Diskrete Konvergenz linearer Operatoren. Part III [M]. In Linear Operators and Aproximation, pp. 196-216. Birkhaeuser, Basel, 1972.

[732] STUMMEL F. Discrete convergence of mappings [M]. In Topics in Numerical Analysis (J. J. H.

Miller, ed.), pp. 285-310. New York: Academic Press, 1973.

[733] STUMMEL F. Approximation methods for eigenvalue problems in elliptic differential equations [M]. In Numerik und Anwendungen von Eigenwertaufgaben und Verzweigungsproblemen (E. Bohl. L. Collatz, and K. P. Hadeler, eds.), pp. 133-165. Birkhaüser, Basel, 1977.

[734] STUMMEL F. Basic compactness properties of nonconforming and hybrid finite element spaces [J]. RAIRO Anal. Numér., 1980(14):81-115.

[735] SWARTZ B, WENDROFF B. The relation between the Galerkin and collocation methods using smooth splines [J]. SIAM J. Numer. Anal., 1974(11):994-996.

[736] SYMM H J, WILKINSON J H. Realistic bounds for a simple eigenvalue and its associated eigenvector[J]. Numer. Math., 1980(35):113-126.

[737] SYMM H J, WILKINSON J H. Error bounds for computed invariant subspaces [M]. Res. Rep. No. 81-02, Angew. Math., Eidg. Techn. Hochschule Zürich, 1981.

[738] SZEGÖ G. Orthogonal Polynomials, Colloquium Publ[M]. No. 233, Amer. Math. Soc., Providence, Rhode Island, 1975.

[739] SZ-NAGY B. Perturbations des transformations

autoadjointes dans l'espace de Hilbert [J]. Comment. Math. Helv., 1946/47 (19): 347-366.

[740] SZ-NAGY B. Perturbations des transformations linéaires fermées [J]. Acta Sci. Math. (Szeged), 1951(14): 125-137.

[741] SZ-NAGY B, FOIAS C. Forme triangulaire d'une contraction et factorisation de la fonction caractéristique[J]. Acta Sci. Math. (Szeged), 1967(28): 201-212.

[742] SZ-NAGY B, FOIAS C. Harmonic analysis of operators on Hilbert space[M]. Akadémiai Kiadó, Budapest and North-Holland Publ, Amsterdam, 1970.

[743] SZYLD D B, WIDLUND O B. Applications of conjugate gradient type methods to eigenvalue calculations[M]. Tech. Rep., Courant Inst., New York Univ, 1978.

[744] TAMME E E. On regular convergence of difference approximations for Dirichlet problem (in Russian) [J]. Eesti NSV Tead. Akad. Toimetised Füüs-Mat., 1977(26): 3-8.

[745] TAYLOR A E. Introduction to Functional Analysis[M]. New York: Wiley, 1958.

[746] TEMPLE G. The computation of characteristic numbers and characteristic functions[J]. Proc. London Math. Soc., 1928(29): 257-280.

Tschebyscheff 逼近定理

[747] TEMPLE G. The accuracy of Rayleigh's method of calculating the natural frequencies of vibrating systems[J]. Proc. Roy. Soc. London Ser. A, 1952(211):204-214.

[748] TEMPLE G, BICKLEY W G. Rayleigh's Principle and Its Applications to Engineering [M]. Constable, London, 1933.

[749] TSUNEMATSU T, TAKEDA T. A new iterative method of solution of a large-scale generalized eigenvalue problem[J]. J. Comput. Phys., 1978(28):287-293.

[750] UEBE G, FISHER J. Computation of the eigenvalues of large econometric models[J]. Comput. and Oper. Res., 1974(1):313-319.

[751] UNDERWOOD R. An iterative block Lanczos method for the solution of large sparse symmetric eigenproblems [M]. Ph. D. Thesis, Rep. STAN-CS-75-496, Comput. Sci. Dept., Stanford Univ., Stanford. california, 1975.

[752] URABE M A. Numerical solution of multi-point boundary value problems in Chebyshev series—Theory and the method [J]. Numer. Math., 1967(9):341-366.

[753] URABE M A. Numerical solution of boundary value problems in Chebyshev series—A method of computation and error estimation[M]. In Conference on the Numerical Solution of Differential E-

quation (J. L. Morris, ed.), Lecture Notes in Mathematics, Vol. 109. pp. 40-86. Springer-Verlag, Berlin and New York, 1969.

[754] URABE M A. A posteriori componentwise error estimation of approximate solutions to non linear equations [M]. In Interval Mathematics (K. Nickel, ed.), Lecture Notes in Computer Science, Vol. 29, pp. 99-117. Springer-Verlag, Berlin and New York, 1975.

[755] VAINIKKO G M. Asymptotic error bounds for projection methods in the eigenvalue problem [J]. U. S. S. R. Computational Math. and Math. Phys., 1964(4):9-36.

[756] VAINIKKO G M. On the convergence of Galerkin method (in Russian) [J]. Tartu Riikl. Ál. Toimetised, 1965a(177):148-152.

[757] VAINIKKO G M. On the stability and convergence of the collocation method [J]. Differential Equations, 1965b(1):186-194.

[758] VAINIKKO G M. The convergence of the collocation method for nonlinear differential equations [J]. U. S. S. R. Computational Math. and Math. Phys., 1966(6):35-42.

[759] VAINIKKO G M. Galerkin's perturbation method and the general theory of approximate methods for nonlinear equations (in Russian) [J]. Ẑ. Vyĉisl. Mat. i Mat. Fiz., 1967a(7):723-751.

[760] VAINIKKO G M. The rate of convergence of approximate methods in the problem of eigenvalues (in Russian)[J]. Ẑ. Vyĉisl. Mat. i Mat. Fiz. , 1967(b):7:977-987.

[761] VAINIKKO G M. Rapidity of convergence of approximation methods in the eigenvalue problem [J]. U. S. S. R. Computational Math. and Math. Phys. 1967c(7):18-32.

[762] VAINIKKO G M. On the convergence speed of the method of moments for ordinary differential equations (in Russian)[J]. Sib. Mat. Z. , 1968 (9):21-28.

[763] VAINIKKO G M. The compact approximation principle in the theory of approximatin methods (in Russian)[J]. Ẑ. Vyĉisl. Mat. i Mat. Fiz. , 1969a(9):739-761.

[764] VAINIKKO G M. A difference method for ordinary differential equations (in Russian) [J]. Ẑ. Vyĉisl. Mat. i Mat. Fiz. , 1969b (9): 1057-1074.

[765] VAINIKKO G M. The connection between mechanical quadrature and finite difference methods (in Russian) [J]. Ẑ. Vyĉisl. Mat. i. Mat. Fiz. , 1969c(9):259-270.

[766] VAINIKKO G M. On the rate of convergence of certain approximation methods of Gaierkin type in an eigenvalue problem[J]. Amer. Math. Soc.

Transl., 1970a(36):249-259.

[767] VAINIKKO G M. On the convergence of collocation method for multidimensional integral equations (in Russian)[J]. Tartu Riikl. Ál. Toimetised, 1970b(253):244-257.

[768] VAINIKKO G M. On the stability of the collocation method (in Russian)[J]. Tartu Riikl. Ál. Toimetised, 1971(281):190-196.

[769] VAINIKKO G M. On the approximation of fixed points of compact operators (in Russian)[J]. Tartu Riikl. Ál. Toimetised, 1974a(342):225-236.

[770] VAINIKKO G M. Discretely compact sequences (in Russian)[J]. Ž. Vyĉisl. Mat. i Mat. Fiz., 1974(14):572-583.

[771] VAINIKKO G M. Convergence of the difference method when seeking the periodic solutions of ordinary differential equations (in Russian)[J]. Ž. Vyĉisl. Mat. i Mat. Fiz., 1975(15):87-100.

[772] VAINIKKO G M. Analysis of Discretization Methods (in Russian)[M]. Univ. of Tartu Publ., Estland, 1976a.

[773] VAINIKKO G M. Funktionalanalysis der Diskretisierungsmethoden [M]. Teubner, Leipzig, 1976b.

[774] VAINIKKO G M. Áber die Konvergenz und Di-

vergenz von Näherungsmethoden bei Eigenwertproblemen[J]. Math. Nachr. , 1977a(78):145-164.

[775] VAINIKKO G M. Áber die Konvergenzbegriffe für lineare Operatoren in der Numerischen Mathematik[J]. Math. Nachr. , 1977b(78):165-183.

[776] VAINIKKO G M. Approximative methods for nonlinear equations (two approaches to the convergence problem)[J]. Nonlinear Anal. , Theory, Method and Appl. , 1978a(2):647-687.

[777] VAINIKKO G M. Foundations of finite difference method for eigenvalue problems[M]. In the Use of Finite Element Method and Finite Difference Method Geophys. (V. Bucha and H. Nedoma, eds.), pp. 173-192. Czech. Acad. Sci. , Prague, 1978b.

[778] VAINIKKO G M, KARMA O O. The convergence of approximate methods for solving linear and nonlinear operator equations (in Russian)[J]. Ẑ. Vyĉisl. Mat. i Mat. Fiz. , 1974a(14):828-837.

[779] VAINIKKO G M, KARMA O O. The convergence rate of approximate methods in the eigenvalue problem when the parameter appears nonlinearyly (in Russian)[J]. Ẑ. Vyĉisl. Mat. i Mat. Fiz. , 1974b(14):1393-1408.

[780] VAINIKKO G M, PEDAS A. On solution of integral equations with logarithmical singularities by

quadrature formulae methods (in Russian) [J]. Tartu Riikl. Ál. Toimetised, 1971 (281) : 201- 210.

[781] VAINIKKO G M, PEDAS A. The properties of solutions of weakly singular integral equations [J]. J. Austral. Math. Soc. Ser. B, 1981 (22) : 424-434.

[782] VAINIKKO G M, TAMME E E. Convergence of difference methods in the periodic soulution problem for equations of elliptic type (in Russian) [J]. Ž. Vyĉisl. Mat. i Mat. Fiz., 1976 (16) : 652-664.

[783] VAINIKKO G M, UBA P. A piecewise polynomial approximation to the solution of an integral equation with weakly singular kernel [J]. J. Austral. Math. Soc. Ser. B, 1981 (22) : 435-442.

[784] VAINIKKO G M, UMANSKII Y B. Regular operators [J]. Funct. Anal. Appl, 1968 (2) : 175- 176.

[785] VANDERGRAFT J S. Generalized Rayleigh method with applications to linding eigenvalues of large matrices [J]. Linear Algebra Appl., 1971 (4) : 363-368.

[786] VAN DOOREN P M. The generalized eigenstructure problem in linear system theory [J]. IEEE Trans. Automat. Control, AC – 26, 1981 : 111- 129.

[787] VAN KEMPEN H P M. On the convergence of the classical lacobi method for real symmetric matrices with non distinet eigenvalues [J]. Numer. Math., 1966(9):11-18,19-22.

[788] VAN LOAN C F. A general matrix eigenvalue algorithm [J]. SIAM J. Numer. Anal., 1975(12):819-834.

[789] VAN VELDHUIZEN M. A refinement process for collocation approximations [J]. Numer. Math., 1976(26):397-407.

[790] VARAH J M. Rigorous machine bounds for the eigensystem of a general complex matrix [J]. Math. Comp., 1968b(22):785-791.

[791] VARAH J M. Rigorous machine bounds for the eigensystem of a general complex matrix [J]. Math. Comp., 1968b(22):793-801.

[792] VARAH J M. Computing invariant subspaces of a general matrix when the eigensystem is poorly conditioned[J]. Math. Comp., 1970(24):137-149.

[793] VARAH J M. Invariant subspace perturbations for a nonnormal matrix[M]. In Information Processing 71, Vol. 2, pp. 1251-1253. North-Holland Publ., Amsterdam, 1972.

[794] VARAH J M. On the numerical solution of ill-conditioned linear systems with applications to ill-posed problems [J]. SIAM J. Numer. Anal.,

1973(10):1257-1267.

[795] VARGA R S. Matrix Iterative Analysis[M]. Prentice-Hall, Englewood Cliffs, New Jersey.

[796] VARGA R S. Minimal Gershgorin sets[J]. Pacific J. Math., 1965(15):719-729.

[797] VOROBYEV Y V. Method of Moments in Applied Mathematics[M]. New York: Gordon & Breach, 1965.

[798] WACHPRESS E L. Iterative solution of elliptic systems and application of the neutron diffusion equations of reactor physics[M]. Prentice-Hall, Englewood Cliffs, New Jersey, 1966.

[799] WAHBA G. Convergence rates for certain approximate solutions to Fredholm integral equations of the first kind[J]. J. Approx. Theory, 1973(7):167-185.

[800] WAHBA G. On the optimal choice of nodes in the collocation projection method for solving linear operator equations[J]. J. Approx. Theory, 1976(16):175-186.

[801] WANG J Y. On the numerical computation of eigenvalues and eigenfunctions of compact integral operators using spline functions[J]. J. Inst. Math. Appl., 1976(18):177-188.

[802] WARD R C. The OR algorithm and Hyman's method on vector computers[J]. Math. Comp., 1976(30):132-142.

[803] WARD R C, GRAY L J. Eigensystem computation for skewsymmetric matrices and a class of symmetric matrices[J]. ACM Trans. Math. Software, 1978a(4):278-285.

[804] WARD R C, AND GRAY L J. Algorithm 530, an algorithm for computing the eigensystem of skew-symmetric matrices and a class of symmetric matrices[J]. ACM Trans. Math. Software, 1978b(4):286-289.

[805] WATKINS D S. Understanding the QR algorithm [J]. SIAM Rev. 1982(24):427-440.

[806] WEINBERGER H F. Error bounds in the Rayleigh-Ritz approximation of eigenvectors[M]. In Part. Diff. Eq. Cont. Mech. (R. Langer, ed.), pp. 39-53. The Univ. of Wisconsin Press, Madison Wisconsin, 1961.

[807] WEINBERGER H F. Variational Methods for Eigenvalue Approxi-mations[M]. SIAM, Philadelphia, Pennsylvania, 1974.

[808] WEINSTEIN A. Etude des spectres des équations aux dérivées partielles de la théorie des plaques élastiques[J]. Mémoire de Sci. Math. 88, Gauther-Vilars, Paris, 1937.

[809] WEINSTEIN A. The intermediate prolbem and the maximum-minimum theory of eigenvalues [J]. J. Math. Mech., 1963(12):235-245.

[810] WEINSTEIN A, STENGER W. Methods of Inter-

mediate Problems for Eigenvalues: Theory and Ramifications[M]. New York: Academic Press, 1972.

[811] WEISS R. The application of implicit Runge-Kutta and collocation methods to boundary value problems[J]. Math. Comp., 1974(28):449-646.

[812] WENDLAND W L. On Galerkin collocation methods for integral equations of elliptic boundary value problems[M]. In Numerical Treatment of Integral Equations (J. Albrecht and L. Collatz, eds.), pp. 244-275. Birkhaeuser, Basel, 1980.

[813] WENDLAND W L, STEPHAN E, HSIAO G C. On the integral equation method for the plane mixed boundary value problem of the laplacian [J]. Math. Meth. in Appl. Sc., 1979(1):265-321.

[814] WIDLUND O. On best error bounds for approximation by piecewise polynomial functions[J]. Numer. Math., 1977(27):327-338.

[815] WIDLUND O. A Lanczos method for a class of non symmetric systems of linear equations[J]. SIAM J. Numer. Anal., 1978(15):801-812.

[816] WIELANDT H. Error bounds for eigenvalues of symmetric integral equations[M]. Proc. Sympos. Appl. Math. 6th., Providence. Amer. Math. Soc. Providence. Rhode Island, 1956.

Tschebyscheff 逼近定理

[817] WIELANDT H. Topics in the Theory of Matrices [M]. Univ of Wisconsin Press, Madison, Wisconsin, 1967.

[818] WILKINSON J H. Note on the quadratic convergence of the cyclic Jacobi process [J]. Numer. Math., 1962(4):296-300.

[819] WILKINSON J H. The Algebraic Eigenvalue Problem [M]. Oxford Univ. Press (Clarendon), London and New York, 1965.

[820] WILKINSON J H. Global convergence of tridiagonal QR algorithm with origin shifts [J]. Linear Algebra Appl, 1968(1):409-420.

[821] WILKINSON J H. Global convergence of QR algorithm [M]. In Information Processing 68, Vol, 1, pp. 130-133. North-Holland Publ., Amsterdam, 1969.

[822] WILKINSON J H, REINSCH C H. Handbook for Automatic Computation [M]. Linear Algebra. Vol. 2. Springer-Verlag, Berlin and New York ,1971.

[823] WING G M. On a method for obtaining bounds on the eigenvalues of certain integral equations [J]. J. Math. Anal. Appl, 1965(11):160-175.

[824] WITTENBRINK K A. High order projection methods of momentand collocation-type for nonlinear boundary value problems [J]. Computing, 1973 (11):255-274.

[825] WOLFE M A. The numerical solution of non-sin-

gular integral and integro-differential equations by iteration with Chebyshev series[J]. Comput. J., 1969(12):193-196.

[826] WRIGHT G C, MILES G A. An economical method for determining the smallest eigenvalues of large linear systems[J]. Internat. J. Numer. Methods Engrg., 1971(3):25-33.

[827] YAMAMOTO T. Componentwise error estimates for approximate solutions of non linear equations [J]. Inform. Process., 1979(2):121-126.

[828] YAMAMOTO T. Error bounds for computed eigenvalues and eigenvectors[J]. Numer. Math., 1980(34):189-199.

[829] YAMAMOTO Y, OHTSUBO H. Subspace iteration accelerated by using Chebyshev polynomials for eigenvalue problems with symmetric matrices [J]. Internat. J. Numer. Methods Engrg., 1976 (10):935-944.

[830] YOSIDA K. Lectures on Differential and Integral Equations [M]. New York: Wiley (Interscience), 1960.

[831] YOSIDA K. Functional Analysis[M]. Springer-Verlag, Berlin and New York, 1965.

[832] YOUNG A. The application of approximate product integration to the numerical solution of integral equations[J]. Proc. Roy. Soc. London Ser. A, 1954(224):561-573.

Tschebyscheff 逼近定理

[833] ZAANEN A C. Linear Analysis[M]. North-Holland Publ., Amsterdam, 1960.

[834] ZABREIKO P P, KOSHELEV A I., KRASNOSELSKII M A, et al. Integral. Equations—A Reference Textbook[M]. Noordhoff Int., Leyden, The Netherlands, 1975.

[835] ZARUBIN A G. The speed of convergence of projection methods for linear equations[J]. U. S. S. R. Computational Math. and Math. Phys., 1979(19):265-272.

[836] ZERBI G, PIESERI L, CABASSI F. Vibrational spectrum of chain molecules with conformational disorder polyethylene[J]. Molecular Phys., 1971(22):241-256.

[837] ZIENKIEWICZ O C. The Finite Element Method [M]. New York: McGraw-Hill, 1977.

[838] ZLÁMAL M. Superconvergence and reduced integration in the finite element method[J]. Math. Comp., 1978(32):663-685.

[839] ZLATEV Z. Use of iterative refinement in the solution of sparse linear systems[J]. SIAM J. Numer. Anal., 1982(19):381-399.

[840] ZWART P B. Multivariate splines with nondegenerate partitions[J]. SIAM J. Numer. Anal., 1973(10):665-673.

编辑手记

国务院参事,中国出版传媒股份有限公司副总经理樊希安说:现在我国每年出书47万种,83亿册,在全世界算得上出版大国,但还不是出版强国.书的品种很多,但读者反映到书店里又买不到好书,满足不了需要.好多书到书店根本没有上架,所谓"见光死",原样退回;还有的"不见光死"——包都没打开,就被原包退回了.相当一部分书跟风出版,重复出版,庸俗,媚俗,格调低下,印刷质量差.出版业也同样存在着供需错位、产品不够优化的问题.也存在一些出版社库存量过大的问题.供给与需求出现了矛盾,出版社应该实现的效益没有实现.造成一些出版社利润下降,甚至亏损.就是从出版企业生存计,供给侧结构性改革也非抓不可了.

Tschebyscheff 逼近问题

那么究竟应该怎么改呢？余以为有一个思路可行,那就是多出大师写的书或写大师的书.数学大师的书更是如此. E. C. Titchmarsh 说:"伟大的数学家已经对人类思想做出了甚至比伟大的文学艺术更加不朽的贡献,因为它是与语言无关的."

"我们生活的时代是大师日益稀缺的时代."一再有学者这样说.究其原因,在这个世界被功利目标充斥之后已经祛魅,大师们的高论已经被浅薄的应试技巧洗刷干净.正如一位文人所说:敬畏才是从一个伟大心灵所写下的伟大作品中学到教益的必备条件.当大师身上神圣的光环一旦退去,我们的敬畏之心已经被蒙昧所遮蔽.因此我们只能从不断地阅读中寻找残留的敬畏之心,从沉寂的经典中淘洗着大师的深邃.

数学大师们得出的定理或建立的理论异于常人,好比中国士大夫的古典高端审美情调,一定是这样的:衣食丰足精神饱满之时,一定要给自己"找不痛快",深宅大院单听梧桐夜雨,满目繁华只看灯火阑珊.切比雪夫的独特之处在于他把数学理论与自然科学技术的实践紧密地结合起来,他曾多次指出:科学在实践中获得了正确的领导地位(《切比雪夫全集》第五集,1951年版第 150 页).

本书的理想读者应该是大中师生,但它可能并不能迎合他们的阅读心理.

平顶山市某中学高三学生曾写出了一条励志标语：

不学习的女人只有两个下场:逛不完的菜市场,穿不完的地摊货.不学习的男人只有两个下场:穿不完的阿迪吊丝,捡不完的破瓶

编辑手记

烂罐.

这完全是以功利的心态、以成功学的俗腔烂调来恐吓那些对学习对读书不上心的学生.就好比用不吃某种食品会对身体有害的恐吓来代替美食家以品尝美味为人生乐趣的品尝活动.

《北京晨报》曾发表过一篇文章指出：

> 功利主义的勃兴，不能让商家负全责，在我们的教育、治理、系统中，乃至对历史的解读、对文化的涵养、对问题的认知中，随处可见功利主义的影子，"有用即真理"甚至已贯彻到生活的许多细节中，当工具理性大大凌驾于价值理性时，"成功学"怎能不勃兴？

其实读书对人生现实处境有所帮助与提升这一点中国古来就有.

宋真宗赵恒曾写了一首劝学诗：

> 富家不用买良田，书中自有千钟粟；
> 安居不用架高堂，书中自有黄金屋；
> 出门莫恨无人随，书中车马多如簇；
> 娶妻莫恨无良媒，书中自有颜如玉；
> 男儿若遂平生志，六经勤向窗前读.

这种功利的倾向目前在出版界也有所反映，所以荣获出版广角2011新锐出版人的刘瑞琳就说："在这个文化真衰败，出版假勃兴的时代，我们所能做的只是——做有尊严的出版、做有底线的出版、做有个性的

Tschebyscheff 逼近问题

出版、做有品质的出版."本书是我们的一次践诺行为.

1937 年,邹韬奋曾写下这样的文字:

"时光过得真快,我这后生小子,不自觉地干了 15 年的编辑.为着做了编辑,曾经亡命过;为着做了编辑,曾经坐过牢;为着做了编辑,始终不外是个穷光蛋;被靠我过活的家庭埋怨得要命,但是我至今'乐此不疲',自愿'老死此乡'."

时代变了,但理想没变.这也是所有有理想的编辑的共同心愿!

本书编译成份较大,所以有些人名、地名的译法可能会与其他书籍不同,包括切比雪夫,以前还被译为契比雪夫,还有译为契伯舍夫.同样的例子文学中更多见.米歇尔·维勒贝克(Michel Houellebecq)是法国最重要,最畅销的作家,其新作《屈服》五天内便销售出十二万册.他同时也是最为国际化的一位大家,几乎所有的作品都被翻译成数十种语言出版.

然而进入中国却出现了问题,由于译者与出版社都各行其事,最终落得一书两名,一人三姓,读者若不小心,还以为是风马牛不相及的作家和作品.事实上米歇尔·乌洛贝克,米榭·韦勒贝克和米歇尔·维勒贝克原本是同一个人.

值得一提的是,切比雪夫在逼近论中给出的公式还可用于数理统计中回归方程的计算,许多搞纯数学的人都瞧不起统计学,即便是号称与统计不分家的概率学家也是如此,比如钟开莱和辛勒共同创办了一系

编辑手记

列讨论概率论难题的讲习班,定期在不同的大学举行讲习班,还编辑了许多学术刊物.有一期讲习班在钟开莱任教的斯坦福大学,时间定在一个周二的下午.辛勒就和钟开莱说,"周二下午斯坦福还有一个统计学大会,很多统计学家肯定两个会议都想来.你不如换个时间吧."钟开莱击掌笑着说,"我就是特意安排在这个时间的!这样所有的统计学家就来不了我的讲习班啦.我最讨厌统计学家了."钟开莱的诙谐言语令人捧腹大笑.

  概率和统计不分家,要做概率上的学问一定要有扎实的统计功底,而钟开莱偏偏又公然宣布自己"最讨厌统计学家."他把自己能施展拳脚的范围缩得很小,就是在这样一个鲜有人探索的领域,他打出了一片天地,成了美国概率论界第一人.他的概率论著作被专业学术论文引用次数最多,据美国数学学会评价,"Chung's writing is literate, elegant, wise, humane. He takes the reader into his confidence, explaining ideas, motivation, and circumstances."如果钟开莱一生研究物理,研究基础数学,也会成为美国物理和基础数学的第一人.

  但在这些纯数学家中切比雪夫是个例外.而且他是举世公认的概率论第一人.

刘培杰

2016 年 6 月 1 日

于哈工大